Flattening the Earth

T0138410

Flattening the Earth

*Two Thousand Years of
Map Projections*

JOHN P. SNYDER

THE UNIVERSITY OF CHICAGO PRESS   *Chicago and London*

The University of Chicago Press, Chicago 60637
The University of Chicago Press, Ltd., London
© 1993 by The University of Chicago
All rights reserved. Published 1993
Paperback edition 1997
Printed in the United States of America
02 01 00 99 98 97   5 4 3 2

ISBN: 0-226-76746-9 (cloth)
ISBN: 0-226-76747-7 (paperback)

Library of Congress Cataloging-in-Publication Data

Snyder, John Parr, 1926–
    Flattening the earth : two thousand years of map projections / John P. Snyder.
      p.   cm.
    Includes bibliographical references and index.
    ISBN 0-226-76746-9 (alk. paper)
    1. Map projection—History.   I. Title.
GA110.S576   1993
526′.8—dc20   92-36750
                                                   CIP

For Jeanne

CONTENTS

ILLUSTRATIONS

PREFACE

ALTHOUGH LATER TRAINED and employed for some thirty years as a chemical engineer, I was drawn to map projections in 1942 as a teenager, continuing a childhood fascination with maps. Eventually I grew tired of using logarithm tables or a slide rule to calculate the necessary rectangular coordinates before I could tediously plot an outline map on a given projection. The advent of pocket calculators in the early 1970s and the completion of a personal project pursuing the history of the mapping of the region where I then lived led to a revival of my interest in map projections in 1976, with a plan to write about Map Projections—History and Application. Soon finding that someone was funded to research and publish the history, I limited my efforts to collecting historical data but publishing various mathematical aspects, to which I added occasional historical notes.

By the time it became clear that the funded project was not going to lead to publication, I felt too busy with the mathematics. This was most fortunate for me, because it led to a second career with the U.S. Geological Survey in Reston, Virginia. The location provided ready access to far more resources than I probably would have found previously. In 1982, when David Woodward invited me to write the first of a set of chapters on the history of map projections for projected volumes in the *History of Cartography* series he was editing with the late Brian Harley, I felt much better prepared to carry out my 1976 intentions.

The education I was receiving while researching my chapters for the series was so enticing that the initial draft of my contribution was completed well before the painstaking editors were able to use or even edit the material. Sensing my restless nature, Dr. Woodward kindly suggested that I approach the University of Chicago Press about publishing the combined chapters as a single work, to be used later in the *History* volumes in a somewhat condensed form. The press has happily agreed to do so. The greater detail available in this current volume still generally provides only a concise exposition of the individual projections, however.

For the current volume, the history of map projection has been arbitrarily divided into four chapters, each covering chronological periods corresponding to plans for volumes 3–6 of the *History of Cartography*. Each chapter is subdivided into two major sections: the earlier projections used during the period, followed by the new projections developed during the period. Types of projections are described in chronological order within each of the two subdivisions.

Chapter 1 discusses the period of the Renaissance, taken as 1470–1669, as well as earlier periods; chapter 2, 1670–1799; chapter 3, 1800–1899; and chapter 4, 1900–1992. The dates for the Renaissance are rather arbitrary, but the first known printed editions of Ptolemy's *Geography*, the first printed world map, and the first printed star map all appeared in the 1470s, while the beginnings of the first geodetic survey of a country and the development of the calculus occurred near 1670 to herald a major transition in mapping precision.

The majority of illustrations in this volume are modern computer plots rather than reproductions of actual maps. Most such maps could not be reduced to printing size for this book and remain legible, and smaller portions of these maps would not sufficiently show the map projections. The homemade computer program used allows a map to be printed in several sections on a dot-matrix printer, maximizing dot density at a size up to 8 by 8 or 8 by 11 inches before reduction for printing. The shoreline file generally used is a reformatted and slightly edited Dahlgren datafile, with about 9,000 points. For figures 1.6 and 1.12 the file is an extraction by MicroDoc (Fred Pospeschil), of the World Data Bank 2 file, with about 125,000 points worldwide. In several cases I have specified map centers identical or nearly so to those used in published maps based on the projection.

I have seen most but not all references listed in the notes. In several cases I've given the source of a reference I've not seen. Also, the number of references given in the text rather than as endnotes results from press format requirements. To emphasize the human aspect of projection history, complete names of persons, when I have found them, are included with references, even if they are not given in any of the papers, and minibiographies, biographical references, and a few portraits have been added to the text.

Over the years I have been kindly helped by many individuals who have supplied me with copies of papers and other valuable assistance in gathering the information included herein. Three have provided very constructive comments on all or parts of the manuscript for this book. I especially thank Tau Rho Alpha, Alden P. Colvocoresses, Richard E. Dahlberg, Carlos B. Hagen, John D. Hill, Derek H. Maling, Mark Monmonier, Clifford J. Mugnier, Arthur H. Robinson, Herbert W. Stoughton, Waldo R. Tobler, Philip M. Voxland, and David Woodward. Gerald I. Evenden, Daniel Strebe, and especially Arthur H. Robinson have contributed corrections to this impression.

For those who have loaned photographs and granted permissions, as credited with the related items, I also express my thanks.

I hope that the flavor of development, description, and usage as well as résumés of the lives of many of the inventors will tempt readers to seek further detail in some of the primary and secondary sources listed.

ONE

Emergence of Map Projections: Classical through Renaissance

FOR ABOUT TWO THOUSAND YEARS, the challenge of trying to represent the round earth on a flat surface has posed mathematical, philosophical, and geographical problems that have attracted inventors of many types. Of course the use of maps predates this period. The contemporary process of mapmaking, however, had a slow beginning because exploration of the earth as a whole is a relatively recent historical development.

Although the true shape of the earth was in doubt for centuries, a number of scholars were convinced the earth was spherical well before Claudius Ptolemy wrote his monumental *Geography* about A.D. 150.[1] Even scholars who considered the earth flat believed the skies were hemispherical. Thus it was soon apparent that preparing a flat map of a surface curving in all directions led to distortion. That distortion may take many forms—shape, area, distance, direction, and interruptions or breaks between portions. In other words, a flat map cannot correctly represent the surface of a sphere.

A globe also has drawbacks in spite of its basic freedom from distortion. A globe is bulky, of small scale, and awkward to measure; less than half its surface can normally be seen at one time. As Ptolemy stated in his *Geography,*

> it is not easy to provide space large enough (on a globe) for all of the details that are to be inscribed thereon; nor can one fix one's eye at the same time on the whole sphere, but one or the other must be moved, that is, the eye or the sphere, if one wishes to see other places.
>
> In the second method wherein the earth is represented as a plane surface there is not this inconvenience. But a certain adjustment is required in representing the earth as a sphere in order that the distances noted therein may be shown on the surface of the globe congruent, as far as possible, with the real distance on the earth. (Translation in Stevenson 1932, 39)[2]

The systematic representation of all or part of the surface of a round body, especially the earth, onto a flat or plane surface is called a *map projection.* Literally an infinite number of map projections are possible, and several hundred have been published. The designer of a map projection tries to minimize or eliminate some of the distortion, at the expense of more distortion of another type, preferably in a region of or off the map where distortion is less important.

1

Historically, one of the first steps in preparing a map is to lay down the graticule or net of meridians and parallels according to the selected map projection. The system of meridians and parallels, or lines of longitude and latitude, respectively, was developed near the beginning of the history of map projections. The history of the development of longitude and latitude is a long and arduous one not described in this work, but the development of map projections as mathematical or geometrical designs continued in spite of the lack of accuracy of terrestrial measurements. The designer could treat meridians and parallels as if they represented precise coordinates, placed as straight lines or smooth curves on the map, even if the related geographical features were subject to misplacement.

In addition to longitude and latitude, terminology in this history involves frequent use of the type of projection. There have been several proposed classifications of map projections.[3] One of the most common classifications, and the one principally used in the following chapters, is based on association, at least conceptually, with a developable surface.[4] Such a surface may be laid flat without distortion. A cylinder or cone may be so developed, but not a globe. The most common projections may be conceptually and, in some cases, geometrically projected onto a cylinder, cone, or plane placed tangent or secant to the globe, hence the categories *cylindrical* and *conic* projections. Because directions, or azimuths, are shown correctly from the center of vertical projections onto a tangent plane, the latter projections are usually called *azimuthal* rather than planar. Many projections do not fall into these categories, leading to additional classes, some with related names like *pseudocylindrical* and *pseudoconic*.[5]

Claudius Ptolemy and a handful of others, within two hundred years before or after the beginning of the Christian era, had provided detailed instructions concerning some methods of map projection, but many of the types of maps dominant before the Renaissance had bases that were philosophical rather than mathematical. These included world maps or *mappaemundi,* as they are called, such as symbolic T-O maps and others on which the landmasses were neatly fitted into a circle.[6] (The T-O design is exemplified in fig. 1.1, which depicts the known continents separated by the Mediterranean Sea shaped like a T and surrounded by an O-shaped ocean.)

FIGURE 1.1. The oldest existing European printed map known, a T-O map by Isidore of Seville, in *Etymologiarum,* 1472. Reproduced from Woodward (1975, 9).

Whether the earth was spherical or flat was not a factor in these maps. They represented no projection as such. For over a thousand years the works of Ptolemy, both maps and technical exposition, were stagnant classics having little meaning as foundations from which to develop other approaches to projection. Occasionally there were other innovations in projection during the centuries prior to the Renaissance, but the subject was quiescent for a millennium under the most generous evaluation.

The transition from the Middle Ages to the Renaissance entailed a basic change in the concept of map projection. During the Renaissance (for our purposes, 1470–1669), interest in map projection began to expand. At last, mapmakers began to build on the work of their predecessors and to innovate, with their work often in turn replaced by still newer projections a few decades later. This development proceeded exponentially from the early fourteenth century to the present time.

The attraction toward scientifically sound map projection was prompted by various factors. Fundamental was the desire for knowledge itself, whether philosophical or scientific. Many of those contributing to the advancement of map construction, and specifically map projections, were renowned philosophers or ordained clergymen. The modern term *Renaissance man* originated from a multiskilled approach to learning often true of the cultural and scientific leaders of the sixteenth and seventeenth centuries. The study of map projections is now frequently termed mathematical cartography, but the development of appropriate mathematics was often contemporary with the development of map projections both during and after the Renaissance. Astronomy was also increasingly mathematical, and star maps utilized projections, so the fields of mathematics, astronomy, and map projections often attracted the same individuals.

The development of map projections was also spurred by pragmatic considerations. Although projections were used to construct star maps of the apparently dome-shaped heavens long before the Middle Ages, the greater geographical awareness during the late Middle Ages and Renaissance led to an increased demand for flat maps of the globe.

The dozen or so different projections developed during the one-and-a-half millennia prior to the Renaissance were augmented by another ten or so general types of new or improved projections developed during the next two centuries. Most were pursuits of the classic goal of reducing the apparent distortion of the map, at least for the particular region being portrayed, but other projections provided a special type of information, such as Gerardus Mercator's projection for navigation. Still others served to remind the map user of the globular shape of the world. Most of the older as well as the newer projections were used for the cartography of the Renaissance.

Delineation of the use of the projections during the Renaissance must be incomplete, but some descriptions are available to us. The modern use of mathematics to prepare maps based on these projections in their ideal form is well known, but the actual techniques used in the fifteenth and sixteenth centuries, when trigonometric and algebraic techniques were still emerging, were often quite different. The draftsman of the Renaissance could work with elementary tools with varying origins. Compasses were known in ancient Egypt (use by a carpenter is mentioned in the Old Testament), and preparing a design with ruler, compasses, and a lead or tin

stylus was mentioned by Theophilus the Monk about 1100.[7] Bronze tools from
A.D. 79 (Pompeii and Herculaneum) include fixed proportional compasses. An extant set of related instruments from about 1618 includes pairs of compasses with a
removable divider point, stylus, fluted pen, and/or crayon-holders, in addition to a
square, a sector, and other items.[8]

The mathematicians, geographers, physicists, clergymen, astronomers, mapmakers, and others who have offered their solutions to the problem of flattening the
globe have done so modestly or arrogantly. No one has found the ideal map projection, unless the requirements set are arbitrarily restrictive. The search continues, as
the ever-escalating volume of literature asserts. Literature about the history of map
projections has been much more limited. Aside from detailed treatises about early
projections, the only lengthy study is an important but unillustrated paper in French
(d'Avezac-Macaya 1863). Over three-fourths of published projections have appeared
since then.

This chapter relies heavily on a few classic nineteenth- and twentieth-century
treatises on the history of map projections, especially the paper by d'Avezac-Macaya
(1863) as well as work by Nordenskiöld (1889), and Keuning (1955). There are some
contemporary descriptions of specific projections; although Keuning states that "no
textbook is known from the XVIth or the XVIIth century, in which they [projections]
are treated," *Geographia generalis,* prepared by the short-lived Bernhard Varenius
(1622–1650), fills much of that role.[9]

In very recent times, computers have been used not only to plot the meridians
and parallels, but also to place geographic information in proper positions on a map
projection. Until this era, the ease of plotting meridians and parallels by geometric
means was of considerable importance, since other geographical data were often
plotted by interpolation between these previously drawn lines. This was especially
true in the years preceding and during the Renaissance. Furthermore, since maps
were treated as works of art portraying the philosophy and beauty of the earth as
well as providing technical information, the very shapes of meridians and parallels
were important in implying the sphericity of the planet.

For many of the projections identified below, the mapmaker seldom gave us a
name or construction technique. Instead, projections have been determined by
measurements on and observations of the map itself.

THE CLASSICAL AND MEDIEVAL LEGACY: MAP PROJECTIONS DEVELOPED BEFORE THE RENAISSANCE

Before meridians and parallels could be used as the framework for a flat map representing the globe, it was necessary to develop an artificial pattern of lines of longitude and latitude (or some comparable arrangement) by which positions on the
earth's surface could be given systematic locations. Eratosthenes (ca. 275–194 B.C.)
had devised a system similar to meridians and parallels, but Hipparchus (ca. 190–after 126 B.C.), an astronomer and mathematician of Rhodes, applied more rigor to
relate astronomical measurements to the determination of *climata* or latitudinal positions. He was the first to use a formalized system of longitude and latitude in which
the meridian and parallel circles were each divided into 360 degrees, each degree
into 60 minutes, and each minute into 60 seconds (the sexagesimal system).[10]

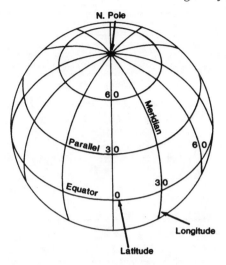

FIGURE 1.2. The system of meridians and parallels on the globe. Reproduced from Snyder (1987b, 9).

With this system, latitude circles are shown (fig. 1.2) on the globe parallel to each other and equally spaced along the surface, the largest circle being the equator (0° latitude) halfway between the poles and other circles, reducing in size to points, representing the poles at 90° north and south latitude. The meridians all make identical circles on the globe, intersecting at equal angles at the two poles and therefore spaced equally along the equator and along any other parallel of latitude. Longitude is counted in degrees up to 180 east (+) and west (−) of a prime meridian which currently passes through Greenwich, but which has been varied through history by nation and by mapmaker. Ptolemy further promulgated the system of meridians and parallels, and it became the standard for spherical coordinates used for the network or graticule on the projected flat map.

The Equirectangular Projection

Probably the projection dominant among those inherited from the pre-Renaissance was a graticule appearing to be rectangular or, in certain cases, square. In some cases it is not actually shown on these old maps, but the graticule is sufficiently implied from markings about the border or from lines drawn for the equator, the Tropic of Cancer, and so forth.

On this cylindrical-type projection, all parallels of latitude are straight, parallel lines, spaced equally for equal changes in latitude. The meridians of longitude are also straight, equidistant, and parallel to each other, and perpendicular to the lines of latitude. The scale is correct along all meridians and along the equator, or else along all meridians and a selected pair of latitude lines equidistant from the equator. The simplicity of construction coupled with valuable scale properties made the projection attractive to mapmakers. If the equator is the parallel of true scale, the graticule is square (provided the intervals of latitude equal those of longitude); common current names of this projection are plate carrée (used by d'Avezac-Macaya in 1863) and plane chart (fig. 1.3).[11]

For a different latitude of true scale, such as 30° N (and S), early chartmakers merely had to know the length of, say, 10° of longitude at this latitude, as well as the

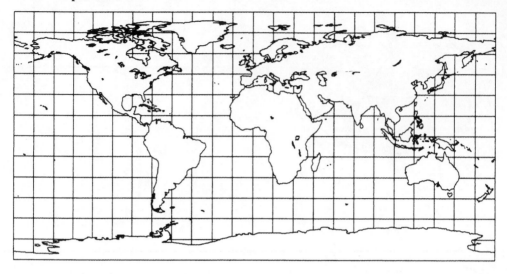

FIGURE 1.3. The plate carrée or plane chart. A modern computer plot of the simplest and one of the oldest map projections. Scale is correct along all meridians and the equator. 15° graticule.

length of any 10° of latitude, to plot the projection accurately. This projection is usually now called equirectangular (fig. 1.4), or rectangular.

This projection was said to have originated with Eratosthenes, who is most famous for his method of determining the earth's circumference, but this theory is disputed by a more careful analysis of Eratosthenes' own descriptions.[12] Ptolemy credited Marinus of Tyre with the invention about A.D. 100, and the latter's name is commonly applied by later scholars to the projection as used during the Renaissance.[13] Marinus chose the parallel of Rhodes (about 36° N) for nearly true scale on a world map by spacing meridians at ⅘ the spacing of parallels.[14]

Ptolemy, however, recommended using the projection only for maps of smaller areas. He further modified the projection to provide correct scale along the central parallel of the actual region shown. All those Greek manuscript maps for his *Geography* which date from the thirteenth century or later, as well as the first Latin manuscript of the atlas, dating from the early fifteenth century, employ Ptolemy's modification.[15] It was used for several of the *tabulae novae* (*tabulae modernae*) or maps added to the Ptolemy group, such as *Europe tabula n*^{ou}, the 1427 map of northern Europe by Claudius Clavus (b. 1388) (fig. 1.5).[16] Martin Waldseemüller (1470–1518) used the square graticule for his 1516 *Carta marina*, and Robert Thorne for his 1527 world map *Orbis vniuersalis descriptio*, while numerous maps of smaller regions appeared on square or rectangular graticules in the late sixteenth- and seventeenth-century atlases and maps of Abraham Ortelius (1527–1598), Gerardus Mercator (1512–1594), Cornelis Wytfliet (d. after 1597), and John Speed (1552–1629).[17]

The portolan charts of the era, portraying the coastlines and ports for sailors, normally have no graticule, but several of the later charts, beginning about 1520,

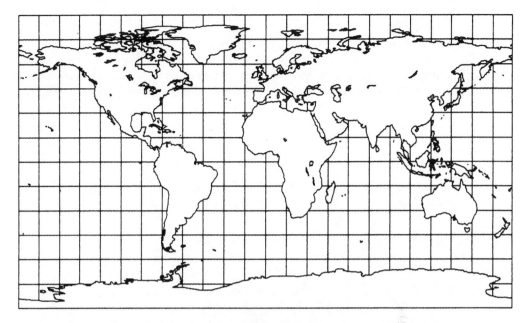

FIGURE 1.4. A modern example of the equirectangular projection, with correct scale along all meridians and, in this case, 30° N and S latitude. 15° graticule.

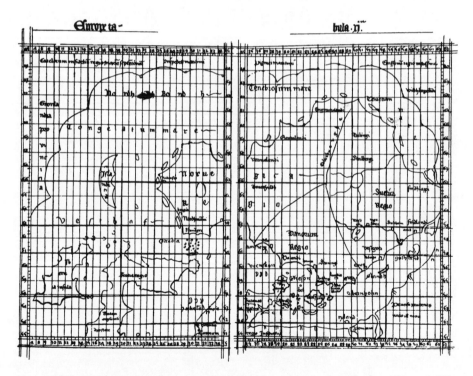

FIGURE 1.5. Northern Europe on an equirectangular projection by Claudius Clavus, 1427. Reproduced from Nordenskiöld (1889, 49).

give clues that the equirectangular projection may have been the basis, at least by then. Whenever the equator and outer meridians are graduated in longitude and latitude, respectively, as on the *Carta uniuersal* portolan chart (or planisphere)[18] of 1529 by Diogo Ribeiro, the almost universally uniform spacing indicates the pre-eminence of this projection for sixteenth-century nautical charts until Mercator's projection became more widely accepted many years after it was presented in 1569.[19]

The Trapezoidal Projection

Rivaling the equirectangular projection for simplicity of construction and prevalence of usage was a projection named the trapezoidal (*trapéziforme*) (fig. 1.6) by d'Avezac-Macaya (1863, 484) because of the shape of its graticule. Used in a rudimentary form for star maps, perhaps by Hipparchus about 150 B.C. and more

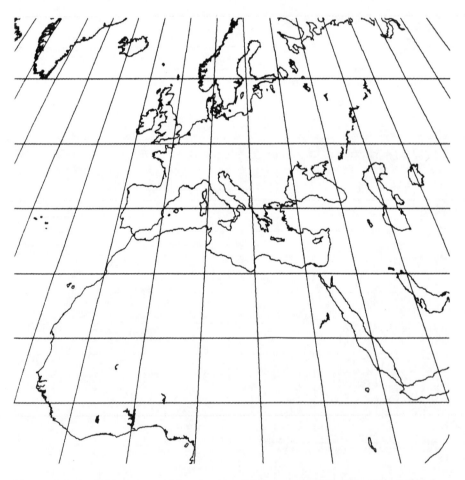

FIGURE 1.6. A modern trapezoidal projection of Europe and North Africa. True scale along 10° and 60° N latitude. 10° graticule. Central meridian 15° E longitude.

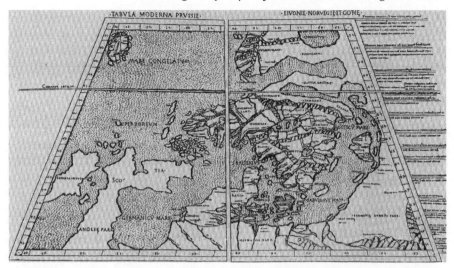

FIGURE 1.7. Northern Europe on a trapezoidal projection without graticule, as shown in a 1507 Rome edition of Ptolemy's *Geography*. Reproduced from Nordenskiöld (1889, 27).

definitely in 1426 by Conrad of Dyffenbach (see fig. 1.20), it was treated by later map historians as an invention of Donnus Nicolaus Germanus (ca. 1420–ca. 1490), who claimed it as such in 1482 and used it more precisely in several manuscript and printed editions of Ptolemy's *Geography*, beginning in 1466.[20]

Nordenskiöld (1889, 86) called it the Donis projection, after a common name applied to Donnus Nicolaus Germanus. Most regional maps in the various editions of Ptolemy by many different cartographers were drawn according to this projection, although the graticule marks are frequently shown only along the edges; the trapezoidal border provides strong confirmation that the projection was used (fig. 1.7). The projection was often the basis for regional maps in other atlases, including those of Ortelius and Mercator. In his 1579 county atlas of England, Christopher Saxton (1542/44–1610/11) regularly used the trapezoidal for county maps, and for his 1583 wall map of England and Wales he repeated the usage (Saxton 1579; Ravenhill 1981).

The projection was constructed by drawing parallels of latitude straight, equidistant, and parallel to each other. A straight central meridian was drawn perpendicularly. The two extreme parallels were marked off true to scale for meridians, starting from the central meridian, and the meridians were then drawn as straight lines connecting the corresponding points.[21] The meridians therefore converged at a common point which was normally not at the pole. The trapezoidal projection thus had advantages over that of Marinus for maps of continents and smaller regions in that two parallels in the region rather than one were correct in scale, and the meridians were shown converging. On the other hand, meridians no longer intersected parallels at right angles on the map, as they do on the globe, except at the central meridian, which was also the only meridian true to scale.

To accommodate the reversal of meridian convergence at the equator, the equator was usually also marked off correctly if it fell between the extreme parallels, and each meridian (other than central) crossing the equator was drawn as two straight segments bent at the equator. The meridians on *Africae tabulae XII*, a map of Africa in the *Geographia* (1588) of Livio Sanuto (1532–1586), although shown only at the edges, each consist of five straight segments bent to approximate changing convergence near latitudes 8° and 24° N and S.[22]

Mercator used a modification of the trapezoidal projection for many of his maps, beginning in 1578. In this form, the parallels of true scale are not the extreme ones, but they are about one-quarter and three-quarters of the distance between the limiting latitudes, to reduce the overall distortion (Keuning 1947). The trapezoidal cannot properly be called a cylindrical projection, but it fits a category often called *pseudocylindrical* (from the Greek *pseudēs*, meaning "false"), another term apparently originated by d'Avezac-Macaya (1863, 483), in which the lines of latitude are straight parallel lines, as they are on cylindrical projections, but the meridians do not remain parallel to each other. (On most pseudocylindricals, the meridians are curved.)

Ptolemy's Three Projections

Claudius Ptolemy was possibly the single most influential individual in the development of cartography in Europe and the Middle East at the dawn of the Renaissance, although he lived thirteen hundred years earlier. He is believed to have lived around A.D. 90 to 170, to have been born in Greece, and to have primarily lived and worked near Alexandria, Egypt (Toomer 1970–80). Details of his life are little known beyond his technical writings, which include an eight-book *Geography* (on mapmaking, quoted earlier) and a thirteen-book *Almagest* (on mathematics and astronomy). These two works were revived in the fifteenth century as the most authoritative existing references on their subjects.

A regular conic projection shows parallels of latitude as concentric circular arcs and meridians as straight, evenly spaced radii of the circles. The angles between the meridians are less than the differences in longitude. Ptolemy, although he made no direct or indirect reference to a cone, introduced two influential projections with concentric, circular arcs for parallels of latitude (like conics), but with meridians that are either broken straight lines or circular arcs. Thus the projections were only coniclike, not conic. They were used intact or with slight modifications for the maps of the *oikoumene* or known world in many of the fifteenth- and sixteenth-century editions of his *Geography*.

On Ptolemy's first projection (fig. 1.8), extending from the parallel of Thule (about 63° N) to the parallel as far south of the equator as Meroë was north (about 16°25′ S), all the parallels are concentric, equidistant, circular arcs. In describing the principles of this projection, Ptolemy declared:

> we shall do well to keep straight lines for our meridians, but to insert our parallels as the arcs of circles, having one and the same center, which we suppose to be the north pole, and from which we draw the straight lines of our meridians, keeping above all else similarity to a sphere in the form and appearance of our plane surface.

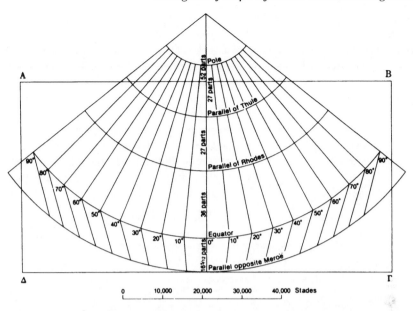

FIGURE 1.8. Reconstruction of the graticule for Ptolemy's first projection. Reproduced from Harley and Woodward (1987, 187). *Copyright University of Chicago Press.*

The meridians must not bend to the parallels, and they must be drawn from the same common pole. Since it is impossible for all of the parallels to keep the proportion that there is in a sphere, it will be quite sufficient to observe this proportion in the parallel circle running through Thule and the equinoctial, in order that the sides of our map which represent latitude may be proportionate to the true and natural sides of the earth. . . .

Now indeed we are not permitted to carry the lines which are to be drawn as meridians through in one straight course to the parallel [opposite to Meroë] but only to the equator . . . ; and with the arc [opposite to Meroë] divided in both directions into ninety parts or segments, equal in size and number to those taken on the parallel of Mero[ë] we can then draw to these marked points the intervening straight lines from those points marked in the equator the course of which will seem deflecting toward the south on the other side of the equator. (Translation in Stevenson 1932, 40, 42–43)

Ptolemy chose to maintain the spacing of meridians along the northern parallel in nearly the proper ratio (taken as 52 to 115) to the spacing along the equator, although neither parallel was drawn to the map scale.[23] Therefore, he held the radii of these two parallels at the same ratio. As a result, the meridians, drawn for only half the globe (180° of longitude), converge at a point about 25° above the North Pole. They are radii of the circular arcs only as far south as the equator. Correct scale is intended along all meridians and is approximately held along the single parallel of Rhodes (about 36° N), used also by Marinus for his projection.

Ptolemy marked the limiting southern parallel (16°25′ S) for meridians with the same spacing as that for Meroë, 16°25′ N, reflecting the actual relationship on the globe; the southern meridians were shown as straight lines from the intersections of northern meridians with the equator to the markings at 16°25′ S. This projection was used for the world map in the Greek manuscript of Ptolemy's maps, in the first printed edition (Bologna, 1477), and in several of the editions of the next thirty years.[24] Also, the regional maps of the 1477 edition were drawn to portions of the same projection, including the bend at the equator, whenever it was crossed.

Ptolemy's second projection, the one he preferred, falls into the class often called *pseudoconic* after d'Avezac-Macaya (1863, 483) introduced this term, with concentric, equidistant, circular arcs for parallels, but with curved rather than straight meridians (fig. 1.9). As Ptolemy outlined the construction,

FIGURE 1.9. Reconstruction of the graticule for Ptolemy's second projection. Reproduced from Harley and Woodward (1987, 187). *Copyright University of Chicago Press.*

We shall be able to make a much greater resemblance to the known world in our map if we see the meridian lines, that we have drawn, in that form in which meridian lines appear on a globe, when the axis of the eyes is imagined as directed upon a motionless globe through a point before the eyes in which occurs the intersection of that meridian and that parallel which divides respectively the longitude and the latitude of the known earth into two equal parts, and also through the center of the globe, so that the extreme parts which lie opposite each other appear and are perceived by the eye in like condition. . . .

That a greater likeness to a sphere is achieved by this method than by the former will be self-evident. When the sphere stands motionless before the eyes, and is not revolved (which necessarily holds true for a plane map), and the eye rests on the middle of the object, one certain meridian which, because of the globe's position, lies at the middle of the plane passing though the axis of the eye, will exhibit the appearance of a straight line, while those on either side appear inflexed with their concave side toward it, and the more so as they lie farther from it, which is also observed here with exact analogy, just as it is also seen that the symmetry of the parallel arcs keeps the proper ratio of one to another, not only in the equatorial line and the parallel of Thule (as was done in the former case), but in the others also, as closely as they can be made—the difficulty of doing this is evident—and that the conformity of latitude as a whole serve toward a true, general longitude ratio, not only in the parallel drawn through Rhodes, but in all of the parallels. . . .

Although for these reasons this method of drawing the map is the better one, yet it is less satisfactory in this respect, that it is not as simple as the other. . . . for me both here and everywhere the better and more difficult scheme is preferable to the one which is poorer and easier, yet both methods are to me retained for the sake of those who, through laziness, are drawn to that certain easier method. (Translation in Stevenson 1932, 43, 45)

The center of the arcs for the parallels is equivalent in distance to about 90° above the North Pole, along the extended straight central meridian. The northern, southern, and middle parallels of the *oikoumene* (respectively about 63° N, 16°25' S, and 23°50' N—the parallel of Syene) are marked off correctly for meridians up to 90° on each side of the central meridian. Each meridian was drawn through one of these sets of three points per meridian with a circular arc, which may be drawn through any three points.

The projection shows the areas of regions in approximately the correct ratio and was named Ptolemy's *homeotheric* (equal-area, that is, having no distortion of area) by d'Avezac-Macaya (1863, 282–83).[25] This is not quite the case, since meridians cannot be circular arcs for this type of projection to be equal-area. If Ptolemy actually knew the correct basis of equal-area construction, although it is not apparent that this is the case, he could have chosen the circular arcs for convenience, much as many other sixteenth- and seventeenth-century projections such as the orthographic, Werner's, and the sinusoidal were crudely drawn with straight segments or circular arcs in place of mathematically sound curves.

The first known use of Ptolemy's second projection was by Donnus Nicolaus Germanus in his manuscript copies of Ptolemy's *Geography* dating from about 1470, but it was the projection for the world map of several printed editions by Donnus Nicolaus Germanus and others, such as those by Henricus Glareanus (1488–1563) in 1510 (a portion of it extended to the North Pole), Bernardo Sylvano in 1511, and Sebastian Münster (1489–1552) in 1540.[26] Neither of Ptolemy's projections was well suited to maps of the full globe, but, especially after Columbus's voyages expanded the *oikoumene* of the Europeans, Ptolemy's projections inspired succeeding ones that were true conics or that exhibited more exacting properties.

Ptolemy also outlined a third, azimuthal-like, new projection while discussing an armillary sphere. His third projection, described in book 7 of his *Geography*, rather than book 1, where the others are discussed, does not appear to have been used. It was designed to show "the hemisphere of the earth in which the *oikoumene* is located," using a modified perspective view of the globe having true scale along the straight central meridian, tapered scale along the straight central parallel of Syene, and with other meridians and parallels "turned concavely toward these straight lines."[27]

Globular Projections

Many planispheres based on so-called globular projections were appealing because they gave the appearance of globelike hemispheres.[28] The first known design was developed and used for star maps by Islamic scholar, especially in mathematics and astronomy, Abū al-Rayḥān Muḥammad ibn Aḥmad al-Bīrūnī (973–after 1050) about 1000 (Keuning 1955, 20; Savage-Smith 1992, 36–37; Tibbetts 1992, 137, 141–42). In 1660, it was reinvented by G. B. Nicolosi, who is usually given credit, and under whose name it will be described. The second design, about 1265, was by Roger Bacon (1214–1294), the English scholar in philosophy and early science. His presumed design resulted in a map of a single hemisphere with a bounding circle equidistantly marked for parallels of latitude.[29] The parallels were drawn as straight lines connecting the corresponding points (fig. 1.10). The central meridian is perpendicular, and the remaining meridians are arcs of circles meeting at the poles and equally spaced along the equator. Two maps, one of each hemisphere and based on Bacon's design, were prepared by Franciscus Monachus (fl. 1524–1555) about 1527.[30]

The next known version of a globular projection was delayed for centuries, but Roger Bacon's work probably remained alive primarily because Pierre d'Ailly (1350–1420) copied some of Bacon's significant work into his own writings, which were familiar to Christopher Columbus.[31] Peter Apian (or Bienewitz) (1495–1552) printed modifications in 1524. In the first, the central meridian instead of the outer ones was marked off equally for the straight parallels. The second was the same as the first except for equally spaced semiellipses for meridians.[32] Apian, mentioned below for his projection innovations of the early sixteenth century, studied mathematics and astronomy at Leipzig and Vienna. He spent his later years at the University of Ingolstadt, south of Nuremberg, as a professor of mathematics, preparing the first tables of trigonometric sines for every minute of arc (Kish 1970–80b).

In 1554 Michele Tramezzino (d. 1579) published (in Venice) a world map in two hemispheres, using Apian's first projection.[33] One or the other of these forms

FIGURE 1.10. Modern reconstruction of Bacon's presumed design for his globular projection. Eastern hemisphere; 10° graticule. Central meridian 70° E longitude.

was used in several editions of Ptolemy by Girolamo Ruscelli (ca. 1504–1566) during the mid-sixteenth century.[34]

Zone maps, in the sense of resembling globular projections, usually consist of hemispheres bounded by circles, but are so-called because meridians are not shown, and the horizontal straight parallels are usually limited to five: the equator, the Arctic and Southern (Antarctic) circles, and the Tropics of Cancer and Capricorn.[35] This is the case with a 1493 zone map by Zacharius Lilius (fl. ca. 1493).[36] Another example, by d'Ailly, includes eight additional parallels of latitude north of the equator.[37] On maps printed in the late fifteenth century, John Eastwood (Eschuid) (d. 1380) and Ambrosius Theodosius Macrobius (fl. A.D. 399–422) included only four of the five classical parallels, omitting the equator.[38] The spacing of these four parallels, while symmetrical about the equator, appears arbitrary on several of the maps, varying considerably from correct spacing along either the circular boundary or a central meridian if it were included.

Nearly a century later, the basic zone-map design was still being used in a 1571 map by Benito Arias Montano (1527–1598) (*Sacrae geographiae tabulam ex antiquissimorum cultor*), which barely retains the classification because of its added features: only the five classical parallels are shown, but they are correctly spaced along the central meridian, and the equator is marked for every 10° meridian, although only the central meridian is drawn (Schwartz and Ehrenberg 1980, 72).

The Earliest Azimuthal Projections

None of the projections discussed so far is still used today, except occasionally the equirectangular. There are four azimuthal (from the Arabic *al* and *sumūt*, the "paths") projections, however, which not only are believed to predate those above, but also are increasingly used in the twentieth century. The terms *azimuthal* and *zenithal* as applied to this class of projections appear to have originated in the latter third of the nineteenth century.[39] Use of these projections in the fifteenth and sixteenth centuries was largely confined to star maps (see below) until Mercator launched the use of the equatorial stereographic so prominently that for the next two centuries it was the standard for eastern and western hemispherical maps.

All directions or azimuths from the center of an azimuthal (or zenithal) projection are correct. As a consequence, the great-circle path, which is the shortest route on the globe, from the center of the projection to any other point is shown as a straight line on the map. An azimuthal projection is projected, geometrically or conceptually, onto a tangent (or occasionally secant) plane from a point along a straight line passing through the point of tangency and the center of the earth. If the point of tangency is at either pole, the projection is said to be in its polar or normal aspect, with concentric circles for latitude lines and straight equally spaced radii as meridians. If the plane is tangent somewhere on the equator, the equatorial (or meridian) aspect results; if tangency occurs between a pole and the equator, an oblique (or horizon) aspect is obtained.[40]

Apparently known in at least one aspect to Egyptian and Greek astronomers and other scholars, as delineated below, were three azimuthal projections that are perspective (that is, in which the sphere may be projected geometrically from some fixed point onto a plane) and one that is not. Each of the four could be constructed in the polar aspect using elementary geometric construction: to draw the necessary concentric circles and straight radii, the only unique information needed was the spacing of the circular parallels, and that was easily described in contemporary terms. The three perspective projections could also be constructed in the equatorial and even oblique aspects by applying geometric relationships without trigonometry or algebra; they were more complicated than the polar aspects, but the geometric concepts required had been known for well over a thousand years, beginning at least with Euclid (ca. 300 B.C.). The actual method of construction is rarely available as a written contemporary description, but some examples are given below.

The Orthographic Projection

The orthographic projection is projected from an infinite distance onto a tangent plane and thus has the classic "globelike" appearance, especially in the oblique aspect (fig. 1.11). Only half the globe is shown at a time, and great distortion occurs

FIGURE 1.11. The oblique orthographic projection. Center 30° N, 70° E; 10° graticule.

near the edges. All meridians and parallels appear as ellipses, circles, or straight lines. Hipparchus used the equatorial aspect in the second century B.C. to determine the places of starrise and starset, while the Roman architect and engineer Marcus Vitruvius Pollio, ca. 14 B.C., used it to construct sundials and to compute sun positions.[41]

The name *analemma*, then applied to a sundial showing latitude and longitude, was generally used until François d'Aiguillon of Antwerp promoted its present name in 1613.[42] Vitruvius, however, apparently originated use of the term *orthographic* (from the Greek *orthos*, or "straight," and *graphē*, or "drawing") for the projection.[43]

It appears in woodcut drawings of terrestrial globes of 1509 (anonymous), 1533 and 1551 (Johannes Schöner), and 1524 and 1551 (Apian), although the shape of meridians is rather crude and more like the circular arcs of a globular projection.[44] On the other hand, in 1515 a carefully drawn oblique orthographic projection of the

Old World was prepared by Johann Stabius (or Stab) (after 1460–1522), a professor of mathematics in Vienna, from a map designed by Albrecht Dürer (1471–1528) of Nuremberg.[45]

Dürer is most famous as an artist, but he closely associated the new art of the Italian Renaissance with the exactness of mathematics (Steck 1970–80). As a result, in his day he was equally prominent as a mathematician, especially for his studies of perspective and geometric proportion, whether of the human body, edifices, or the globular world. Thus he was interested in the orthographic projection and polyhedral globes (mentioned below). Schöner (1477–1547), a printer and globemaker of Nuremberg, published his own studies of mathematics and astronomy (Rosen 1970–80). The oblique orthographic also appears as an inset for the world map *Universi orbis seu terreniglo* of 1578 by Gerard de Jode (1509–1591).

Possibly some of these globelike views were actually scenographic projections, to use an obsolete term also proposed by Vitruvius, that is, perspective projections from some finite point away from the earth or globe;[46] some equatorially centered insets by Nikolaus Visscher in 1660 rather clearly fit this description. This is generally unlikely, however, given both the apparent hemispherical coverage (the scenographic would be less than hemispheric) and the fact that a scenographic or general perspective view is considerably more difficult to construct geometrically than the orthographic projection. Generally the orthographic was little used for geographic maps.

The Gnomonic Projection

The gnomonic or central projection, another early perspective azimuthal, is projected from the center of the earth onto a tangent plane and was originally called horologium or horoscope (clock or hour watcher) because of its relationship to sundials, which have gnomons. Thales of Miletus (ca. 624–547 B.C.) used the projection in the oblique form.[47] The angles between the hour markings on a sundial designed for a particular latitude are identical with the angles between the meridians on a gnomonic projection centered at the same latitude, counting each 15° of longitude from the central meridian as one more hour from noon.

To read time, a gnomonic map centered north of the equator is first rotated so that the North Pole is south of the center of projection rather than north. Then the triangular gnomon of the sundial is placed with its base along the central meridian and its shadow-casting back parallel to the earth's polar axis and touching the North Pole on the map. The local sun time is indicated by the meridian along which the shadow of the sun then falls.

The origin of the use of the more obvious projection name *central* is uncertain. The term *gnomic* has been used since at least 1749 and *gnomonic* since at least 1836 (Emerson 1749; DeMorgan 1836). The chief value of the gnomonic projection lies in the fact that it shows all great-circle arcs, not just those passing through the center as on other azimuthal projections, as straight lines. This includes all meridians and the equator. All other parallels of latitude are shown as ellipses, parabolas, hyperbolas, or (on the polar aspect) circles.

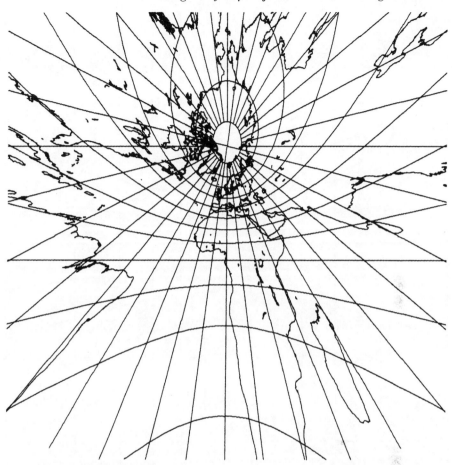

FIGURE 1.12. Simplified reconstruction of an oblique gnomonic projection with the range and center used by Ritter (1610). Center 45° N, 10° E; 10° graticule. All great-circle paths are straight lines, including all meridians and the equator. All other parallels are conic sections.

Use of the projection for maps during the Renaissance was rare, appearing to postdate 1600. One early geographic example is oblique, centered near Nuremberg, with a 5° graticule, and was published in 1610 (fig. 1.12).[48] Johannes Kepler (1571–1630) employed the equatorial aspect (fig. 1.13) for a star map of 1606, and various aspects were used for constellations in a 1612 atlas by Christoph Grienberger and for some star charts in 1619 by his colleague Orazio Grassi.[49] Kepler's scientific brilliance is well known to most technical students of the solar system because he developed the three fundamental Kepler's Laws of planetary motion, published in the early seventeenth century after Kepler's study of the records of the planets' positions by Tycho Brahe (1546–1601) (Gingerich 1970–80).

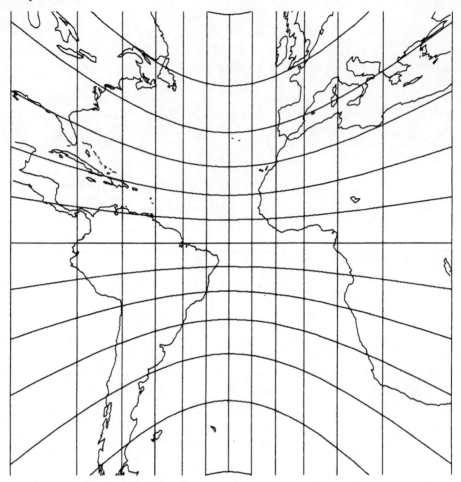

FIGURE 1.13. The equatorial gnomonic projection. Center 30° W longitude; 10° graticule.

The Stereographic Projection

The third major perspective azimuthal is the stereographic (from the Greek *stereos*, or "solid"), so named by d'Aiguillon in 1613 in place of the earlier name *plani-sphere*.[50] It is projected onto a plane from the point on the globe exactly opposite the point of tangency (or of the center of the circle along which the plane is secant, if this is the type of contact). Therefore, not all the world can be shown from one projection center.

The projection is conformal, so that all small (actually infinitesimal) elements are shown without shape or angular distortion, although larger regions are distorted in shape. This feature was apparently first realized by Edmond Halley (1656–1742), famous for a comet, mathematics, and thematic maps. He published a proof of con-formality (1695–96), including the statement "In the stereographic projection, the angles, under which the circles intersect each other, are in all cases equal to the spherical angles they represent; which is perhaps as valuable a property of this projection as that of all the circles of the sphere on it appearing circles [see below]; but this, not being commonly known, must not be assumed without a demonstration," which Halley then provided.[51]

On any aspect of the stereographic, all meridians and parallels and in fact all great circles and small circles are shown as circles, except for any meridian or great circle passing through the center and the parallel opposite in sign to the central parallel. These are shown as straight lines, which are circles of infinite radius. Thus the commonly shown ecliptic, a great circle which is the apparent path of the sun through the stars, is plotted as a circular arc (or, if it passes through the center of the projection, a straight line).

On the polar aspect, the meridians and parallels are the straight lines (intersecting at the central pole) and the concentric circles, respectively, that appear on all polar azimuthals, but the spacing of the circles gradually increases away from the pole (fig. 1.14). The radii of these circles are proportional to the trigonometric tangent of half the colatitude (90° minus the latitude, or the angular distance of the latitude from the pole). In 1650, without displaying a diagram, Varenius described the geometric construction of the polar aspect, named only as the "first easy Mode, the Eye being placed in the Axis," as follows, utilizing the perspective nature of the stereographic projection:

> In any Plain or paper let the middle *point* P, be taken for the Pole, and from that as from a *Center*, let the great or small *Periphery* [circle] be drawn (as we desire to have our Maps great or small) which we shall have for the *Æquator*. These two may be taken at pleasure, but the other *points* and *Peripheries* shall be found from them. Let the *Æquator* be divided into 360 *deg.* and *streight lines* being drawn through the *Center* and the beginning of every *deg.* these shall be the Meridians, from which that which is drawn at the beginning of the first *degree* from these 360, shall be taken for the first, so the rest of the *lines* shall shew the rest of the Meridians and *Longitudes* of the Earth from the first Meridian. Now the Parallels of *Latitude* must be described. There are four *Quadrants*, or quarters of the *Æquator*, the first 0, 90: the second 90, 108 [*sic*; should be 180]: the third 180, 270: and the fourth 270, 0. Let those be noted for the more easy appellation with the letters AB, BC, CD, DA, and let one be taken from these, for *Example*, BC, from every one of whose *degrees* as also from the 20 [*sic*; should be 23] *deg.* 30 *min.* and the 66 *deg.* 30 *min*, let occult *streight lines* be drawn to the *point* D, (the term of the *Diameter* BD) or let the *Rule* be only applyed to D, and brought round through every degree of the *Quadrant* BC: and let the 23 *deg.* 30 *min.* and the 66 *deg.* 30 *min.* in which these *streight lines* cut the *Semidiamiter* PC, be noted, and from P as from a *Center*, and the *Peripheries* be described through every *point* taken in PC. These *Peripheries* shall be the Parallels of the *Latitudes* unto which in the first, and opposite Meridian, *viz.* AP, and CP, the numbers may be ascribed from the *Æquator* towards P, *to wit*, 1, 2, 3, 4, even to 90. (1650; 1693 translation, 319, 320)

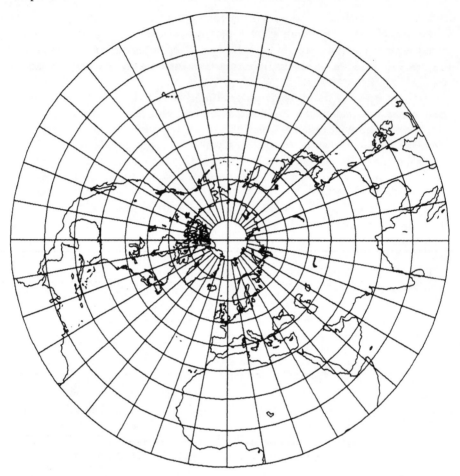

FIGURE 1.14. The polar stereographic projection. Northern hemisphere. 10° graticule. The first conformal projection, preserving local shapes.

Known to Hipparchus, Ptolemy, and probably earlier to the Egyptians, the polar stereographic was used for numerous star maps, including those first printed, *Imagines coeli septentrionales cum duodecim imaginibus zodiaci* (fig. 1.15) and *Imagines coeli meridionales,* by Dürer in 1515.[52] His graticule is limited to ecliptic longitude lines radiating from the central ecliptic poles, but star positions indicate the stereographic. Johannes Honter (1498–1549) in 1532 (*Imagines constellationum borealium; Imagines constellationum australium*) included some small circles of the celestial graticule.[53]

In 1507 Gualterius Lud (1448–1547) of St. Dié prepared the earliest existing world map based on the polar stereographic, presumably (not seen by this writer) as two hemispheres.[54] Gregor Reisch (ca. 1470–1525) from 1512 and Apian from 1524 were among others to use the polar aspect.[55] In 1596 John Blagrave (ca. 1550–1611),

FIGURE 1.15. Albrecht Dürer's use of the polar stereographic projection for his 1515 star map of the northern ecliptic hemisphere. Star positions confirm the projection, even though parallels are missing. *National Gallery of Art, Washington, D.C., Rosenwald Collection.*

a mathematician, surveyor, and instrument maker of Reading, England, on his map *Nova orbis terrarum descriptio optice proiecta,* used the north polar stereographic to the equator but appended the southern hemisphere in four sections according to an uncertain projection neither conformal nor azimuthal, which placed the world map in a square.[56]

The oblique stereographic (fig. 1.16) was used astronomically by Theon of Alexandria (fl. A.D. 364) in the fourth century.[57] It was recommended for geographical maps as an original projection by Stabius, and promoted as the fourth of four "new" projections by Johannes Werner (1468–1522) of Nuremberg in his 1514 revision

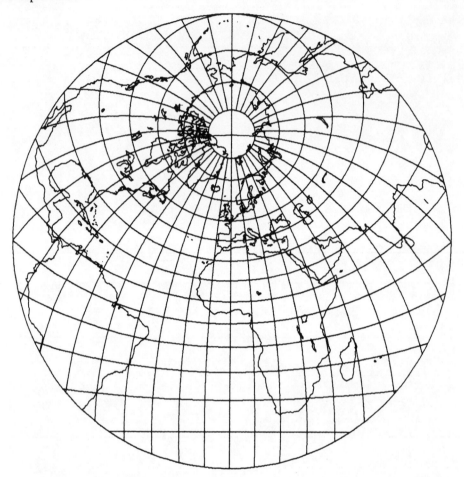

FIGURE 1.16. The oblique stereographic projection, a hemisphere centered at 40° N, 0° E. 10° graticule. All meridians and parallels are circular arcs or straight lines, intersecting at right angles.

and translation into Latin of part of Ptolemy's *Geography*.[58] Werner included a 10° graticule of this projection centered on 50° N, near the latitude of Nuremberg (fig. 1.17), and recommended it for men of distinction as an ideal representation of the globe.[59] Werner's three other projections were heart-shaped and actually original; they are discussed later.

The construction of the oblique aspect was the most complex of those described by Varenius in 1650 (fig. 1.18):

> But to describe such a *Map*, let us take *London* to possess the *Center* of the Map: we take his *Latitude*, or the *Elevation* of the *Pole*, to be the 51½ *degree*, the Eye is placed in the *point* opposite to the *Vertex*, or in the *Nadir* of the place: the *Table*, or *Glass* is the Plain of the *Horizon*, or another Parallel

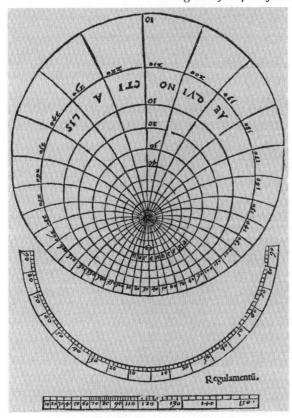

FIGURE 1.17. A partial 10°
graticule by Werner (1514) using the oblique stereographic
projection centered at 50° N
latitude. Reproduced from
Nordenskiöld (1889, 92).

to it; if you please to represent a larger portion than the *Hemisphere*, which is
more commodious in this Method, *to wit*, that the Plain at least may pass
through the depressed *Pole*.

Therefore in the Plain, let the *Center* E, be taken for *London*, and the
described *Periphery* ABCD, which sheweth the *Horizon*, must be divided
into four *quarters*, and every one of these into 90 *degrres* [*sic*]: let the *Diameter* BD, be the Meridian *line*: B the *North Pole* [*sic*]: D the *South Diameter.*
And the *line* of the rising and setting *Æquinoctial*, sheweth the *primary vertical*. A, the *Occident*, C, the *Oriental Cardo*, or sheweth the place which is
distant 90 *degrees* in the *primary vertical point*. All the *vertical points* are
represented in *streight lines*, drawn through the *Center* E, to every *degree* of
the *Horizon*. But to shun confusion it is better to omit them, and to adjoyn a
Circumductile Rule to the *Pixil* affixed in E.

Then let BD be divided into 180 *degrees*, as in the former Mode, by drawing Right Lines from A, to every *degree* of the *Semiperiphery* BCD, That
point in EB, which sheweth the 52 *deg.* of the *Arch* BC, shall be the projecture of the *Arctick Pole*: Let the *point* in ED, be noted with the letter P, which
representeth the 52 *deg.* of the *Arch* DC, (by accounting from C, to D) shall

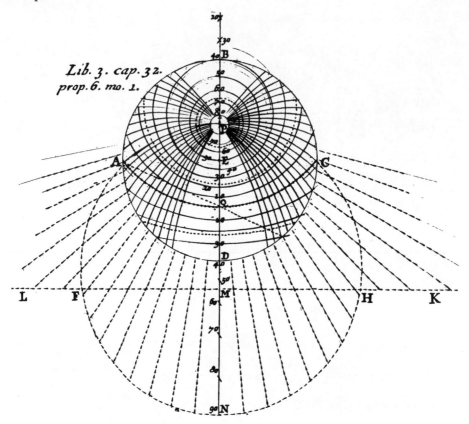

FIGURE 1.18. Diagram for construction of the oblique stereographic projection, added by Blome to his translation of Varenius. Reproduced from Varenius (1650; 1693 trans., facing 290). *Library of Congress.*

be the projecture of the intersection of the *Æquator,* and the Meridian of *London.* Let the letter Q, be noted, and from that towards the letter P, let the numbers of the *degrees,* 1, 2, 3, &c. be ascribed. Also from Q, towards D, and from B, towards P, *viz.* 52, 53, 54, 55, &c.

Then the *points* being taken from P, of the equal *degrees, viz.* 99 and 99, also 88, and 88; let these be described about these parts as the *Diameters* of the *Peripheries* of the *Circles,* which shall represent the Parallels or *Circles* of *Latitude,* and the *Tropicks,* and *Polary Circles* with the *Æquator.*

To describe the Meridians, first, let a *Periphery* be described through the *points* APC: that shall shew the Meridian, which is 90 *degrees* absent from *London.* His *Center* shall be M, in BD, (protracted into the *point* N, which sheweth the *Antarctick Pole*). Let PN, the *Diameter,* be drawn through M, Parallel to AC, which is FH; protracted from both parts in K,L. Moreover let the *Circle* PHNF, be divided into 360 *deg.* and Right *lines* from the *point* P, to every *deg.* (or only by application of the Rule) which shall cut the *line*

KFHL. The *Circles* must be described through every *point* of the Section, and both the *Poles* P,N, as through three given *points* which shall represent all the Meridians: the *Centers* of the *Arches* to be described are seated in the same KL, *viz.* those which are found by the former Section, but to be taken with this condition that the most remote *Center* at L, be chosen for the nearest Meridian from BDN, towards A, and for the second, the second from this. (1650; 1693 translation, 330–31)

In spite of Stabius and Werner, the oblique stereographic was seldom the basis for maps during the sixteenth and seventeenth centuries. It was used by Jacques de Vaulx (ca. 1555–1597) of France in a manuscript atlas of 1583 for a pair of hemispheres centered on Paris and its opposite point, by John Blagrave in 1596 for an elaborate star map, and by cartographers such as John Speed (e.g., in 1627) a few decades later for maps of Europe and Asia.[60]

The equatorial aspect of the stereographic was employed by the Arab astronomer az-Zarqalī (Arzachel) (d. 1100) of Toledo in the eleventh century on an astrolabe design.[61] It apparently did not appear as a world map until the 1583 atlas by de Vaulx, although claimed differently by Keuning.[62]

It was *Orbis terrae compendiosa descriptio,* the world map of 1587 in two hemispheres by Rumold Mercator (1546/48–1614, son of the famous Gerardus) (one of them reconstructed in fig. 1.19), inserted with revisions into the Mercator *Atlas* of 1595, that launched the equatorial aspect of the stereographic into a two-century tenure as the standard projection for eastern and western hemispheres.[63] Later users often chose the same central meridians, but there were variations. About 1595, Jodocus Hondius (1563–1612) shifted the central meridians 90°, centralizing the Atlantic and Pacific oceans in each hemisphere, to show circumnavigational routes by Francis Drake (1539/41–1596) and Thomas Cavendish (1560–1592).[64] Philipp Eckebrecht (1594–1667) prepared a map in 1630 for Kepler's astronomical work with one hemisphere split down the central meridian, half placed to either side of the intact hemisphere centered on Europe and Africa.[65] The New World, South America, Africa, and the East Indies also appeared separately on the equatorial stereographic in seventeenth-century atlases by Mercator, Speed, the Blaeu family, John Ogilby (1600–1676), and others.

To construct the equatorial aspect involved the equivalent of straightedge, compasses, and dividers. Using Varenius's 1650 description, for this "second Mode, the Eye being placed in the plain of the Æquator," a circle was drawn for the outer meridians of the hemisphere, and perpendicular diameters provided the equator and the central meridian, with the poles at each end of the latter (1650; 1693 translation, 322–23). Each of the two northern quadrants of the outer meridians was divided into 90 equally spaced degrees, and straight lines were drawn from these points to the South Pole. The intersections of these lines with the equator provided, with the two poles, three points through which a single circular arc could be drawn for each meridian, using Euclidean methods, as Varenius pointed out. Points marked along the central meridian at the same spacing as those along the equator provided third points for parallels of latitude as circular arcs through the equidistant points on the outer meridians. Varenius added, however, that "the operation will be less ob-

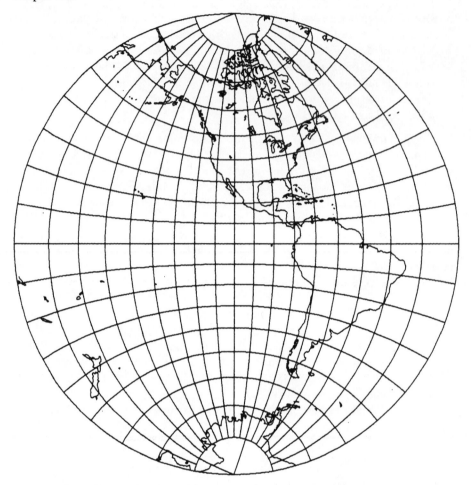

FIGURE 1.19. Modern outline reconstruction of the equatorial stereographic projection as used in Mercator atlases beginning in 1595 for the western hemisphere. 10° graticule. Central meridian 110° W longitude.

noxious to error" if a "*Canon* [or Table] of [trigonometric] *Tangents*" were used to mark these intersections in proportion to the tangents of half the angles from the center, "especially in great Maps" (322–23).

An elaborate adaptation of the equatorial stereographic projection occurred in the 1670s when John Adams (fl. 1670–1690) prepared a large map of England at a stated scale of one degree of latitude per foot (about 1:4,400,000). After making measurements and calculations, Ravenhill (1978) concludes that the stereographic mentioned in Adams's Latin note to the reader as a starting point is used to determine a radius of the globe from the spacing of the limiting parallels of 50° N and 56° N. The parallel of 53° is then plotted as it would be on the equatorial stereographic projection, but other parallels and the meridians are placed more nearly as they would be on a Bonne projection (described later).

The Azimuthal Equidistant Projection

The fourth ancient azimuthal projection, one that is not a geometric perspective, is most commonly called the azimuthal equidistant. On it all distances and directions from the center of the projection are shown correctly. The equatorial and oblique aspects, much more complicated to construct, apparently were not used with graticules until the nineteenth century, but they have been considered by some to be the de facto projections for some early charts of small portions of the coasts.[66] This usage, however, requires that all bearings and distances to coastal points be measured from a single point; on extended coastal charts, such as portolans, they were measured from coastal point to coastal point.

The easily constructed polar aspect is said to have been used for star maps by Egyptians in some holy books, but it made its earliest extant appearance on an incomplete and rudimentary star map of 1426 by Conrad of Dyffenbach (fig. 1.20). The aspect had appeared several times by the mid-sixteenth century; for example, it was used by Dürer about 1515.[67] The polar form consists simply of equally spaced latitude circles centered about the pole, with radiating equally spaced, straight meridians. The projection was used for two terrestrial hemispheres by Glareanus about 1510, and by several others between 1511 and 1524, including Giovanni Vespucci, who in 1523 showed the southern hemisphere split into two semicircular halves placed tangent to the northern hemisphere.[68] Mercator used the projection as an inset for north polar areas on his famous navigational map of 1569.[69]

The projection is named for Guillaume Postel (1510–1581) in France and Russia (the Postel projection), but he did not use it until 1581, on a map of the northern hemisphere with insets of the southern.[70] A 1582 map by Michael Lok (fl. 1582–1615) is based on a portion of the polar projection bounded by two meridians 160° apart.[71] The azimuthal equidistant became a nearly exclusive standard projection for polar terrestrial maps for centuries to come, although it was only half as popular for polar celestial maps as the stereographic.[72]

Apparently the first to prepare a north polar azimuthal equidistant map of the entire world was Urbano Monte of Milan with a 1603 map of sixty-four sheets. A similar world map was prepared by Louis de Mayerne Turquet in 1648; this inspired others and led to the Cassinis' construction of a world map on the floor of the Paris Observatory in the 1680s.[73]

NEW PROJECTIONS OF THE RENAISSANCE

New Conic Projections

The opportunities for exploration and the increasing geographical knowledge that resulted from Columbus's voyages to what was shortly to be called America led to extensive revisions of map projections suitable for displaying the known world. Perhaps the first to adapt Ptolemy's projections to post-Columbian mapping was Giovanni Matteo Contarini (d. 1507). On his only known map, a world map of 1506, Contarini modified Ptolemy's first coniclike projection (fig. 1.8) in three ways: he doubled the span of meridians from 180° to the full 360°; he extended latitudes to the North Pole (shown as a circular arc); and he continued the meridians unbroken to about 35° S without a bend at the equator.[74] He thus presented a rudimentary

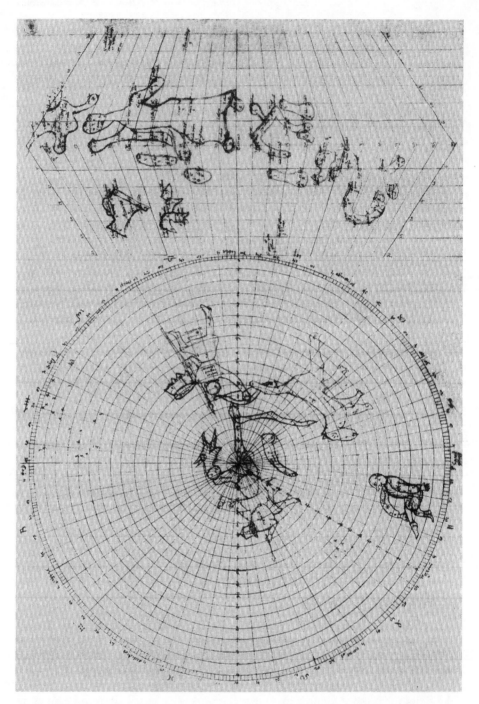

FIGURE 1.20. Polar azimuthal equidistant projection as used by Conrad of Dyffenbach in 1426 for a partial star map of the northern ecliptic hemisphere, with trapezoidal projection above. Reproduced from Uhden (1937, facing 8), original in the Vatican Library. *Library of Congress.*

simple or equidistant conic projection having the same scale relationships north of the equator as those of Ptolemy.

In 1507–8, Johannes Ruysch (d. 1533) generally followed the Contarini modifications, but placed the North Pole at the center of the circular arcs of latitude and equally spaced the parallels from this pole to the map limits at 38° S (reconstructed in fig. 1.21).[75] The equator is nearly but not quite at the same scale as the meridians.

The scale distortion along the southern latitude lines is substantial in both the Contarini and Ruysch versions, but this is also true of modern polar azimuthal equidistant world maps with the projection center at the North Pole. Wilhelm Schickard (1592–1635), a German astronomer and mathematician of Tübingen, Germany, who designed and built a working model of the first modern mechanical calculator and published various astronomical ephemerides, was apparently the first of a half-dozen seventeenth- and eighteenth-century mapmakers to use the conic projection for star maps.[76]

This projection has the advantage of providing less overall shape distortion in the regions being shown than would be the case for a polar azimuthal equidistant projection; the disadvantage is the discontinuity at the cut along the element of the cone. Centered at each pole, but both extending to 35° N declination, not the equator, Schickard's star maps of 1623 consist of cones cut along the meridian of a solstice

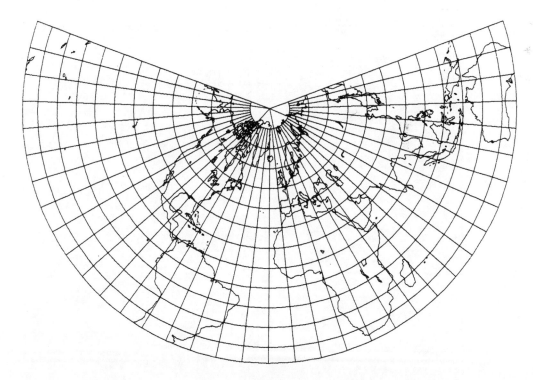

FIGURE 1.21. Modern outline reconstruction of equidistant conic projection as used by Ruysch, adapting Ptolemy's first projection in 1507–8 for his full world map (north of 38° S). Effective standard parallels 90° N, 3°40′ S. 10° graticule.

(but without a graticule), the pole at the center and apex, and a circumference of about 0.84 of a complete circle. A similar type of conic projection was used for maps of smaller regions by Rumold Mercator, Henry Briggs (1561–1630), Willem Janszoon Blaeu (1571–1638), and others in the late sixteenth and early seventeenth century.[77]

Briggs, who used a conic projection for a well-known map of North America showing California as an island, was a professor of mathematics in London and later at Oxford. His most notable contribution was the modification of John Napier's invention of logarithms to produce "common" or "Briggian" logarithms to the base 10.[78] He computed a table of thirty thousand logarithms (1 to 20,000 and 90,000 to 100,000). The table was completed by Adrian Vlacq and published in 1628, together with Briggs's tables of logarithms of trigonometric sines and tangents.

Some maps, appearing to follow the pattern of conic projections, do not comply on closer inspection. For example, *Carta da navegar de Nicolo et Antonio Zeni Furono in Tramontana Lano.M.CCC.LXXX.* (fig. 1.22) is a 1558 map of northern regions accompanying the Zenos' annals as published by descendant Niccoló Zeno (1515–1565). It is based on a coniclike projection in which parallels are concentric circular arcs and meridians are straight, but the meridians do not converge at the center of the circular arcs. Therefore, the meridians do not intersect the parallels at

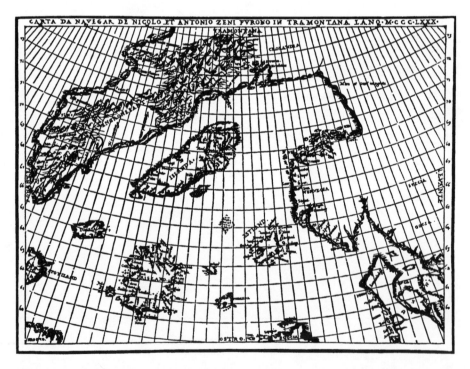

FIGURE 1.22. The Zenos' use of a modified equidistant conic projection for a 1558 map of the North Sea. Reproduced from Nordenskiöld (1889, 53). Meridians are not true radii of the concentric circular arcs of parallels.

right angles, and the graticule is a sort of conic version of the trapezoidal projection. Maps of Europe and some smaller regions by Mercator, Hondius, and Speed also show these modifications, which were apparently a rudimentary prototype of the equidistant conic with two standard parallels.[79] Equidistant parallels were used on almost all maps based on conic projections until the nineteenth century, and, as the simple or equidistant conic projection with meridians drawn as true radii of the latitude circles, conic projections are still the basis for many regional maps.

Ptolemy's second projection (fig. 1.9), with the curved meridians, was adapted to mapping of the whole world in two parallel channels. One channel consisted of moderate alterations, with a strong resemblance to Ptolemy's prototypes, such as the projection for Henricus Martellus Germanus's pre-Columbian world map of ca. 1489.[80] As on Ptolemy's projection, parallels of latitude are concentric equidistant circular arcs centered about 90° above the North Pole along the extended central meridian.

The parallels extend, however, from the North Pole to 40° S, instead of from about 63° N to 16° S as on Ptolemy's projection. The meridians are almost certainly (from measurement) circular arcs, each in three sections (instead of Ptolemy's one), with bends at the equator and the Arctic Circle. South of the Arctic Circle the meridian arcs are concave with respect to the central meridian; north, they are convex.

The breaks at the equator of the Henricus Martellus Germanus map are relatively gentle; on Waldseemüller's post-Columbian world map of 1507 (the one that includes the name *America* for the first time), the bend at the equator is much more abrupt for corresponding meridians.[81] Keuning surmises that this abruptness would not have occurred if the outer woodblocks used for printing the map were larger. With minor alterations, Apian copied Waldseemüller's projection and map at reduced scale in 1520, with a version by 1551 showing four parallels and no meridians except for the outer limits, which are smoothed curves no longer having the earlier bends; in 1561 Johannes Honter produced a similar world map with smooth curves but with a complete 10° graticule extending to the South Pole.[82]

Small hemispherical insets by Waldseemüller on his 1507 map were constructed on a similar projection, but with single arcs for the meridians from the Arctic Circle to 40° S, the limits of these insets. These insets were the bases for hemispheres drawn by Johannes ze Stobnicza (d. 1530) in his *Introductio in Ptholomei cosmographiam* (Cracow, 1512).[83] During the same period, in addition to his 1511 pre-Columbian world map on Ptolemy's second projection, Sylvano included the New World with a second world map (reconstructed in outline form in fig. 1.23), also in his 1511 edition of Ptolemy's *Geography*. On this map Sylvano extended the meridians of Ptolemy's projection to 320° of the globe, spacing meridians essentially equally along parallels and thus using noncircular meridians more closely resembling the Stabius-Werner approach below, while markedly decreasing the radii of the latitude circles to place their center about 10° above the North Pole.[84]

The second channel for modification of Ptolemy's second projection involved greater changes. About 1500, Stabius invented a series of three attractive heart-shaped (cordiform) projections, in addition to believing he had invented a fourth, the oblique stereographic, discussed above.[85] They were further publicized by Werner, who was an ordained priest with a pastorate. His pastoral duties were limited,

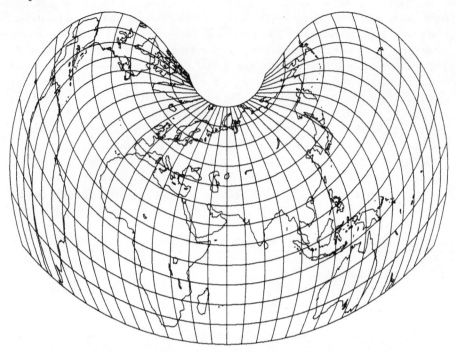

FIGURE 1.23. Modern outline reconstruction of the "Bonne" projection as used by Sylvano, adapting Ptolemy's second projection in 1511 for his world map (limited to 320° of longitude), centered at about 60° E longitude, and with latitudes between about 80° N and 40° S. Standard parallel about 47°; 10° graticule, but Sylvano used various *climata* instead of degrees of latitude.

however, and he spent much of his time studying astronomy, mathematics, and geography (Folkerts 1970–80). He related spherical trigonometry to the solution of various celestial and terrestrial problems, and designed several instruments to compute astronomical positions. His important treatises, prepared between 1505 and 1522, dealt especially with spherical triangles and conic sections.

In advancing Stabius's projections, Werner in 1514 prepared two 10° graticules (figs. 1.24, 1.25) at Nuremberg, including them with his edition of Ptolemy's *Geography*.[86] Ostensibly the graticules were for Stabius's first two projections, but the second graticule is constructed like the third projection rather than the second. All of the first three projections are equal-area (although this feature was apparently unknown to the inventors) and differ from Ptolemy's second projection in that (1) the full globe may be portrayed completely on the second, and almost completely on the third, (2) the North Pole is made the center of the concentric equidistant arcs of circles representing parallels, (3) each parallel is marked off equidistantly for meridians in direct proportion to the distances along the equator, and (4) the equator is marked off equidistantly, but not necessarily in proper ratio to distances along the central meridian, which is constructed at the correct scale.

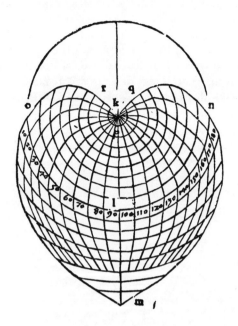

FIGURE 1.24. 10° graticule by Werner (1514), presenting one hemisphere of his first projection. Reproduced from Nordenskiöld (1889, 88).

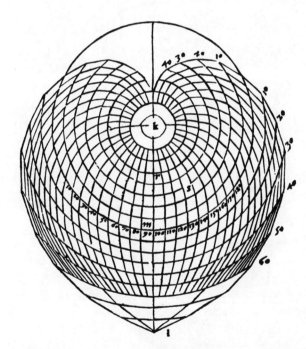

FIGURE 1.25. 10° graticule by Werner (1514), presenting his third projection (full world). Reproduced from Nordenskiöld (1889, 88).

On Stabius's and Werner's first projection, a semicircle representing the equator is divided into 180° of longitude. Thus the ratio of the length of a degree of longitude on the equator to a degree of latitude on the central meridians is π/2; while this projection was arbitrarily limited to a hemisphere, it could encompass 720°/π or 229°11′ of longitude without overlapping.

On the second projection (fig. 1.26), the circle representing the equator (having the same radius as the equator in the other two projections) is marked off for meridians at the true scale of the central meridian. Thus the ratio is 1.0 instead of π/2, and the projection can be used for displaying the entire globe.

The third projection was designed so that the chord rather than the arc connecting two points 90° apart in longitude on the equatorial circle is the same length

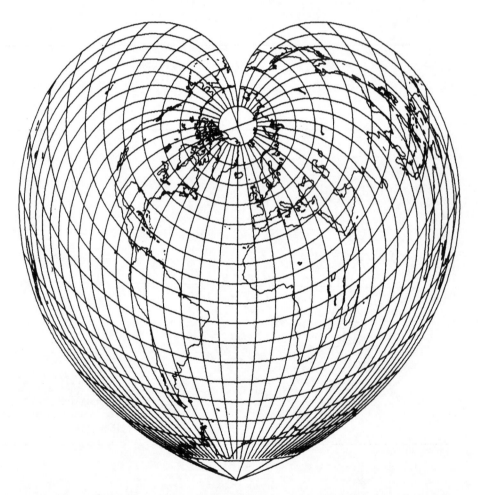

FIGURE 1.26. Modern outline reconstruction of a 1530 world map by Apian (n. 88) based on Werner's second projection, now called the Werner projection. 10° graticule. Central meridian 10° W longitude.

as 90° in latitude along the central meridian. This results in a ratio of π/3 instead of π/2 or 1.0, but is hardly distinguishable from the second, except that the 60th parallel north forms a complete circle, while north of it the 180th meridians east and west of the central meridian slightly overlap by up to 16°.

The first Stabius-Werner projection was apparently not used, while the third was apparently initially employed by Oronce Finé (1494–1556) for a world map of 1536, followed by Giovanni Paolo (*Cosmographia universalis ab Orontio olim descripta*) in 1556 and by Hadji Ahmed (fl. 1534–1560) for a map (*Perfect and Complete Engraving and Description of the Entire World,* in Turkish) of 1560.[87] The second, however, became known as the Werner (or Stab-Werner) projection and first appeared on a world map by Apian, *Tabula orbis cogniti universalior* (Ingolstadt, 1530).[88] Asia was mapped on this Werner projection in the 1570 Ortelius atlas *Theatrum,* and both Asia and Africa so appear in Mercator atlases from 1595 to at least 1609.

The (second) Werner projection was modified in 1531 to produce a double cordiform world map by Finé.[89] Separate northern and southern hemispheres were drawn according to the Werner projection and centered about the respective poles; the two circular arcs representing the equator were made tangent at the central meridian. After other editions, Finé's double cordiform projection was used by Mercator for his renowned world map of 1538, the first to name North America (reconstructed in fig. 1.27).[90]

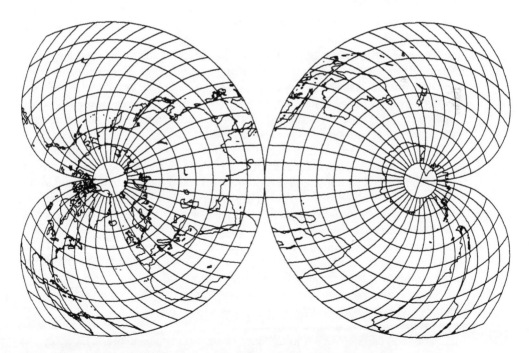

FIGURE 1.27. Modern outline reconstruction of the double cordiform world map by Mercator in 1538, using his central meridian of about 70° E longitude. 10° graticule.

There was occasional later use, such as a map of the northern hemisphere by Christian Sgrooten (1532–1608) in 1592, but the cordiform projections all but disappeared by the eighteenth century in favor of Bonne's projection. This more general adaptation of Werner's projection provides less angular distortion for a map of a continent, but also closely resembles the progressive styles of Sylvano, Apian, and Honter.[91]

The Norman navigator Guillaume Le Testu (ca. 1509–1572) used a flatter Ptolemy-like version of the double cordiform in a world map of 1566.[92] The parallels were still concentric equidistant circular arcs, but they were centered about 440° beyond the North and South Poles along the central meridian, extending this meridian as a straight line with the same spacing in degrees as those of latitude. The meridians 90° from the central meridian form a circle broken only by the interruption at the equator, which is equally marked off for 180° of longitude in each direction. Remaining meridians are smooth connecting arcs, not necessarily circular.

Earlier, in 1556, Le Testu had shown cartographic dexterity with six different existing or new projections in a manuscript sea atlas.[93] The types include Finé's octant (projection of an eighth of a globe), Apian's globular, an Apian-like oval, another modified Werner, a rudimentary attempt at an oblique perspective, and a four-lobed star. The last projection, on which petal lobes, bounded by circular arcs of meridians 90° apart, were joined at the North Pole, was among the first of a general style repeatedly used up to the present in different forms.

Oval Projections

The oval type of projection resembles a worldwide version of some of the globular hemispheres. It was produced with several variations. Common to nearly all oval projections are equidistant horizontal lines for parallels and curved meridians equidistant at the equator. While medieval *mappaemundi* were bounded in the early twelfth century by curves that could be called rudimentary ovals, they should not be called oval map projections because they lack either graticules of meridians and parallels or geographic arrangements that would imply agreement with the above definition even if a graticule were superimposed.

One of the earliest existing, if not the earliest, oval maps to show a graticule is a detailed map by Francesco Rosselli (b. 1447/48) from the first decade of the sixteenth century.[94] The equator and central meridian are both evenly divided in this 10° graticule, and the ratio of their lengths is very close to 2:1. Rosselli's meridians are near ellipses or ovals connected at the poles and cutting each parallel approximately equidistantly. Previously it has been stated that the 2:1 ratio first appeared on a world map (fig. 1.28) in a *Libro* of 1528 by Benedetto Bordone (1460–1539), who has usually been considered the inventor of the oval type, and whose meridians are smoother arcs than those of Rosselli's (Keuning 1955, 21).

Typus orbis terrarum, the world map of Abraham Ortelius in his 1570 *Theatrum,* is probably the most artistic example of an oval projection, but the projection is nearly identical with the one Battista Agnese (fl. 1535–1564) used on several maps of about 1540 (reconstructed in fig. 1.29).[95] Its ratio is also 2:1, but its outer oval consists of two semicircular meridians with flat poles half the length of the equator. Meridians are all circular arcs equidistant along the equator; if greater than 90° of longitude from the center, they are semicircles of equal radii; if less than 90°, the

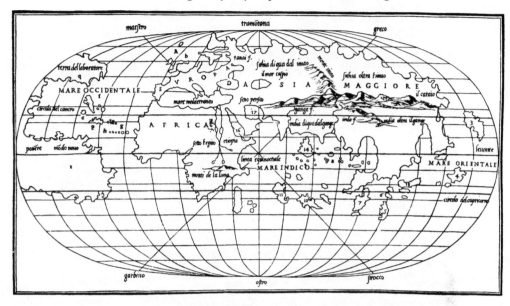

FIGURE 1.28. Bordone's 1528 oval projection of the world. Reproduced from Nordenskiöld (1889, pl. 39[2]).

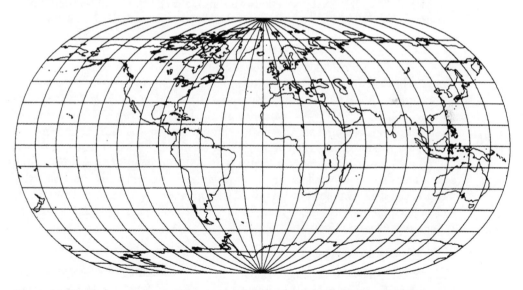

FIGURE 1.29. Modern outline reconstruction of the oval projection used by Agnese about 1540 and Ortelius in 1570 for world maps. 15° graticule. Central meridian 15° W longitude.

arcs are adjusted to pass through the two ends of the central meridian. The inner hemisphere thus has the same design as that of Apian's globular projection of 1524 described earlier (see n. 32). Ortelius's map of the Americas was based on a portion of the same projection.[96]

Several other world maps of the period use the 2:1 ratio, but on others the ra-

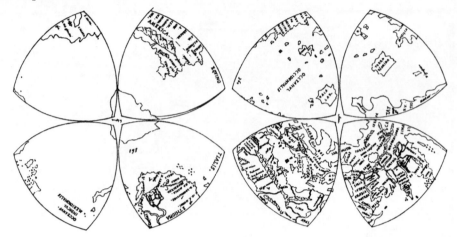

FIGURE 1.30. Leonardo da Vinci's octant world map of about 1514. Reproduced from Nordenskiöld (1889, 77).

tio varies to as low as 1.33:1. The latter was used for a map of 1544 attributed to Sebastian Cabot (ca. 1475–1557), the son of Venetian explorer John Cabot (ca. 1450–1498/99).[97] Oval projections were rarely used after 1600, although some modern pseudocylindrical projections (especially Max Eckert's third projection, described in chapter 4) are reminiscent of them.

Octant maps were little more than novelties of the era. The surface of the globe may be conveniently divided into eight equilateral spherical triangles, each section bounded by the equator and two meridians 90° apart. One of the first uses of spherical octants was made by the prolific Florentine artist and scientific genius Leonardo da Vinci (1452–1519). On his *mappamundi* of about 1514 (fig. 1.30), the sides of the octant maps are circular arcs each centered on the opposite vertex, and the octants are clustered in groups of four about each pole.[98] Intermediate meridians and parallels are not shown. In a graticule in *Sphaera mundi* (Paris, 1551), Finé added meridians and parallels (fig. 1.31) as circular arcs within each octant, symmetrical about a straight central meridian and apparently equidistant at their termini and centers; the projection was used for maps of 1556 by Le Testu and of 1616 by Daniel Angelocrater (1569–1635).[99]

Globelike Projections

The globular projections of Roger Bacon and Apian were further modified in 1643 by Georges Fournier, a Jesuit priest of Caen, France.[100] For his first projection, Fournier marked the outer and central meridians of the circular hemisphere at equal intervals and connected them with circular arcs for parallels of latitude. For the meridians, he drew ellipses passing through the poles and equidistant marks on the equator. The meridians are the same on his second projection, but parallels are straight lines connecting the same points on the outer meridians.

The projection most often identified as globular, and so named by Aaron Arrowsmith (1750–1833) of England in 1794 as if it were originated by Philippe de La

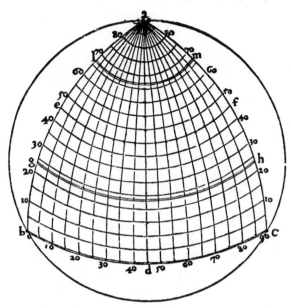

FIGURE 1.31. 5° graticule by Finé in 1551, projecting one octant of the globe. Reproduced from Nordenskiöld (1889, 94).

Hire (1640–1718), was modified from Fournier's first projection by Sicilian-born Giovanni Battista Nicolosi (1610–1670), who was only reinventing al-Bīrūnī's projection of about 1000. In 1660 Nicolosi, a chaplain in Rome, changed the elliptical meridians to circular arcs, equidistant only at the equator (fig. 1.32), and used the projection for a set of maps of the eastern and western hemispheres and five conti-

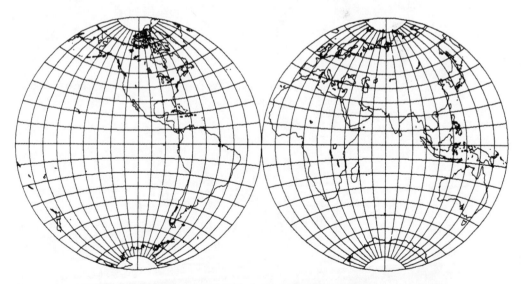

FIGURE 1.32. Modern western and eastern hemispheres, using the "Nicolosi" globular projection. All meridians and parallels are circular arcs or straight lines. 10° graticule. Central meridians 110° W and 70° E, respectively.

nents.[101] This all-circular projection, neither equal-area nor conformal, became standard for eastern and western hemispheres during the nineteenth century after the equatorial stereographic popularized by Mercator finally fell into disuse.

Gore maps, consisting of strips bounded by two meridians usually 10° to 30° apart and either bounded by the equator or extending pole to pole, were primarily drawn as sections for globes, beginning in the early sixteenth century. A number of gore maps were used in atlases and thus became flat map projections in their own right. Waldseemüller prepared a world map consisting of twelve 30°-wide gores, with a 10° graticule, in 1507 (reconstructed in fig. 1.33), the year his modified conic world map was printed (Schwartz and Ehrenberg 1980, 26).

Glareanus described one type of construction of a set of globe gores (fig. 1.34) in *Poetae laureati de geographia liber unus*, first published in 1527 (Nordenskiöld

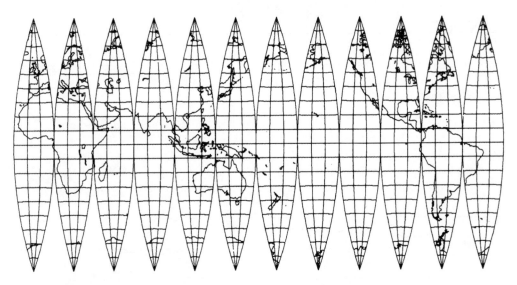

FIGURE 1.33. Modern outline reconstruction of Waldseemüller's 1507 world map of twelve gores, each 30° of longitude wide. 10° graticule. Interrupted Nicolosi globular projection is used as a close approximation.

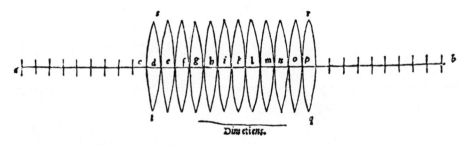

FIGURE 1.34. Diagram by Glareanus in 1527 to illustrate his method of preparing gore maps. Reproduced from Nordenskiöld (1889, 74).

1889, 74–75). According to his description, a straight line two and a half times the length of the equator of the proposed globe is divided into thirty equal parts. The middle twelve parts constitute the equator itself. Circular arcs are struck through each end of these twelve parts, with the other point of the *circinus* or compasses ten parts in distance along the original line, in a direction such that the arcs intersect to form twelve gores 30° wide in longitude. This construction makes the meridians 4–6% longer than their true length relative to the equator, but formation into a globe would alter this disparity.

How Glareanus drew meridians and parallels within the gores is not clear. From the appearance of contemporary gore maps, other meridians, if shown, are circular arcs equally spaced along the equator, except for a straight central meridian (if shown) on each gore. The central meridian of each gore (whether shown or not) is evenly divided for latitudes at the same spacing used to set off meridians of the same degree interval along the equator. The parallels of latitude are apparently nonconcentric, circular arcs with radii that appear to be the same as the length of the element of a cone tangent to the globe at the particular latitude. If so, this is a prototype of the polyconic projection of the nineteenth century. The latter, however, does not use circular arcs for meridians, and the concept of tangent cones does not seem to appear in the literature of the sixteenth century.

Some other gore maps are joined at the poles instead of the equator, such as sets of gores 10° of longitude wide in northern and southern hemispheres on a 1555 map by Antonio Floriano and a 1542 map by Alonso de Santa Cruz (1505–1567).[102] One map with much larger gores, ca. 1550, has four 60° gores plus one of 120° with horizontal straight parallels.[103] At first glance, the map resembles an interrupted sinusoidal, but the meridians are circular arcs, not sinusoids.

From time to time, cartographers have also been intrigued by the concept of projecting the globe onto a polyhedron as a compromise between a flat sheet and the round globe. Dürer (1538) seems to have been the first when he proposed projection onto a regular tetrahedron (four triangles), dodecahedron (twelve pentagons), icosahedron (twenty triangles), and other solids. He prepared diagrams (fig. 1.35) but apparently did not construct maps based on them (see Fisher and Miller 1944, 92–97).

Mercator's Projection for Navigators

Unquestionably the most famous projection is the one simply named for the inventor Gerardus Mercator (fig. 1.36). Mercator was born Gerhard Kremer in Rupelmonde, Flanders (then in the Netherlands), in 1512, but he latinized the name (*kramer* is Dutch for "peddler" and *mercator* is Latin for "merchant") when he entered the University of Louvain (also then in the Netherlands) in 1530.[104] There he soon developed interests in philosophy, theology, mathematics, geography, astronomy, engraving, and calligraphy. He prepared a globe in 1536 and his first published map (Palestine) in 1537. In 1538, he was the first to name North America as such, on his double cordiform map mentioned earlier (see n. 90).

In 1544, Mercator was one of dozens of Louvain inhabitants arrested and imprisoned for heresy. Five were executed, but after seven months Mercator, who was strongly supported by Louvain authorities, was cleared and released. In 1552 he

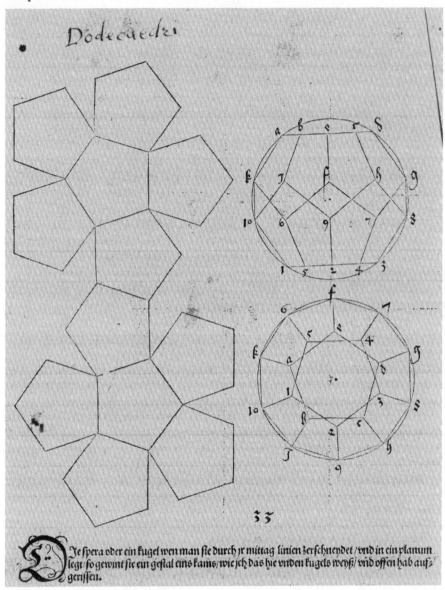

FIGURE 1.35. Dürer's (1538) diagram showing one flattened and two assembled dodecahedra as possible substitutes for a globe. *Library of Congress.*

moved to Duisburg, Germany, where he presented his cylindrical projection in 1569 and prepared numerous maps and terrestrial and celestial globes. Many of these maps were published a few at a time during his lifetime and then as a bound collection, the *Atlas sive cosmographicae meditationes de fabrica mundi et fabricati figura,* by his son Rumold the year after Mercator's death in Duisburg from a stroke

FIGURE 1.36. Gerardus Mercator (1512–1594), prominent Flemish mapmaker and inventor of the most famous map projection, devised as an aid to navigation. *U.S. Geological Survey.*

and cerebral hemorrhage in 1594. This was the first time a book of maps was titled with the name of the mythological Greek Titan Atlas, who was condemned to carry the heavens, earth, and pillars separating them on his back and shoulders.

The famous 1569 projection appeared as a world map of eighteen sheets mounted on twenty-one sections totaling about 1.3 meters by 2 meters in size. Like the equirectangular projection, it is a regular cylindrical projection, with equidistant, straight meridians, and with parallels of latitude that are straight, parallel, and perpendicular to the meridians (fig. 1.37). Unlike those of earlier cylindrical projections, the parallels are spaced more widely as the poles are approached. In fact, the spacing is directly proportional to the increasing scale along the parallels, or as the trigonometric secant of the latitude, and the poles cannot be shown at all because they are infinitely far from the equator on the map. As a result of this design, the Mercator is a conformal map projection, the second one developed (the stereographic was first).

Mercator's chief purpose in developing the projection was navigational. All lines of constant bearing (loxodromes or rhumb lines) are shown straight. The projection thus became valuable to sailors, who could follow a single compass setting (adjusted for magnetic declination, or the variation of true from magnetic north) based on the bearing or azimuth of the straight line connecting the point of departure and destination on the map. Mercator entitled his map *Nova et aucta orbis terrae descriptio*

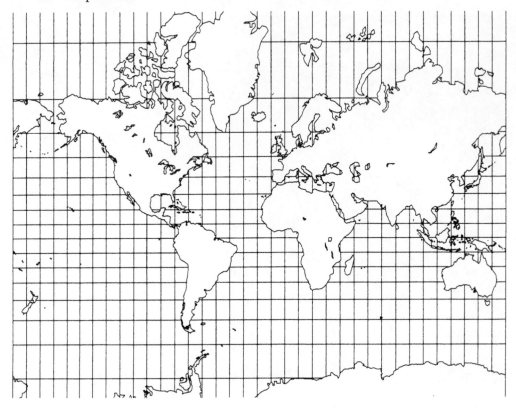

FIGURE 1.37. Modern outline reconstruction of the 1569 map on which Mercator introduced his navigational projection. Center, north-south range and 10° graticule reflect Mercator's choice.

ad usum navigantium emendate accommodata (A new and enlarged description of the earth with corrections for use in navigation). Placing his description of the nature of the projection over much of his portrayal of North America, he tells us:

> In this mapping of the world we have had three ends in view: first, so to spread out the surface of the globe into a plane that the places shall everywhere be properly located, not only with respect to their true direction and distance, one from another, but also in accordance with their due longitude and latitude; and further, that the shape of the lands, as they appear on the globe, shall be preserved as far as possible. For this there was needed a new arrangement and placing of meridians, so that they shall become parallels, for the maps hitherto produced by geographers are, on account of the curving and the bending of the meridians, unsuitable for navigation, and in the high latitudes the contour and position of localities, on account of the oblique cutting of meridians and parallels, are so strangely distorted that they cannot be recognized, and there can be no proper judging of distances. In the maritime

charts for navigators, the degrees of longitude, measured along the parallels increase toward the pole beyond the proportion they have on the globe, for they always remain equal to the degrees at the equator, while the degrees of latitude do not increase at all, so that here, too, it is inevitable, that the shape of the lands is enormously distorted, and that not only longitudes and latitudes, but also directions and distances are far from correct. Great mistakes result for this reason; but the greatest is this, that if a mapping of three places on one side of the equator is made in the form of a triangle, and any intervening place is properly located, with correct direction and distance in respect to the corners of the triangle, it is impossible that the corner points agree in the same way. Taking all this into consideration, we have somewhat increased the degrees of latitude toward each pole, in proportion to the increase of the parallels beyond the ratio they really have to the equator. In this way, we have reached this result, that in whatever way one maps out two, three, or more places, provided that of these four things, difference in longitude, difference in latitude, distance, and direction, he keeps any two in any one place with reference to any other, everything will be correct in comparison of one place with another; and no error will be found to have been made anywhere, such as must inevitably be made in the ordinary sailing charts, in many ways, particularly in the higher latitudes. (Translation in Fite and Freeman 1926, 76–78)[105]

Mercator probably determined the spacing graphically, but how he spaced his parallels is subject to conjecture. Nordenskiöld (1889, 96) says that "Mercator seems to have calculated the length of the intervals between every tenth degree of the parallel" by using the reciprocal of the cosine of the midlatitude of the interval. This can be done graphically by constructing a right triangle with one angle equal to the midlatitude of the particular interval and the adjacent side equal in length to the distance between two meridians 10° apart. The hypotenuse of this triangle is then the approximated length of the distance between the two parallels 5° above and below the midlatitude. Nordenskiöld (1889, 96) found discrepancies of about 2% in comparing values calculated in this manner with those measured from a reproduction of the map available to him, but Keuning (1955, 18) found a better fit using a different reproduction and concluded that Nordenskiöld's hypothesis was correct.[106] McKinney (1969, 472), on the other hand, suspects that Mercator's method "may have been an empirical one of transferring rhumb lines from a globe to a chart."

Edward Wright (1561–1615) later studied the projection, apparently independently, with such care that a number of writers, including Halley, have credited Wright rather than Mercator with the invention.[107] Many of Wright's earlier years were spent as a student and lecturer at Cambridge University, and interrupted by occasional trips to sea. He was considered one of the most eminent mathematicians of the time. Then about 1596 he moved to London, where he became a navigational adviser to the East India Company. He also translated John Napier's pioneer work on logarithms from Latin to English.[108] In his *Certaine Errors in Navigation*, Wright mentions Mercator derisively: "by occasion of that Mappe of Mercator, I first thought of correcting so many, and grosse errors, and absurdities, as I haue alreadie touched. . . . But the way how this should be done, I learned neither of Mercator, nor of any man else" (Wright 1610, 12 of unpaginated preface). His own development of the mathematics for the projection with straight rhumb lines is described as follows:

> at euery point of latitude in this planisphaere, a part of the meridian keepeth the same proportion to the like part of the parallel, that the like partes of the meridian, and parallel haue each to other in the globe, without any explicable error. . . . we haue an easie way laid open for the making of a table (by helpe of the Canon of Triangles) whereby the meridians of the Mariners Chart may most easily and truly be diuided into parts, in due proportion from the aequinoctial toward either pole.
>
> For (supposing each distance of each point of latitude, or of each parallel from other, to conteine so many parts as the Secans of the latitude of each point or parallel conteineth) by perpetuall addition of the Secantes answerable to be latitudes of each point or parallel vnto the summe compounded of all the former secantes, beginning with the secans of the first parallels latitude, and thereto adding the secans ofthe [sic] second parallels latitude, and to the summe of both these adioyning the secans of the third parallels latitude, and so forth in all the rest, we may make a table which shall truly shew the sections and points of latitude in the meridians of the nautical planisphere; by which sections, the parallels are to bee drawne.
>
> As in the table following, we make the distance of each parallel from other, to be one minute. (Wright 1610, 15, 17–18)

His subsequent "A Table of Latitudes of euery minute of a meridian of the Sea Chart, in such parts whereof each minute of the aequinoctial containeth 10,000," cumulatively from the equator and with first differences, thereby gave the spacing from the equator with far greater accuracy than did Mercator (Wright 1610, 20, 21–112).

Erhardt Etzlaub (ca. 1460–1532) of Nuremberg used a similar projection for a small map limited to Europe and North Africa on the cover of some sundials constructed in 1511 and 1513 (fig. 1.38).[109] It extended in 1° intervals of latitude from the equator to 67° N with increasing spacing, but the principle remained obscure until Mercator's independent development.

While usage of the projection began slowly, Peter Plancius (1552–1622) apparently used it for sea charts ("caerten met wassende breedten") for which the Dutch government granted him a patent in 1594. One world map was prepared about 1597 (*Typus totus orbis terrarum*) and others by 1599 by Jodocus Hondius.[110] In 1599 another world map, *A true hydrographical description of so much of the world as hath been hetherto discouered,* was published by Richard Hakluyt, but the cartographer was probably Wright.[111] The first sea atlas of the world was also the first atlas on the Mercator projection, by the Englishman Robert Dudley in 1646–47.[112]

Not until later did the Mercator projection as a world map become all too commonly used for geographic purposes; in addition, it has always been an important navigational tool. When using the projection for geographic purposes, unintended by Mercator, one should not overlook the importance of the fact that Europe is portrayed about twice its true area in proportion to countries near the equator.

FIGURE 1.38. Europe and North Africa. Etzlaub's 1511–13 map for the cover of a sundial, using the spreading latitudes as found in the later Mercator projection. South is at the top. Reproduced from Bagrow (1964, 150).

The equations for the precise modern-day Mercator projection when based on the sphere are

$$x = R \lambda; \text{ and}$$
$$y = R \ln \tan (\pi/4 + \phi/2),$$

where ϕ is the latitude (+ if north, − if south), λ is the longitude (+ if easterly, − if westerly) from the chosen central meridian, which is also the longitude of the y axis for rectangular coordinates, while the x axis lies along the equator. Coordinate x is positive easterly and y is positive northerly. The radius of the earth taken as a sphere is R measured at the scale of the map. In these and subsequent formulas, angles are measured in radians (degrees multiplied by $\pi/180°$). A radian is a more "natural" unit for angular measurement, because it is the angle ($180°/\pi$ or about $57°18'$) enclosing an arc of a circle equal in length to the radius of the circle. The above formula for y is equal to R times the integral $\int_0^\phi \sec \phi \, d\phi$, closely achieved by Wright with "perpetuall addition of the Secantes" for each "minute of a meridian," although he provided no equations as such. The term ln refers to the natural logarithm, or logarithm to the base e (2.71828 . . .).[113]

The Sinusoidal Projection

The last of the major new projections of the Renaissance was the sinusoidal, to use another of the terms first applied by d'Avezac-Macaya (1863, 481). It combines some important qualities with simple construction: Areas are shown correctly, and the scale along the central meridian and along every parallel of latitude is correct. Parallels of latitude are shown straight, parallel, and equidistant. After a straight central meridian is constructed perpendicular to the parallels, the other meridians are marked off along each parallel at their true distances. The meridian curves connecting these points therefore appear as sinusoids, or as sine (or cosine) curves (fig. 1.39). These are mathematically defined curves, but the base Latin word is *sinus,* a bend or curve. The projection is in effect an equatorially centered modification of the pole-centered Werner projection, and it is also pseudocylindrical.

The origin of this projection has been variously reported, with a resulting variety of names based on presumed inventors. Apparently the first to show it was Je[h]an Cossin of Dieppe, who used it for a world map dated 1570, with carefully drawn sinusoidal meridians.[114] Some maps of the period look sinusoidal, but the meridians on closer inspection are circular arcs, not sinusoids.[115] The projection was, however, used by Hondius for maps of South America and Africa in some of his editions of Mercator's atlases of 1606 to 1609. This presumably led to one of the names, the Mercator equal-area projection (Deetz and Adams 1934, 161).

Nicolas Sanson d'Abbeville (1600–1667) of France used the projection beginning about 1650 for maps of several continents, while John Flamsteed (1646–1719), the first astronomer royal of England, used it, as he broadly stated in a letter to Isaac Newton, "for drawing the charts of the constellations, which I do after a new manner (too long here to be described) so as the appearance to the naked eye is less distorted than by any projection I have yet seen."[116] Thus developed the common name Sanson-Flamsteed, while still others refer to the projection as the Mercator-Sanson (Reignier 1957, 242).

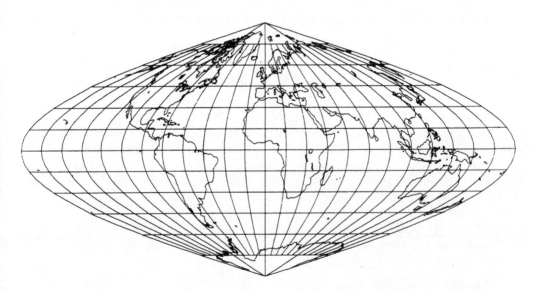

FIGURE 1.39a. A sinusoidal projection of the world. All parallels and the central meridian are at correct scale; there is also no distortion of the area scale. Central meridian 0°; 15° graticule.

FIGURE 1.39b. South America centered on the sinusoidal projection at longitude 60° W. A common choice of projection for this continent and Africa from the seventeenth to the twentieth centuries. 10° graticule.

The early sinusoidal meridians often appeared as a set of short straight lines with rounded bends, but this is undoubtedly due to the lack of drafting instruments for lines of continuously changing curvature.

CONCLUSIONS

The first map projections had little supporting information. The subject was basically mathematical, but although geometry and the beginnings of algebra and trigonometry had been developed by the time of Ptolemy, these were only minimally applied to map projections until after the Renaissance. Projections serve as tools to help display geographical data, but the placement of lands over the face of the earth was little known except as a philosophical concept amplified by very limited exploration until the escalating maritime investigation of the Renaissance.

Rudimentary in application or not, a few map projections found their way into extant pre-Renaissance literature. These projections are often known to us only through verbal descriptions and secondhand references, but they served as important, simply constructed beginnings. The equirectangular projection of Marinus, with its straight meridians and parallels, was the simplest and, with Ptolemy's two projections using circular arcs, appeared during the second century A.D. Four azimuthal projections, at least in polar form, were developed in Greece or Egypt a few centuries earlier.

After the surge of interest by Hipparchus, Marinus, Ptolemy, and other early Greeks and Egyptians, the development of techniques for map projection waned. Centuries of stagnation resulted from religious intolerance, wars, and other disorders, leading to the destruction of much early work and the stifling of innovation during the intellectual drought of the Dark and Middle Ages.

Over a thousand years after Ptolemy, only the trapezoidal and crude globular projections were added to the work of Ptolemy and his predecessors to complete the short list of projections available at the beginning of the Renaissance (fig. 1.40). In 1470 the information contained on many maps was still so inaccurate that use of rudimentary map projections was often of little consequence. During the next two centuries, however, notable improvement in both arenas occurred almost hand in hand. The Mercator, sinusoidal, and Werner projections developed during this period preserve important characteristics of conformality, area, or particular scales, although some of these properties were probably not even known, just as the earlier stereographic projection was not known to be conformal.

Many early maps, especially those dating from the Renaissance, visually indicate the use of a given projection or its prototype. Only occasionally are the reasons for its use and the means of construction available, aside from inference. Ease of construction remained significant, as it has since, but this attraction gradually decreased

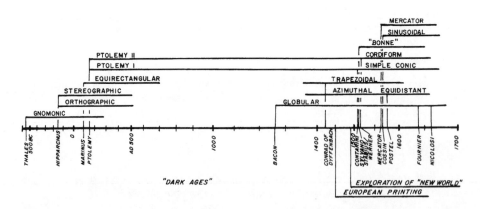

FIGURE 1.40. Logarithmic time line, 600 B.C. to A.D. 1700, showing origin of important projections, inventors or promoters, and some related events. Vertical positions of projections are not significant. Extensions of horizontal lines to left or right of vertical lines imply indefinite extent.

in importance as other characteristics became increasingly desirable with the years. The trapezoidal and equirectangular graticules of meridians and parallels constructed entirely with straight lines gave way only slowly to those employing a mixture of circular arcs and straight lines, especially the coniclike projections of Ptolemy and of those who improved on his designs as well as azimuthal, oval, and globular types.

On Mercator's new projection for navigation, the property of straight rhumb lines was central to its use rather than the rectilinear graticule; the mathematical spacing between the parallels was a complicating factor, but Wright advanced beyond Mercator's initial approach at resolution. More complex curves entered with the sinusoidal and cordiform projections, including Werner's, but these were offset by a reduction in some of the distortion. By the end of the Renaissance, some sixteen different projections, depending on the method of counting, could be listed (table 1.1).

Modern reasons for using a projection could be quite different from those given before or during the Renaissance by artisans who knew no calculus and sometimes little of the applicable contemporary mathematics. It remained for the dramatic development of the calculus in the late seventeenth century to answer the need for more mathematically based map projections. This need was created by greater precision in marine and topographic surveys, and by more precise measurements of the earth's shape, but it could build on a small but firm Greek, Egyptian, and central European foundation constructed during a millennium and a half.

TABLE 1.1. Map Projections Used during the Renaissance

Modern Names, (Contemporary Names), and Earliest Dates	Figures	Likely Advantages for Contemporary Users
Equirectangular (Marinus) ca. A.D. 100	1.3, 1.4, 1.5	Very simple straight-line graticule. Scale correct along one parallel and all meridians (also along parallel of opposite sign).
Trapezoidal, Donis ca. 150 B.C. (?); A.D. 1426	1.6, 1.7	Simple straight-line graticule. Scale correct along two parallels in region (but only along one meridian).
Ptolemy's first ca. A.D. 150	1.8	Straight converging meridians; equidistant concentric circular arcs for parallels. Proposed by established authority, Claudius Ptolemy.
Ptolemy's second, and moderate alterations ca. A.D. 150	1.9, 1.23	Scale correct along three parallels. Proposed by established authority, Claudius Ptolemy.

(continued)

TABLE 1.1. *Continued*

Modern Names, (Contemporary Names), and Earliest Dates	Figures	Likely Advantages for Contemporary Users
Globular, various types ca. 1000; ca. 1265; 1524; 1643; 1660	1.10, 1.32	Approximate globelike appearance with circular arcs and straight lines for most graticules.
Orthographic (analemma, orthographic) ca. 150 B.C.	1.11	True globelike appearance.
Gnomonic, central (horologium, horoscope) ca. 550 B.C.	1.12, 1.13	Projection from center of globe. All great-circle paths shown straight.
Stereographic (planisphere, stereographic) ca. 150 B.C.	1.14–17, 1.19	Projection from opposite surface of globe. All meridians and parallels are circular arcs or straight.
Polar azimuthal equidistant B.C. (?); A.D. 1426	1.20	Simple construction with straight meridians at true angles; equidistant concentric circles for parallels
Simple or equidistant conic 1506	1.21, 1.22	Simple construction with straight meridians at reduced angles for true scale along some parallel; equidistant concentric circular arcs for parallels.
Werner, cordiform ca. 1500	1.24–27	Equidistant concentric circular arcs for parallels, marked off at true scale for meridians. Two are suitable for world maps.
Oval ca. 1510	1.28, 1.29	Equidistant parallel lines for parallels; curved meridians. Full world shown.
Octant maps 1514	1.30, 1.31	World shown in eight equal sections with reduced distortion.
Gore maps 1507	1.33	World shown in numerous touching gores, with reduced distortion. Used for globes.
Mercator 1511, 1569, 1599	1.37, 1.38	Rhumb lines shown straight.
Sinusoidal, Sanson-Flamsteed 1570	1.39	Equidistant parallel lines for parallels, marked off at true scale for meridians.

Note: Map projections are listed in the order described in chapter 1.

TWO

Map Projections in an Age of Mathematical Enlightenment, 1670–1799

B Y THE END of the Renaissance, more than sixteen map projections had
been developed in varied attempts to resolve the problem of portraying
the round earth on a flat sheet. This legacy includes relatively simple approaches such as showing meridians and parallels as a square grid (the plate carrée),
as arcs of circles (the "Nicolosi" globular), or as a geometric perspective centered on
a pole (the orthographic and stereographic). There were also more sophisticated
approaches, notably the Mercator, designed to enable navigators to set a constant
bearing.

The field of mathematics, however, was still too elementary to permit much
development of map projections to fit certain characteristics. There is even considerable doubt that some of the early projections with little distortion of area were
known to have this feature at the time, because they had other features more obvious to the designer, such as true scale along all parallels of latitude (on the Werner,
Sylvano's pseudoconic, and sinusoidal). In any case, the construction of the graticule
of meridians and parallels was often so crude, with large circular arcs rather than
correct curves, that such a projection as drawn did not accurately retain equality of
area, even if land could be correctly mapped relative to these lines.

About half of these sixteen or so projections continued in use after the Renaissance: the rectilinear equirectangular and trapezoidal, the globular, the azimuthals
(especially the stereographic and polar azimuthal equidistant), the Mercator, the
sinusoidal, and the projection that became known as the Bonne. The simple conic
projection, dating to Ptolemy, continued to receive refinements, reducing its overall
distortion with more carefully chosen parallels of true scale.

The post-Renaissance century would have seen fewer than ten new map projections had it not been for the work of J. H. Lambert, who nearly doubled this number
by applying a branch of mathematics rarely used in cartography until his work of
1772—the calculus. Developed in the late seventeenth century, the calculus gave
mathematicians one of their greatest tools. With its first significant application to
map projections by Lambert, mapping likewise advanced immeasurably, although
Lambert's three major projections (the conformal conic, the transverse Mercator, **55**

and the azimuthal equal-area) were not appreciated as his contributions for a century.

Of more pervasive importance was the fact that cartography and the shape of the earth itself were subjected to definitive mathematical analysis. Not long preceding the development of the calculus was another mathematical tool, tables of logarithms. They were first developed and calculated in the early seventeenth century and for 350 years were regularly used to reduce the effort involved in computations. A third major relatively contemporary activity affecting mapmaking was a more accurate measurement of the size of the earth, via the Cassinis' geodetic survey of France; the resulting conclusion was that the earth more closely approximates a slightly flattened ellipsoid of revolution rather than a perfect sphere. This added both to the precision of mapping and to its mathematical complexity. It is a tribute to Isaac Newton's genius that he was a central figure in both the development of the calculus and the determination of the shape of the earth.

These monumental advances overshadow other steps in the field of eighteenth-century map projections. But as the eighteenth century began, in addition to work on the simple conic there were also less significant contributions on azimuthal perspective projections with reduced error as well as on globular projections.

While the history of map projections during this period (and later) has received less attention than that of the Renaissance, descriptive contemporary literature is more readily available. Therefore, although the period did not attract a Nordenskiöld or Keuning (but d'Avezac-Macaya treated it as thoroughly as he did the years through the Renaissance), it is possible to reconstruct the developments during the eighteenth century perhaps more accurately than those of the earlier periods.[1]

EIGHTEENTH-CENTURY USE OF EARLIER MAP PROJECTIONS

The Equirectangular Projection

Popular though the equirectangular type of projection (or plain chart or plane chart) (see figs. 1.3, 1.4) may have been during the Renaissance, by the eighteenth century many projections within the reach of cartographers were more attractive. The popularity of the equirectangular therefore declined. But this projection, with its equally spaced straight lines for parallels and equally spaced straight meridians perpendicular to the parallels, was used by Matthias Seutter (1678–1757) and others for several regional maps, generally consisting of one or two nations.[2] Some sailing charts, such as those by Louis Renard and Joshua Ottens, were based on this projection in spite of the availability of the superior Mercator projection.[3] Many of the charts in *The English Pilot. The Fourth Book,* published in dozens of editions from 1689 to 1794, were examples of the equirectangular projection which also continued to show the numerous compass roses that made the old portolan charts so distinctive.[4] Nevertheless, each *Pilot* contained one or two smaller-scale maps based on the Mercator.

The equirectangular projection can be seen in a handful of contemporary star maps, chiefly of the zodiac region.[5] In some cases the straight center line is the celestial equator, while the ecliptic is curved, and in other cases the reverse is true. In 1718 John Senex (d. 1740) produced an important map of the zodiac, centered on the ecliptic, but with a 2° graticule of right ascension and declination superimposed on a 15′ graticule of longitude and latitude relative to the ecliptic.[6] This may be the first map on the oblique equirectangular projection (the oblique plate carrée in this case).

The Trapezoidal Projection

The trapezoidal projection (fig. 1.6) is almost as easily constructed as the equirectangular. All meridians and parallels are straight lines, but the meridians converge, giving it reduced linear scale distortion along the parallels with correct scale along two of them. The projection was often used for terrestrial maps into the eighteenth century, rivaling in popularity new projections as it did the equirectangular during the Renaissance.[7] It was only slightly used, however, for celestial maps, generally for individual constellations, and by Charles Messier (1730–1817) in 1759 to show part of the path of a comet.[8]

The trapezoidal holds a special distinction during this period because it was used for two well-known maps of the era: One is Philippe de La Hire's (1640–1718) outline *Carte de France, corrigée par ordre du Roy sur les observations de M$^{ss.}$ de l'Académie des Sciences*, first published in 1693, which showed how much the presumed territory had shrunk in size as a result of Jean Picard's surveys.[9] This map led King Louis XIV to reflect that he lost more from the survey than from battle. The second map is John Mitchell's 1755 *A Map of the British Colonies in North America*, used to delineate the boundaries of the newly independent United States for the peace treaty of 1783.[10] This was actually the second edition of his map of the same year, but the title of the first edition referred to *British and French Dominions*. There were serious boundary disputes adjacent to Maine and the Lake of the Woods resulting from map inaccuracies, but they cannot be blamed on the projection.

The trapezoidal projection was commonly used for maps of countries in atlases by the Sanson, De l'Isle, and Homann families, and others.[11] It was increasingly in competition with the simple conic projection, often within the same atlas. By the twentieth century, the latter had become dominant for this usage and the trapezoidal was extinct.

The Azimuthal Projections

The Stereographic Projection

Like the projections discussed above, the stereographic projection was relatively easily constructed. Except for a straight central meridian and one straight parallel of latitude, the graticule consists entirely of circles or circular arcs. It had reached maturity during the Renaissance, but just before the eighteenth century its conformality (correctness of local angles and shapes) was realized by Halley, and this was an important reason for continued usage. The polar stereographic was no longer used only for hemispheres; now it was the basis of insets of polar regions. In 1673 Pietro Todeshi, in his edition of Willem Janszoon Blaeu's (1571–1638) map of America (on the equatorial stereographic), included an inset of the "unknown southern land," but on the south polar stereographic instead of the more common azimuthal equidistant.[12] Blaeu had used both the north and south polar stereographics as insets on his Mercator-projection world map years earlier.[13]

Complete polar stereographic hemispheres (figs. 1.14, 1.15) still appeared: Guillaume De l'Isle (1675–1726) in 1730, and with Philippe Buache (1700–1773) in 1763, used this projection for northern and southern hemispheres of the earth, and dozens of cartographers of the seventeenth and eighteenth centuries used it for celestial hemispheres.[14] The equatorial aspect (fig. 1.19) dominated, however, for terrestrial maps based on the stereographic. The eastern and western hemispheres were thus shown by the Sansons in 1675, Edward Wells (1667–1727) in 1701, Senex in 1721, Johan Gabriel Doppelmayr (1671?–1750) in 1733, and De l'Isle in the same atlases with the polar hemispheres.[15] Individual continents, shown on the equatorial stereographic by Mercator and others previously, continued to be mapped on the same aspect numerous times.[16] In 1750, Johann Baptist Homann (1663–1724) so showed Europe, Asia, Africa, and America.[17]

Seutter used the equatorial stereographic for the map of a fictitious continent, "Schlaraffenland," a country of idleness and luxury.[18] Since it was claimed to extend from 360° to 540° in longitude, and between 60° N and S in latitude, the projection was a reasonable choice.

The oblique stereographic (fig. 1.16) was still rarely used, although the construction (figs. 1.17, 1.18) of the graticule was only a little more difficult than that of other stereographic aspects. Homann (1750) employed the oblique aspect for the region in which a partial eclipse of the moon was visible, centering his map at about 18° S, and showing somewhat more than one hemisphere.[19] In 1778 Noel André used this aspect for celestial hemispheres centered at the two ecliptic poles at declinations 66½° N and S.[20] The oblique aspect also appeared on a map central to a world-tour game. It was dedicated by Thomas Jefferys in 1770 to the Prince of Wales[21] and is a composite of two projections: The stereographic, centered on London, continues south to about the Tropic of Capricorn. A nonstandard projection was used for the portions of South America and Africa south of there.

The Azimuthal Equidistant Projection

Use of the azimuthal equidistant projection was essentially unchanged during the eighteenth century. The equatorial and oblique aspects remained generally too complicated to construct, since the curves for meridians and parallels do not follow standard geometric curves, but Bradock Mead (alias John Green) in 1717 and, with additional details, Lambert in 1772 told how to prepare these aspects.[22] Mead suggested plotting from tables of bearings and distances, or from the globe itself, to produce the oblique or "Horizontal Projection, with Azimuth Lines, to shew the Bearing and Distance of all Places within the Map from London in the center." After deriving the equations for the equatorial aspect, Lambert prepared a table of polar coordinates (as azimuth and angular distance from the chosen projection center) and showed how the principle could be applied to other azimuthal projections (including oblique aspects as well) by using the appropriate trigonometric function of this angular distance.

The polar aspect (fig. 1.20), however, with equally spaced meridians and parallels, is among the simplest projections to construct and is easily understood, even in terms of Renaissance mathematics. It continued as the standard for most pole-centered terrestrial regions and hemispheres (rare stereographic exceptions are mentioned above), and it was heavily used for celestial hemispheres. Still the polar stereographic was employed twice as often for the same purpose.[23]

Perhaps the most grandiose seventeenth- or eighteenth-century use of the polar azimuthal equidistant was for a world map seven meters in diameter and laid about 1680 on one of the floors of the Paris Observatory under the direction of Jean Dominique Cassini (1625–1712), the Italian-born astronomer who revolutionized geodesy, surveying, and cartography.[24]

Conventional usage of the same aspect (normally for a hemisphere) continued in atlases such as those by the Sansons (1675). The progress of Buache's theories about a south polar continent was shown in successive editions of maps of the terrestrial southern hemisphere as far north as 20° S, again using the polar azimuthal equidistant.[25] In the first map of 1739 he showed only a few hundred miles of incomplete coastlines in two sections; on another, in 1754, two complete unconnected land masses are shown, but at the South Pole itself appears a *Mer glaciale* (Icy Sea).

Other Old Azimuthal Projections

The long-known orthographic and gnomonic projections (figs. 1.11, 1.12, 1.13) continued to receive only slight attention. Mead's 1717 book, published in English without his name, described the construction of "fifteen different" projections in much the style of Varenius (1650).[26] Claiming that Varenius had "committed many Mistakes . . . and omitted some of the Projections," Mead, also omitting some, included the polar, equatorial, and oblique "Stereographick," the polar and equatorial "Orthographick," and the oblique azimuthal equidistant among his choices, but pointedly left out the gnomonic projection, "whose Meridians are strait Lines, and Parallels Hyperbolas, as being as useless as troublesome" (1717, preface, 84). Homann, for example, in about 1750 represented globes in his atlas with a crude oblique orthographic on which both poles are shown, but parallels of latitude are curved circular arcs.[27]

In his last year Ignace-Gaston Pardies (1636–1673), a Jesuit mathematician in Paris, issued a set of six celestial maps based on the gnomonic, projecting the sky geometrically onto the six faces of a cube from its center, perhaps the first application of Dürer's proposal in 1538 to use polyhedra as the basis for maps of the sphere.[28] The two celestial poles (of the equator, not the ecliptic), the two equinoxes, and the two points midway on the equator are centered on the faces. Thus there are four equatorial and two polar gnomonic aspects. These maps with their colorful constellations inspired similar sets by Jonas Moore (1617–1679) (dated 1681), P. Grimaldi in 1711 in Chinese, Homann for Doppelmayr published in 1742, with earlier maps, and Georg Friedrich Kordenbusch (1731–1802) in 1789.[29]

Since both the orthographic and gnomonic are perspective projections, the cartographers drawing the equatorial aspects were able to use purely geometric means to plot points along the straight parallels in the first case for the elliptical meridians and along the straight meridians in the second case for the hyperbolic parallels of latitude.[30]

The Mercator Projection

The Mercator projection (fig. 1.37), with its straight loxodromes or rhumb lines, gradually became standard for much of the maritime mapping. Maps of the major oceans were thus cast. Edmond Halley's 1686 map of trade winds and his landmark

A New and Correct Chart Shewing the Variations of the Compass in the Western and Southern [North and South Atlantic] *Oceans* in 1701, the first printed map to show lines of magnetic declination (or any other isolines in the usual sense), are Mercator charts.[31] The explorer James Cook (1728–1779) carried a Mercator-based *Chart of the South Pacifick Ocean* published in 1767 by Alexander Dalrymple of the East India Company.[32] The oldest known chart of the Gulf Stream is based on the Mercator, namely, *A New and Exact Chart* (1769 or 1770) of the North Atlantic by Benjamin Franklin and Timothy Folger.[33]

At times this projection focused on land, for example, a map of Siberia, *Carte des pays traversés par le cap^ne. Beerings depuis la Ville de Tobolsk jusqu' à Kamschatka* by Jean Baptiste d'Anville in 1737, showing Vitus Bering's travels, and *A Map of the British Empire in America* by Henry Popple in 1733.[34] Nonnavigational use of the Mercator was only beginning, but it was to become all too common, and led to commonly held but badly distorted impressions of the shapes and areas of landmasses.

The Sinusoidal Projection

Although Jean Cossin and especially Jodocus Hondius had employed the sinusoidal projection (fig. 1.39) in the late sixteenth and early seventeenth centuries, the Sansons used it so frequently between 1650 and 1700 that it is commonly given their name. In both Sanson volumes (1675, 1683), for example, the sinusoidal was the basis (even if crudely) for maps of Africa, Asia, Europe, and North America, and, in the 1683 work, for South America as well. De l'Isle and later his son-in-law Buache used it during the eighteenth century not only for several continents, but also for regions such as Mexico and Florida, the future Canada and United States, and sections of Africa and South America.[35] A few English mapmakers, such as Wells in 1700, employed the sinusoidal occasionally, and mappers of the heavens in France (especially Messier in numerous maps from 1762 to 1790), England (notably Flamsteed), and Germany (John Elert Bode [1747–1826]) used it, but the projection was made most famous by the French.[36]

Flamsteed (1729), in his star atlas published posthumously, superimposed a graticule relative to the ecliptic on his equatorially centered base map, and thus made use of the oblique sinusoidal projection as well, as did J. Fortin in 1776 and Bode in 1782.[37]

The "Bonne" Projection

The French geographer and hydrographic engineer Rigobert Bonne lived from 1727 to 1795. His name is almost universally applied to an equal-area projection that becomes the Werner and the sinusoidal projections at its polar and equatorial extremes, respectively. The "Bonne" projection (fig. 2.1), however, was used before Bonne's birth.

There are those who attribute the Bonne projection to Apian in his 1520 world map, but this is essentially a copy of Waldseemüller's 1507 modification of Ptolemy's second projection. Both Waldseemüller and Apian have bends in the meridians at the equator, and the meridians are shown as circular arcs in segments rather than as continuously changing curves. At best these are rudimentary forms of the Bonne

FIGURE 2.1. A Bonne projection of Europe, with standard parallel 50° N, central meridian 20° E. 10° graticule. There is no distortion along the central meridian and the standard parallel; scale is also true along all parallels, and correct area scale is maintained throughout.

projection. Sylvano's world map of 1511 (fig. 1.23) and Johannes Honter's world map of 1561 more closely approach the Bonne, since the meridians are spaced along parallels at almost the correct intervals. De l'Isle used the Bonne principle for a map of Africa in his atlases beginning no later than 1700, and for Vincenzo Maria Coronelli (1650–1718) it served as the projection for a map of the regions of present-day Indonesia and northern Australia in 1696.[38] By 1763 De l'Isle and Buache were showing Europe and Asia on this projection, a practice followed by many mapmakers into the late twentieth century.[39] Bonne himself used the projection extensively in his 1752 maritime atlas of the coast of France (Reignier 1957, 164).

The actual philosophy involved in the first regional use of the Bonne projection is not clear, aside from its eventual development from Ptolemy's second projection, but it is a logical transition from the use of Werner's projection. The latter was originally a projection for world maps, but was later used in atlases by Ortelius and Mercator for Asia and Africa. The curvature of the central parallel is corrected with the Bonne, and thus the distortion of shapes in temperate and tropical zones can be much reduced.

Like the Werner and the equidistant conic projections, the Bonne projection is plotted with concentric, equally spaced circular arcs for parallels. The center of these arcs is chosen, like that of normal conic projections with one standard parallel, so that the arc for the chosen central parallel has a radius that equals the length of the element of a cone tangent to the globe at this parallel. This radius thus equals the radius of the globe at the scale of the map times the trigonometric cotangent of this latitude. The chosen central meridian is drawn straight as a radius of this circular arc. It is marked off at true distances from the standard parallel to the other latitudes, and arcs are drawn for each from the same center. Once the parallels are thus constructed, the plotting of other meridians is identical in principle with the plotting of meridians on either the Werner or sinusoidal: each parallel is marked off at true distances for the meridians, starting with the central meridian, and the corresponding points are connected to obtain the meridians, which are complex curves. With the concentric circular arcs for parallels and the curved meridians, the Werner and Bonne projections fall into the pseudoconic classification.

On the Bonne, all areas are in correct proportion, the scale is correct along the central meridian and along all parallels, and there is no distortion of shape and angles along the central meridian and central parallel. Distortion of shape and angles increases, however, with the distance from the central meridian and central parallel. The formulas for rectangular and polar coordinates may be expressed as follows for modern computation:

$$x = \rho \sin \theta, \text{ and}$$
$$y = \rho_0 - \rho \cos \theta,$$

where $\rho = R (\cot \phi_1 + \phi_1 - \phi)$ and

$$\theta = \lambda \cos \phi/(\cot \phi_1 + \phi_1 - \phi).$$

ϕ, λ, x, y, and R are as defined on page 49, and ϕ_1 is the central parallel (+ if north, − if south). If ϕ_1 is made the North Pole or $\pi/2$ radians, the formulas simplify to those for the Werner projection.

Polar coordinates are ρ and θ, with the radius ρ measured from the center of the circular arcs representing the parallels and θ measured counterclockwise from the central meridian, although this orientation does not need to be understood to use the formulas. Radius ρ_0 is calculated from the formula for ρ, but using the latitude of origin ϕ_0 in place of ϕ and ρ_0 in place of ρ. This latitude ϕ_0 may be placed at the bottom, center, or elsewhere on the map without affecting the appearance of the map.

THE NEW PROJECTIONS

Map Projections as an Emerging Mathematical Science

Mathematical formulas are now regularly used for older projections, but these equations were generally not developed at the time, even in rudimentary form. (A few modern projections as well were designed from a geometric standpoint, with only partial formulas. Complete plotting equations were not developed until it became desirable to plot them using a computer.) Mead's (1717) construction procedures regularly used straightedges, compasses, and dividers. For relatively small-scale maps of the world or of a hemisphere, geometric construction was often both practical and sufficient. This was not the case for mapping resulting from detailed surveys, however.

High-precision, large-scale mapping was launched chiefly by Jean Dominique Cassini and Jean Picard (1620–1682) in 1669 with the first geodetic survey of an entire country, France.[40] The foundation for this survey was laid in large part by the development of triangulation by Gemma Frisius (1508–1555) and Tycho Brahe during the sixteenth century and its first successful application in 1615 by Willebrord Snellius (1580–1626) (J. R. Smith 1986, 57–66). Proceeding with the prevailing assumption that the earth is a nearly perfect sphere, Picard and later Cassini's son Jacques (1677–1756) completed a survey of the meridian through Paris by 1720. It was found that the length of each degree of latitude was not the same, as it would be on a sphere, and based on the measurements Jacques Cassini claimed the earth to be closer to a prolate spheroid or ellipsoid of revolution, elongated at the poles.

On the other hand, Englishman Isaac Newton (1642–1727) and others insisted that the earth must be essentially an oblate ellipsoid, flattened at the poles, due to the effects of centrifugal force. To settle the controversy, beginning in 1735 the French Academy of Sciences sent survey expeditions to measure meridians in Peru (now Ecuador) and Lapland. By 1738 they had confirmed Newton's prediction of oblateness, and a complication was added to all future precision mapping, with renewed measurements made every few years (up to the present) to establish more accurate dimensions for the figure of the earth.

In addition to knowing the shape of the earth, it was also necessary to measure correctly the coordinates of points on its surface. With the development of optics, the ability to measure latitude with more precisely constructed astronomical telescopes accelerated in the seventeenth and eighteenth centuries (L. Brown 1949, 180–207). Accurate longitude across the oceans, so long elusive with the resulting loss of men and wealth in shipwrecks, was finally measurable with the aid of John Harrison's chronometer in the mid-eighteenth century (L. Brown 1949, 208–40; Quill 1966).

The scientific enlightenment that nurtured the improvement in instrumentation was of course at least as much at work in mathematics. The need for better map projections to support the greater accuracy in measurement of positions and of the shape of the earth was answered with previously unavailable mathematical techniques. Geometric construction could be used for several important projections such as the stereographic and the Bonne, but such techniques were not useful for the oblique azimuthal equidistant and Mercator projections, for projections based on the ellipsoidal earth, and for several of the graticules about to be developed by Lambert. To calculate rectangular coordinates of sufficient accuracy for graticule intersections, it was often necessary to multiply several decimal places of trigonometric functions for every point. Perhaps the most important invention before the twentieth century to reduce the tedium but still obtain accurate results was the logarithm.

John Napier (or Neper, 1550–1617) of Scotland invented logarithms chiefly to aid in trigonometric calculations. His explanation and tables in 1614 were in a form somewhat different from what are now called Napierian or "natural" logarithms.[41] Mathematically, a logarithm (log) of a number is the power to which another number, called the base, must be raised to equal the first number, that is, if $y = \log_B x$, then $x = B^y$, where B is the base. For natural logarithms, $B = 2.7182818 \ldots$, a convenient base universally called e and used in several conformal map projection equations, such as those for the Mercator, and in the calculus.

Napier's work immediately attracted the attention of mathematicians Wright and Briggs, mentioned earlier for their contributions to cartography. Wright, who saw the value of logarithms for navigation, translated the work from Latin into English, and Briggs worked with Napier to use a base of 10, tables for which Briggs calculated and first published in 1617. This led to a much simpler arrangement of tables, because multiplying any number by 10, 100, or 1,000 does not change the decimal part of the Briggian or common logarithms. To multiply two numbers, one can look up the decimal part of their respective logarithms, interpolating between entries, then attach the simple integer part, add the two logarithms, and look up the number that has the sum as its own logarithm, again using interpolation. Related operations apply to division and to finding powers and roots.

Logarithms provided an important calculating tool until the advent of twentieth-century pocket calculators and computers, but of much greater analytical significance was the invention of the calculus. The calculus is a branch of mathematics concerned with small elements of functions. There are two principal types, differential and integral. In differential calculus, the amount by which one variable changes when some other related variable changes a very small, actually infinitesimal, amount is determined. Integral calculus is the reverse: a desired sequence of infinitesimal changes in one variable with respect to another is summed or integrated to determine what function will produce these changes or to determine how much the value of one variable changes when the other variable is changed by a given finite quantity.

When applied to an existing map projection, differential calculus may be used to determine, for example, the scale factor in any direction at any given point. The scale factor is the ratio of a very small (infinitesimal) distance measured on the map at this point to the true distance on the earth reduced in size to the stated scale of the map. When measured in various directions, the scale factor may be used to determine instantaneous distortion in area, shape, or angles. Integral calculus may be used to attempt to convert a desired pattern of scale or shape distortion into formulas for the entire projection. If the integration cannot be done precisely, it can usually be accomplished by numerical approximation.

The calculus was invented almost simultaneously by Englishman Isaac Newton and Gottfried Wilhelm Leibniz (1646–1716) of Germany. This fact was to lead to a highly acrimonious rivalry between them. Newton developed the calculus first, in 1665–66, calling it the "theory of fluxions," but Leibniz was the first to publish it, in 1684–86, and the latter presented the special symbols still in use. Their work so nearly coincided in time largely because of the rapid advancement in algebra and geometry, especially under René Descartes (1596–1650), and the development of an elementary form of the calculus by Bonaventura Cavalieri in 1635 at Bologna (Struik 1948, 131–58).

Although Antoine Parent (1666–1718) (see below) applied principles of the calculus in deriving some of his minimum-error perspective projections, it appears the calculus was first formally used for cartography by Lambert (1772) in developing most of his important new projections. One of Lambert's concerns was for new projections in which all local angles remain the same shape on the map as they are on the globe. Only two projections, the stereographic and the Mercator, had this characteristic before he added three more. The property is called conformal or orthomorphic (true-shape) in English, although Karl Friedrich Gauss in Germany and Adrien Germain in France introduced these respective terms; the principal German word is *winkeltreu* (angle-true) (Gauss 1843; Germain 1866, 31). The mathematical treatment of the concept is often associated with complex algebra and with the Cauchy-Riemann equations, named for two nineteenth-century mathematicians, Augustin-Louis Cauchy (1789–1857) of France and Bernhard Riemann (1826–1866) of Germany. The two added important rigor to the development of these equations providing conditions for conformality.

The Cauchy-Riemann equations had actually been presented by Jean Le Rond D'Alembert (1717–1783) in a 1752 paper on resistance of fluids. While Lambert developed his conformal projections by taking other approaches, both Joseph Louis Lagrange (1736–1813) and Leonhard Euler (1707–1783) employed D'Alembert's solutions, although giving him minimal credit. As with the calculus and numerous other mathematical innovations of the seventeenth and eighteenth centuries, the full impact of D'Alembert's work was only felt during subsequent centuries.[42]

Perspective Projections with Low Error

Analytical mathematics was prominent in the development of new map projections even as the eighteenth century began. It has already been pointed out that the orthographic, gnomonic, and stereographic projections are direct perspective projections of the sphere onto a tangent or secant plane. In addition to these projections from infinity, from the center, and from the opposite surface of the sphere, respectively, perspective projections may be prepared with the point of projection at any other location between the center of the sphere and infinity, on the same side of the center of the sphere as the desired map center or on the opposite side (with reversal of the geometric map so as to appear "normal").

Several other locations for this point have been formally proposed in attempts to minimize distortion on the resulting map, beginning with French mathematician-physicists La Hire and Parent in 1701 and 1702, respectively, and extending into the nineteenth century. All these projections have the same properties as the general group of azimuthals, namely, that all azimuths from the center are correct, and that the scale pattern is solely a function of distance from the center.

As geometric attempts to minimize error, these special projections were not intended to prepare a view from space, because they are actually views of the surface from a point on the opposite side of the center. La Hire chose as his point of perspective a point located at 1.707 or $[1 + (\sqrt{2})/2]$ times the length of the radius from the center, in a direction diametrically opposite the center of the map, which was

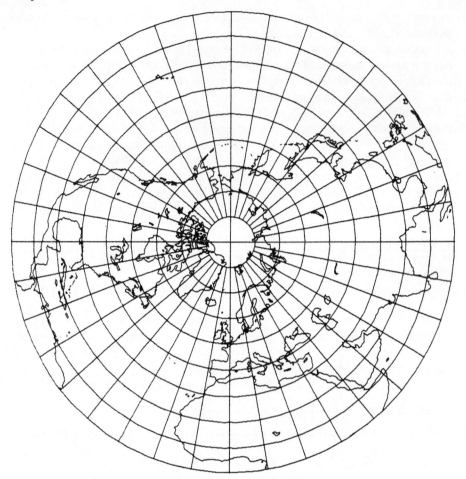

FIGURE 2.2. A polar La Hire projection of the northern hemisphere. This perspective projection retains the 45° N latitude at half the radius of the equator. 10° graticule.

normally the North Pole to simplify construction (fig. 2.2) (Maurer 1935; La Hire 1704). With La Hire's selection, the radius of the circle representing the equator on the north polar aspect is exactly twice the radius of the circle representing the parallel at 45° N. The other parallels are not uniformly spaced, however.

Parent proposed three low-error perspective projections.[43] First, he placed the point at 1.594 times the radius, solved from a lengthy equation to minimize the "entire change of length increase" proceeding 90° out from the center. His second proposal involved a distance of 1.732 or $\sqrt{3}$, so that the greatest error in the scale factor along a meridian is as small as possible, with the plane of projection passing through the center of the globe. His third, specifying certain area stipulations for a central plane, led to a distance of 2.105 times the radius.

Both La Hire and Parent were born and died in Paris. La Hire wrote several treatises on conic sections in the late seventeenth century (Taton 1970–80). He was wary of the new calculus, however, and much preferred analytic geometry in developing his work. He studied astronomy, geodesy, and other natural sciences with equal vigor. Parent was similarly eclectic in science, but considered the calculus a useful tool (Briggs 1970–80). His uncompromising, Spartan nature both offended colleagues and led to significant contributions.

None of these early French perspective projections has been used significantly, but they helped set the stage for increasing reliance on mathematical analysis rather than elementary geometry to determine map projection parameters.

Occasionally perspective projection was used strictly as visual art: In 1740 Matthias Seutter included small perspectives of most of the eastern and western hemispheres as viewed about one earth-diameter above the equator (Portinaro and Knirsch 1987, 234). About 1760, Philippe Buache appears to have used a near-stereographic azimuthal perspective, centered on the North Pole, for a map of mountain chains extending to 50° S. His perspective center is about one-fourth of the globe radius in space above the South Pole.[44]

Some Modified Globular Projections

Mead (1717), in the process of describing the construction of several map projections already in use, such as azimuthals as previously mentioned, the "plain Chart" in its square form (the plate carrée), the "Projection of an eighth Part of the Globe" or octant, the sinusoidal, the trapezoidal, and "Wright's (commonly call'd Mercator's) Chart," also discussed some original or little seen projections. One is a very close approximation to the equidistant conic and is described shortly. Two others are simple modifications to the first globular projection of Fournier (1643), also modified by Nicolosi (1660).[45]

In each case, Mead called for construction of a circle with horizontal and vertical diameters representing the equator and central meridian, respectively. The two diameters are each equally divided into 180 degrees, and the outer circle equally into 360 degrees. The parallels of latitude are circular arcs joining corresponding points on the outer circle (the 90th meridian east and west of center on the Fournier and Nicolosi globulars) with points along the central meridian.

So far this projection is identical to its predecessors, which Mead did not credit, but he then, for one modification, marked off each parallel from the central meridian with true distance for each meridian and connected the points so established with longitude lines.[46] The meridians are thus not circular arcs, and the projection is similar to the later polyconic projection of the nineteenth and twentieth centuries, except that the radii of the parallels are determined differently.

For the second modification described by Mead, the meridians are merely straight lines connecting the poles at the ends of the vertical diameter with the equally spaced points along the equator, where a sharp break occurs in passing from northern to southern latitudes.[47] Mead (1717, 78) stressed that due to distortion, "it is neither usual nor convenient to represent a Hemisphere after this Method. But it is often us'd in exhibiting one of the Quarters, and particular Kingdoms." This writer has not seen examples of its use.

The Improved Simple or Equidistant Conic Projection

Claudius Ptolemy was responsible for the earliest forms of the simple conic projection through his "first" projection, but it was modified so much by Contarini and Ruysch in the early sixteenth century that its later use barely merits the appellation "Ptolemy's projection" often applied by modern textbooks. A major improvement was provided by Joseph Nicolas De l'Isle (1688–1768) (or Delisle, or de L'isle, depending on the map) of an illustrious French family of mapmakers, and brother of the more famous Guillaume. He developed an arrangement with two standard parallels along which the scale is held true, but with details that make it different from the modern equidistant or simple conic with two standard parallels.

On De l'Isle's projection, now obsolete, the parallels are concentric and equidistant circular arcs, with the central parallel at a radius equal to the cotangent of this latitude times the radius of the sphere at the scale of the map (Schott 1882, 291). So far, this is the same design as that of the simple conic with one standard parallel. The two standard parallels are generally placed ¼ and ¾ of the latitude range from the south to north limits of the map. These are marked off at true scale with respect to meridians, which are then connected to the respective points with straight lines. Thus the meridians are not quite radii of the arcs of circles representing the parallels of latitude; therefore, the meridians generally do not quite intersect the parallels at right angles.

De l'Isle first used his projection with standard parallels 47½° and 62½° N (and latitude limits of 40° and 70° N) on a large map of Russia he produced in atlas form in 1745 while working at St. Petersburg.[48] (De l'Isle used the trapezoidal projection for the larger-scale maps in the same atlas.) Mercator is also credited by some with originating the use of two standard parallels, although he did not use this approach for his large 1554 map of Europe as claimed, since it has curved meridians.[49] The meridians fail to intersect parallels at right angles on several of the earlier conic maps by Mercator, Hondius, and Speed and, into the eighteenth century, by, for example, Herman Moll (fl. 1680–1732) and Emanuel Bowen.[50]

In any case, the "true" simple or equidistant conic, with concentric circular parallels of latitude and straight converging meridians intersecting parallels at right angles, whether with one or two standard parallels, was commonly used for regional maps by many cartographers. These included De l'Isle in 1730 and in later revisions of his atlas by Buache, Lewis Evans (ca. 1700–1756) in America in 1749 and 1755, William Faden (1750–1836) in the 1770s, and Messier in plotting the path of a 1758 comet.[51] The geographical regions ranged from the middle-Atlantic colonies of North America to the entire northern Pacific Ocean. Schickard's earlier star maps were revised and again published in 1687, displaying two polar celestial hemispheres as conics with a cone constant of ⅔ (the ratio of the angle between meridians on the map to the angle of the globe), the cones cut along the right ascension or meridian of the summer solstice.[52] His design inspired Johann Jacob Zimmermann (1644–1693) in 1692 and Christoph Cellarius (1638–1707) in 1705 to construct similar plots.[53] The projection has remained a staple for atlas coverage of small temperate-zone countries to the present day (fig. 2.3).

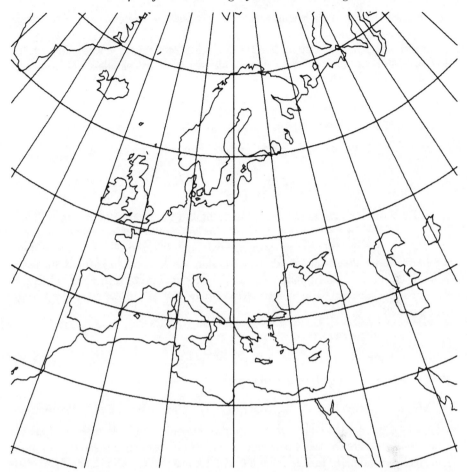

FIGURE 2.3. Europe using the equidistant or simple conic projection with standard parallels 20° and 70° N, central meridian 20° E, and 10° graticule. Scale is true along the standard parallels and all meridians.

One of Mead's (1717, 106–17) descriptions of new and existing projections involved a rather laborious construction of a "new and more exact Method than any of the former, for projecting particular Maps" that is a very close approximation to the simple conic with two standard parallels. The method consists of drawing a trapezoid 1° of longitude wide at the north and south limiting latitudes, symmetrical about the central meridian and with a height equal to the number of degrees of latitude spanned by the map. The central meridian and the limiting latitudes are correct in scale. The parallels of latitude are straight lines, parallel to each other and extended to the edges of the trapezoid. This trapezoid is then repeated with abutting sides. The resulting graticule has straight converging equally spaced meridians, while the parallels are short straight segments of lines appearing to connect as equidistant concentric circular arcs.

For modern computation, the formulas for rectangular coordinates of the simple conic projection for the sphere with two standard parallels can be represented thus:

$$x = \rho \sin \theta, \text{ and}$$
$$y = \rho_0 - \rho \cos \theta,$$

where $\rho = R\,[(\phi_2 \cos \phi_1 - \phi_1 \cos \phi_2)/(\cos \phi_1 - \cos \phi_2) - \phi]$,

$$\theta = n\,\lambda, \text{ and}$$
$$n = (\cos \phi_1 - \cos \phi_2)/(\phi_2 - \phi_1).$$

Symbols are the same as those for the Mercator and Bonne projections (pp. 49, 62), except that ϕ_1 and ϕ_2 are the two standard parallels, and n is the cone constant.

This projection with only one standard parallel is theoretically the same as if the two standard parallels ϕ_1 and ϕ_2 above are equal, but the third and fifth of the above formulas become indeterminate because they degenerate to zero divided by zero. The equations, however, can be rewritten as follows, if latitude ϕ_1 is the single standard parallel:

$$\rho = R\,(\cot \phi_1 + \phi_1 - \phi);$$
$$n = \sin \phi_1.$$

The formulas for x, y, and θ are unchanged.

Murdoch's and Euler's Approaches to the Equidistant Conic Projection

Selection of standard parallels on the equidistant conic by simple rules, such as De l'Isle's ¼ and ¾ (above), were convenient, but others were challenged to find more analytically sound bases for the selection of parameters. Two who pursued the challenges were a Scottish-born clergyman-mathematician, Patrick Murdoch, and a Swiss, Euler, who was perhaps the most prolific eighteenth-century mathematician.

Murdoch (ca. 1700–1774) was both an ordained Anglican priest (by 1738) and a fellow of the Royal Society (beginning in 1745) (Maling 1983). Born in Dumfries, Scotland, and schooled in Edinburgh, he served a number of parishes as rector of Stradishall for several years in Suffolk County, England, retaining the rectorship until his death. In 1748 he entered Cambridge and became a doctor of divinity in 1757.

Beginning in 1756, he began to devote most of his efforts toward geography, but he had already published a book in 1741 listing distances of parallels along the meridian of the Mercator projection for the ellipsoid (or spheroid). While the flattening of the sphere was 50% too great based on his conclusions from information available at the time, this was probably the first table applying Mercator's and Edward Wright's sixteenth-century principles to the newly established oblate ellipsoid and, according to Maling (1983, 114), "almost certainly led to the production of the first map or chart ever to be compiled for the spheroid." Murdoch revised the table about 1760, applying more rigorous mathematics and a more accurate oblateness of the earth ellipsoid.

Murdoch published papers on astronomy and optics as well as a translation in six volumes of a then recent German textbook by Anton Büsching, published in English as *A New System of Geography* in 1762. Of special interest here is Murdoch's description of three variations of the simple conic projection, presented in 1758.[54]

In Murdoch's concept, a secant cone cuts the sphere at two latitudes corresponding to the upper and lower limits of the area being mapped (not the standard parallels). The tip of this cone is then, in effect, moved slightly away from the sphere along the polar axis, so that the cone intersects the sphere at two other latitudes somewhat closer together than the two limiting latitudes, and between them. The new position is chosen so that the total area of the projection equals the total area of the corresponding part of the globe. The projection is not equal-area in the standard sense of achieving the correct area for each small portion.

As on other true conics, the parallels of Murdoch's conics are concentric arcs of circles, and meridians are equally spaced radii of the circles. On his first and third projections, parallels are equidistant; on his second they are geometrically projected from the center of the sphere onto the cone. The first and third projections, as well as Euler's (below), are true equidistant conics, with different approaches to reducing the variation of scale over the map. The two standard parallels in these cases may only be determined by trial and error or iteration, but Murdoch did not attempt to calculate their actual values. He was proposing (in his first and third projections) maps on which the "intersections of the meridians and parallels will be rectangular," the "distances north and south will be exact; and any meridian will serve as a scale," two parallels will be correctly scaled, and other measurements will not be far wrong for a map of limited extent (Murdoch 1758; 1809 reprint, 216–17). For his first projection, in his own words:

> Let E*l*LP, fig. [2.4], be the quadrant of a meridian of a given sphere, its centre C, and its pole P; EL, E*l*, the latitudes of two places in that meridian, EM their middle latitude. Draw LN, *ln*, cosines of the latitudes, the sine of the middle latitude MF, and its cotangent MT. Then writing unity for the radius, if in CM we take $Cx = Nn/(Ll \times MF \times MT)$, and through *x* draw *x*R, *xr*, equal each to half the arc L*l*, and perpendicular to CM; the conical surface generated by the line R*r*, while the figure revolves on the axis of the sphere, will be equal to the surface of the zone described in the same time by the arc L*l*; as will easily appear by comparing that conical surface with the zone, as measured by Archimedes. And lastly, if from the point *t*, in which R*r* produced meets the axis, we take the angle C*t*V in proportion to the longitude of the proposed map, as MF the sine of the middle latitude is to radius, and draw the parallels and meridians as in the figure, the whole space SOQV will be the proposed part of the conical surface expanded into a plane; in which the places may now be inserted according to their known longitudes and latitudes. (Murdoch 1758; 1809 reprint, 216)

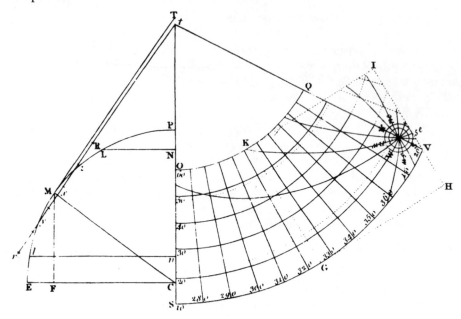

FIGURE 2.4. A diagram prepared by Murdoch for construction of his first equidistant conic projection. Reproduced from Murdoch (1758; 1809 reprint, fig. 1 for p. 216). *U.S. Geological Survey Library.*

In modern notation, his equation for Cx can be simplified to $Cx = (\sin \delta)/\delta$, where δ is defined below, and the formulas for ρ and n become

$$\rho = R\,[(\cot \phi_0 \sin \delta)/\delta + \phi_0 - \phi], \text{ and}$$

$$n = \sin \phi_0,$$

where $\phi_0 = (\phi_2 + \phi_1)/2$, and

$$\delta = (\phi_2 - \phi_1)/2$$

n is the cone constant, and ϕ_1 and ϕ_2 are the limiting parallels, not the standard parallels. Symbols ρ, n, and R, are as defined for the simple conic projection (above), for which the equations for x, y, and θ also apply to Murdoch's projections.

For Murdoch's second and third projections, he briefly described their construction, still using figure 2.4, and overlooking the fact that the change in Cx also affects cone constant n as well:

> If it be required to draw a map, in which the superficies of a given zone shall be equal to the zone of the sphere, while at the same time the projection from the centre is strictly geometrical [his second projection]; take Cx to CM, as a geometrical mean between CM and Nn, is to the like mean between the cosine of the middle latitude, and twice the tangent of the semidifference of latitudes; and project on the conic surface generated by xt. . . . Artists may use

either rule; or, in most cases, they need only make Cx to CM as the arc ML is to its tangent [his third projection], and finish the map; either by a projection, or, as in the first method, by dividing that part of xt which is intercepted by the secants through L and l, into equal degrees of latitude. (Murdoch 1758; 1809 reprint, 218)

Thus Cx becomes $\sqrt{\cos \delta}$ and $\delta/\tan \delta$, respectively, but the formulas, corrected for the changes in the cone constant (for the second projection from this writer's calculations, and for the third by Young 1920, 40), may be expressed as follows:

Murdoch's second projection:

$$\rho = R \left[\cot \phi_0 \sqrt{\cos \delta} + \tan (\phi_0 - \phi) \right];$$
$$n = \sin \phi_0 \sqrt{\cos \delta}.$$

For his third projection:

$$\rho = R (\delta \cot \delta \cot \phi_0 + \phi_0 - \phi);$$
$$n = (\sin \phi_0 \sin \delta \tan \delta)/\delta^2.$$

Symbols are as before. Murdoch's third projection provides a remarkably low-error choice of parameters for mapping a given region with the equidistant conic, although it is not a minimum-error projection as is sometimes stated. In spite of the brief description of the third as quoted above, Murdoch provided a more complete geometric description in his English translation of Büsching, where many of the new maps were engraved by Thomas Kitchin according, apparently, to this third projection sans, of course, Young's much later correction.[55]

The actual standard parallels may be determined for either Murdoch's first or third projections (the second, nonequidistant conic projection is omitted here), as the two values of α satisfying the following equation (using the respective equations above to calculate n):

$$\cos \alpha + n\alpha = (\cos \phi_0 \sin \delta)/\delta + n\phi_0.$$

In Euler's 1777 investigation of the simple conic projection, he designed a projection now given his name, one in which the errors of distance between any two meridians along the two extreme parallels are equal and opposite in sign to the error in the corresponding distance along the central parallel (Euler 1777; Hinks 1912, 82–84). The resulting formulas given here for this projection may be compared with those for Murdoch's and De l'Isle's (the symbols are the same):

$$\rho = R \left[(\delta/2) \cot (\delta/2) \cot \phi_0 + \phi_0 - \phi \right];$$
$$n = (\sin \phi_0 \sin \delta)/\delta.$$

Euler spent his professional life first in St. Petersburg, then in Berlin, and again in St. Petersburg, studying finally under the auspices of Empress Catherine (Struik 1948, 167–68). He produced over seven hundred books and papers covering mathematics, celestial mechanics, and hydraulics. Another of his three 1777 works provided the first formal proof that the surface of a sphere cannot be transformed to a plane without distortion of some sort.[56] His total blindness beginning in 1766, when he returned to the Russian capital, made his achievements all the more remarkable.

Colles's Perspective Conic Projection

Another conic projection involving a mapmaker who was not an academic was that of Christopher Colles (1739–1816), an Irish-born engineer who emigrated to Philadelphia in 1771, developed water supply systems and engines, and lectured on geography, physics, and mechanics.[57] Unable to sell many of his ideas, he died almost penniless in New York City. He is best known among historians of the eastern United States for his also commercially unsuccessful *A Survey of the Roads of the United States of America*, published in 1789 with a set of eighty-three strip maps extending from New York to Virginia.

In 1794 Colles published the first portions of an aborted *Geographical Ledger*, with five large maps and a description of a conic projection equivalent to a perspective projection from the center of the sphere onto a cone secant at latitudes of about 31°32' and 58°28' N.[58] These values were the result of relating the cone to limiting parallels at the Tropic of Cancer and the Arctic Circle in a geometric manner described in detail by Colles. For the polar and equatorial regions, perspective azimuthal and cylindrical projections were also described, secant at 73°27' and 16°33', respectively, but non-U.S. maps were never included. Colles (1794, v) carefully computed the error in scale and said that the "most erroneous parts will be found to differ only 1-22th part of the truth, which is a degree of accuracy far exceeding any extensive map upon any other mode of projection." There was apparently no further use of the projection.

Cassini and His Transverse Equidistant Cylindrical Projection

The most famous Cassini was born Giovanni Domenico Cassini in Italy; he changed his given names to Jean Dominique not long after being hired in 1669 for astronomical research in Paris, and shortly thereafter for the survey of France (L. A. Brown 1949, 215–24). His grandson César François Cassini de Thury (1714–1784), the third of four generations involved, was also director of the Paris Observatory and of the survey. In 1745 the latter devised a new map projection which, with some modifications, now bears the Cassini name.[59] This projection was used for the official topographic maps of France until replaced by the Bonne projection in 1803 (Steers 1970, 242).

The well-known plate carrée (fig. 1.3), or the equirectangular projection with the equator at true scale, is conceptually based on a cylinder wrapped around the globe and tangent to the equator. The parallels of latitude (ϕ) as well as the meridians of longitude (λ) are equally spaced straight lines. Therefore, the spherical coordinates may be converted to rectangular coordinates simply as $x = R\lambda$ and $y = R\phi$. If Cassini's projection is applied to the sphere (fig. 2.5), it may be conceptually developed by first rotating this cylinder 90° so it is tangent to a selected meridian instead of the equator. If the graticule of meridians and parallels is also rotated on the globe so that the tangent meridian becomes the equator of the rotated graticule, and then these are made, say, dashed lines, these transformed meridians and parallels would be plotted on the Cassini with dashed straight lines exactly the same as

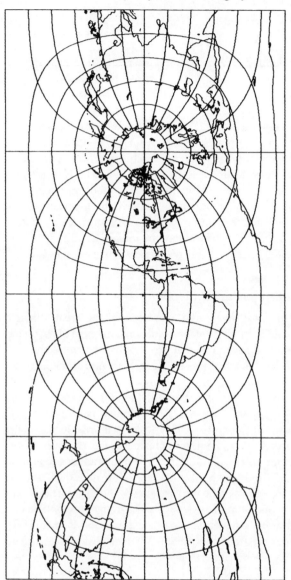

FIGURE 2.5. A Cassini projection of the world, central meridians 90° W and E, 15° graticule. A transverse plate carrée, it has correct scale along the central meridians and along all straight lines perpendicular to them on the map.

the regular equally spaced meridians and equally spaced parallels are plotted on the plate carrée. The original graticule of the globe is then plotted as the curved meridians and parallels on the Cassini by placing them between or across the dashed lines on the map just as they fall between or across them on the globe. Positions on the Cassini projection are then located in direct proportion to the distance from the tangent meridian, along great-circle paths perpendicular to and spaced at true distances along this meridian. The computation may be done more directly with these formulas, which obscure the simplicity of the basic construction:

$x = R$ arcsin (cos ϕ sin λ), and

$y = R$ arctan (tan ϕ/cos λ),

where R, ϕ, λ, x, and y are as defined earlier. The arctan function must be applied in what is sometimes called an arctan$_2$ form, in which π radians must be added to the normal result if the denominator of the argument (the cosine term in this case) is negative.

The projection as originated by Cassini and used for France was applied only to the ellipsoid. Instead of meridians and parallels (except for the central meridian), Cassini substituted a system of rectangular grid coordinates with the central meridian (that of the Paris Observatory) as one axis (T. Craig 1882, 80). Thus squares were shown, much as are rectangular coordinate grids on many modern topographic maps. Use of the Cassini projection for nineteenth- and twentieth-century topographic mapping of several European countries, including German states and Great Britain, involved more practical corrections developed by Johann G. von Soldner in Germany about 1810.[60] For such mapping the projection is called the Cassini-Soldner.

Lambert's Cornucopia of Important Projections

About a century after the calculus was invented, it was used for the greatest single advance in the direct development of modern map projections. In one paper, Johann Heinrich Lambert (1772) presented seven original projections, several of them of considerable later importance. In the order presented in his paper, they are now generally known by the following names (Lambert did not suggest any names): (1) Lambert conformal conic, or conical orthomorphic; (2) Lagrange, named for the mathematician who described its general case in 1779; (3) transverse Mercator; (4) (Lambert) cylindrical equal-area; (5) transverse cylindrical equal-area; (6) Lambert azimuthal (or zenithal) equal-area; and (7) Lambert conical equal-area, or isospherical stenoteric. Of the seven, numbers (1), (3), and (6) are among the most common twentieth-century projections, (2) and (4) are of moderate interest, and (5) and (7) are rarely used. Although his projections came into use slowly, their importance was soon recognized by other leading mathematicians, such as Lagrange (see below) and Gauss (in the 1820s), as well as by projection historian d'Avezac-Macaya (1863, 438–39) ("a treatise [by Lambert] of great importance").[61] As Albert Wangerin said in an 1894 edition of Lambert's work on projections,

> The importance of Lambert's work consists mainly in the fact that he was the first to make general investigations upon the subject of map projection. His predecessors limited themselves to the investigations of a single method of projection, especially the perspective, but Lambert considered the problem of the representation of a sphere upon a plane from a higher standpoint and stated certain general conditions that the representation was to fulfill, the most important of these being the preservation of angles or conformality, and equal surface or equivalence. These two properties, of course, can not be attained in the same projection.

Although Lambert has not fully developed the theory of these two methods of representation, yet he was the first to express clearly the ideas regarding them. The former—conformality—has become of the greatest importance to pure mathematics as well as the natural sciences, but both of them are of great significance to the cartographer. It is no more than just, therefore, to date the beginning of a new epoch in the science of map projection from the appearance of Lambert's work. Not only is his work of importance for the generality of his ideas but he has also succeeded remarkably well in the results that he has attained. (Translation from Deetz and Adams 1934, 78)

Lambert (fig. 2.6) was born in 1728 in Mulhausen, Alsace, the town then being a part of Switzerland, and after the age of twelve was largely self-taught, especially in mathematics, astronomy, physics, and mechanics (Maurer 1931). His published works began to appear in 1755, and in ten years he was in Berlin with titles and grants from Frederick the Great and was ranked with the leading mathematicians of Europe. His published works in both French and German included studies of comets, the planets, convergence of series, probability, geometry, photometry and illumination, meteorology, and mechanical devices. Thus one of the most outstanding contributors to map projections was involved in cartography only as one of a great many interests. He died after a stroke in Berlin in 1777.

FIGURE 2.6. Johann Heinrich Lambert (1728–1777), inventor of some of the most important map projections and an outstanding mathematician, astronomer, and physicist. *U.S. Geological Survey.*

The Lambert Conformal Conic Projection

Lambert developed the conformal conic as the oblique aspect of a family containing the previously known polar stereographic and normal (equatorial) Mercator projections. As he stated,

> Stereographic representations of the spherical surface, as well as Mercator's nautical charts, have the peculiarity that all angles maintain the sizes that they have on the surface of the globe. This yields the greatest similarity that any plane figure can have with one drawn on the surface of a sphere. The question has not been asked whether this property occurs only in the two methods of representation mentioned or whether these two representations, so different in appearances, can be made to approach each other through intermediate stages. . . . if there are stages intermediate to these two representations, they must be sought by allowing the angle of intersection of the meridians to be arbitrarily larger or smaller than its value on the surface of the sphere. This is the way in which I shall now proceed. (Lambert 1772; 1972 trans., 28)

Lambert used the calculus with geometry, algebra, and trigonometry to develop both the spherical and ellipsoidal forms of the conformal conic projection with two standard parallels, and he included a small map of Europe as an example (Lambert 1772; 1972 trans., 28–38, 80–90). The ellipsoidal formulas for the limiting cases of the Mercator and polar stereographic required a few modifications which Lambert also delineated.

One of the earliest uses of the Lambert conformal conic projection occurred in 1777 for two polar celestial hemispheres. The cartographer was C. B. Funke (1736–1786), and the cones, cut along the meridian or right ascension of the vernal equinox, each formed two-thirds of a complete circle.[62] Lambert had stated that his new projection

> yields conical globes, on which it has long been the custom to represent hemispheres of the heavens. Here we have achieved such a representation in which all angles preserve their true magnitude, and consequently the star images have the largest possible similarity to the heavens. Zimmermann's well known star cones do not have this advantage. (Lambert 1772; 1972 trans., 37)[63]

Nevertheless, Lambert's conformal conic projection was hardly used between its introduction and its revival by France for battle maps during the First World War, except for being independently derived during the nineteenth century. It is now standard for much official mapping throughout the world, however, sharing importance in this respect only with Lambert's transverse Mercator.

In basic construction (fig. 2.7) the conformal conic resembles the simple or equidistant conic, each having concentric circular arcs for parallels of latitude and straight equally spaced radii as meridians. On the conformal conic, however, the parallels are spaced so that conformality is achieved; that is, each small (actually infinitesimal) shape is correctly shown so that, as mentioned earlier, all angles are correct about any given point. To obtain conformality, Lambert set up an equation in which a differential length of a parallel of latitude was required to bear the same

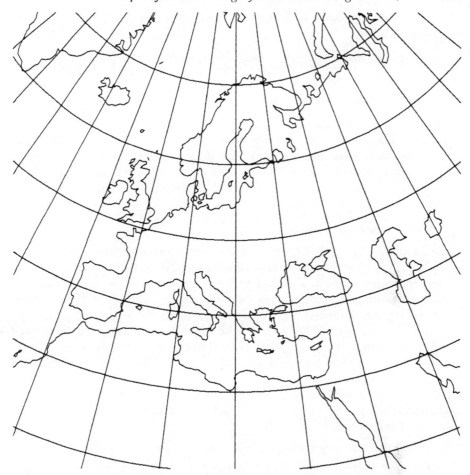

FIGURE 2.7. Europe, using the Lambert conformal conic projection, with standard parallels 20° and 70° N, central meridian 20° E, and 10° graticule. Similar to fig. 2.3, but scale is only true along the standard parallels, and parallels gradually increase in spacing away from the central parallel. However, all very small shapes are correct.

proportion to a differential length of the meridian at the same point on the conic projection as the corresponding elements on the globe. By integrating this differential equation over the entire map, he obtained the necessary equations for plotting the projection.

As a result, the parallels are gradually spaced farther apart as the distance from the central parallels increases. Either one or two parallels may be made standard, or true to scale. If these parallels are in the northern hemisphere, the meridians converge at the North Pole, and the South Pole cannot be shown. For standard parallels in the southern hemisphere, the reverse is true. There is no geometric means of spacing the parallels, and the formulas even for the spherical form are relatively complicated. For the projection with two standard parallels, the equations may be expressed as follows:

$$\rho = R \cos \phi_1 \tan^n(\pi/4 + \phi_1/2)/[n \tan^n(\pi/4 + \phi/2)];$$
$$n = [\ln \cos \phi_1 - \ln \cos \phi_2]/[\ln \tan (\pi/4 + \phi_2/2)$$
$$- \ln \tan (\pi/4 + \phi_1/2)].$$

If there is only one standard parallel, the formula for ρ is the same, but $n = \sin \phi_1$. Symbols and formulas for θ and for conversion of polar to rectangular coordinates are identical with those given previously for the simple conic projection. Lambert used different symbols, with colatitudes (90° − the latitude) and common logarithms (to base 10) instead of natural logarithms (ln) to base e or 2.71828, but his spherical formulas are equivalent to these.

The "Lagrange" Projection

Like the Bonne, Sanson-Flamsteed, and some other projections, the Lagrange projection was promoted but not invented by its namesake. Instead, it was derived by Lambert as a projection in which the entire earth may be shown conformally in a circle, except for "singular" points at the poles, where conformality fails. Furthermore, all the meridians and parallels are shown as circular arcs. Lambert actually derived a general case for the sphere, all forms of which are equivalent to plotting each meridian to coincide with a meridian of the equatorial aspect of the stereographic projection which is equal in value to the original longitude (each measured from the straight central meridian) multiplied by a constant (Lambert 1772; 1972 trans., 41–43). Each parallel of latitude may be plotted to coincide with a parallel of latitude on the equatorial stereographic which bears a more complicated relation to the given latitude:

$$\tan (\pi/4 + \phi'/2) = [\tan (\pi/4 + \phi/2)/$$
$$\tan (\pi/4 + \alpha/2)]^n,$$

where ϕ is the latitude of the new projection, α is the latitude shown as a straight line, ϕ' is the latitude of the equatorial stereographic, and n is the constant multiplier used above for meridians.

For the stereographic itself, $n = 1.0$ and α is equal and opposite in sign to the latitude of the center of the projection (0° for the equatorial aspect). After deriving the equations in equivalent forms, Lambert showed as his example the projection we call the Lagrange, in which $n = \frac{1}{2}$ and $\alpha = 0$, the equator. For this case (fig. 2.8), the formulas, converted to rectangular coordinates, simplify to the following (although Lambert did not pursue this step):

$$x = 4R \cos \phi' \sin (\lambda/2)/[1 + \cos \phi' \cos (\lambda/2)], \text{ and}$$
$$y = 4R \sin \phi'/[1 + \cos \phi' \cos (\lambda/2)],$$

where $\sin \phi' = \tan (\phi/2)$, and x, y, R, and λ are as defined earlier.

The "Lagrange" projection is rarely used for published maps, other than in outline form as part of map projection studies. Lagrange himself (fig. 2.9) was another leading mathematician of the period highly respected by Lambert, who communi-

FIGURE 2.8. The "Lagrange" projection of the world. Reproduced from Lambert (1772; 1894 ed., fig. 11). 10° graticule. The projection is conformal, and all meridians and parallels are circular arcs. *Library of Congress.*

FIGURE 2.9. Joseph Louis Lagrange (1736–1813), a leading mathematician who extensively analyzed conformal projections with graticules of circular arcs.

cated with him about conformal projections (Lambert 1772; 1972 trans., 49). Born in Turin of Italian-French ancestry, he taught in Turin after completing school, until he was invited to Berlin by Frederick the Great in 1766 (Struik 1948, 188–90). When Frederick died in 1786, Lagrange moved on to Paris. While still in Berlin, he published an extended study of Lambert's class of conformal projections in which all meridians and parallels are circular arcs.[64] In his general case, Lagrange varied not only the meridian spacing and the parallel of latitude which is shown as a straight line; he also incorporated the effect of the ellipsoid. Both the calculus and complex algebra (in which he consistently used the numerical $\sqrt{-1}$ rather than the present-day standard symbol i for the imaginary base) were used in his derivations, whereas Lambert did not use complex algebra.

While the formulas for plotting the general Lambert/Lagrange class of projections may be given in several forms, the following form seems the most compact for the spherical version (Lagrange did not use this form of the equations): With n and α as described under Lambert's form (above),

$$x = 4Rm \ (\sin \ n\lambda)/C, \text{ and}$$
$$y = 2R \ (m^2 - 1)/C,$$

where $m = \{[\cos \ \phi/(1 - \sin \ \phi)]/[\cos \ \alpha/(1 - \sin \ \alpha)]\}^n$,

$$C = m^2 + 2m \ \cos \ n\lambda + 1,$$

and the origin of the axes is at the central meridian and α. At the poles, these formulas are indeterminate, but $x = 0$ and $y = 2R$ for the North Pole and $-2R$ for the South Pole.

With other values for his equivalents of constants n and α, Lagrange could achieve various scale relationships, such as a projection minimizing the rate at which the scale changed near any chosen place. For his example of the latter, he chose Berlin, with $n = 1.17$, $\alpha = 18°30'$ S latitude, and the central meridian through Berlin (fig. 2.10).[65]

The Transverse Mercator Projection

Lambert's third new projection is now known by several names, but the most general is the transverse Mercator, a term first applied by Germain (1866, 347). Lambert provided formulas only for the spherical version (1772; 1972 trans., 57–65). Other names, such as the Gauss-Krüger, refer to mathematicians who developed ellipsoidal corrections in the nineteenth and twentieth centuries (Gauss 1825; Krüger 1912). It is the latter form that is primarily now used throughout the world for large-scale topographic and planimetric maps, although as recently as 1911 it was "seldom used" (Close and Clarke 1911, 663).

Lambert described this projection as an improvement over the sinusoidal, for which he described the construction, but he made no mention of the equal-area characteristic of the latter, if indeed he realized its existence (1772; 1972 trans., 57–58). Then he said,

> We will here preserve two characteristics of this design; first that the middle meridian shall be a straight line, equally sub-divided, and then that the equator shall be a straight line, perpendicularly cutting the meridian. To these

FIGURE 2.10. A plot of Lagrange's projection minimizing the scale change near Berlin. 10° graticule. A special case of his conformal projections with circular graticules.

requirements we however also add that all meridians shall penetrate all parallels at right angles, and that the degrees of latitude everywhere are in correct proportion to the degrees of longitude. These latter two conditions have as a consequence that the objective is contained within the much more general version [his analysis of conformal projections] investigated previously. (1972 trans., 58)

The transverse Mercator (fig. 2.11) is the projection obtained by conceptually rotating the cylinder of the Mercator projection 90° so that it is wrapped around a meridian instead of around the equator. The projection is completed by placing a set of transformed meridians and transformed parallels on the globe in exactly the same manner as described for the Cassini projection, but plotting them on the transverse Mercator with, say, dashed straight lines like the equally spaced meridians and

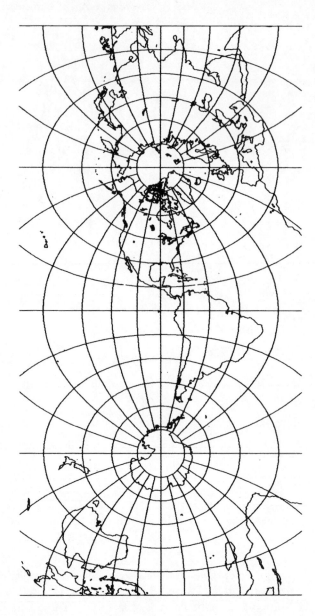

FIGURE 2.11. A transverse Mercator projection of most of the world as a sphere, with central meridians 90° W and E. 15° graticule. Conformal like its parent Mercator projection, it maintains constant scale along the central meridian.

unequally spaced parallels of the regular Mercator (rather than the square pattern of the plate carrée). The actual meridians and parallels may then be placed on the map relative to the dashed lines, just as they fall on the globe. Like the Cassini, the transverse Mercator is not a geometric projection. This transformation may be stated in the formulas

$$x = (R/2) \ln [(1 + \cos \phi \sin \lambda)/(1 - \cos \phi \sin \lambda)], \text{ and}$$
$$y = R \arctan (\tan \phi/\cos \lambda),$$

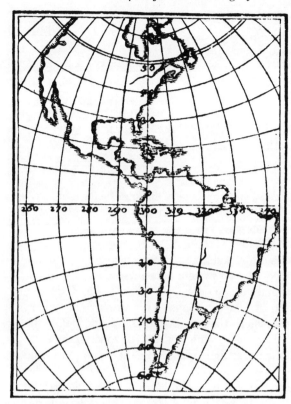

FIGURE 2.12. The Americas as drawn on the transverse Mercator projection when Lambert first presented the projection. 10° graticule. It became the leading projection in the twentieth century for large-scale maps. Reproduced from Lambert (1772; 1894 ed., fig. 13). *Library of Congress.*

where symbols are as defined earlier, and the axes are the same as those for the Cassini. The formula for y is identical with that for the spherical Cassini. These formulas are almost identical with those presented by Lambert (he used colatitude with the symbol ϵ instead of latitude, with some rearrangement of terms) (Lambert 1772; 1972 trans., 61, 62).

Lambert illustrated this projection with a small map of North and South America (fig. 2.12). Because the central meridian is mapped at a constant scale (like the equator of the normal Mercator), the projection is especially suited for regions that have a greater expanse north-south than east-west.

Normal and Transverse Cylindrical Equal-Area Projections

Having defined three new conformal projections derived from the well-established stereographic and Mercator, Lambert turned to equal-area or equivalent representations. He moved briefly in another direction, however, to derive the equations for constructing the equatorial and oblique azimuthal equidistant projections, as noted earlier.

The simplest equivalent projection of the earth is Lambert's normal cylindrical equal-area (fig. 2.13),

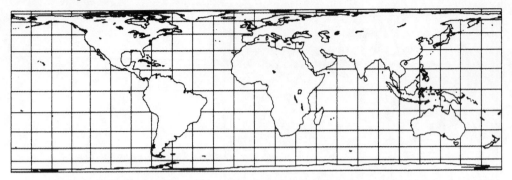

FIGURE 2.13. A modern cylindrical equal-area projection of the world, in the form Lambert devised in 1772, with the equator free of all distortion. The polar regions have extreme shape and scale distortion.

in which the meridional circles are straight parallel lines, which intersect the equator, and its parallels, which are also straight, at right angles. This case is very easy. The equator is divided into 360 equal parts, one for each degree, and the meridians are now straight lines cutting the equator at right angles, and the parallels are inserted where their sines fall above and below the equator. . . . The entire procedure rests on the fact that the zones of the earth from the equator to the poles increase their spatial content as the sine of the latitude. This requires that the degrees of latitude become noticeabl[y] smaller towards the pole, and endlessly small at the pole. However, since the first 30 degrees of latitude are not very different, a map of Africa, or of other countries lying near the equator, comes off well (see Fig. [2.14]). (Lambert 1772; 1972 trans., 71–72)

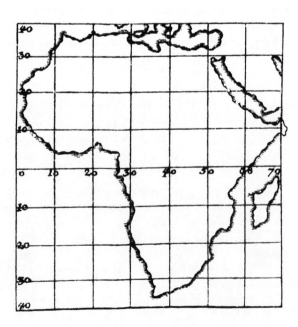

FIGURE 2.14. Lambert's map of Africa using his cylindrical equal-area projection, with no distortion along the equator; 10° graticule. Reproduced from Lambert (1772; 1894 ed., fig. 17). *Library of Congress.*

He illustrated with just a graticule for one hemisphere as well as the larger outline map of Africa. Because of the sine function, the parallels may also be constructed geometrically, from the sphere orthographically onto a tangent cylinder, and the formulas are simply

$$x = R\lambda, \text{ and}$$
$$y = R \sin \phi,$$

although Lambert did not formally list them.

Lambert then rotated the conceptual cylinder as he had done for the transverse Mercator, to obtain the corresponding equal-area projection for "countries, such as America, for example, which have their greatest dimension in a North-South direction" (1972 trans., 72). Applying the same principles of transformation, the formulas for the transverse cylindrical equal-area (fig. 2.15) may be shown as follows, using earlier symbols:

$$x = R \cos \phi \sin \lambda;$$
$$y = R \arctan (\tan \phi / \cos \lambda).$$

The formulas for x in the normal plate carrée, Mercator, and cylindrical equal-area projections are all identical, as are the formulas for y in the corresponding transverse aspects. Lambert described the manner of construction, using a map of Asia, but again he did not list formulas (1972 trans., 73).

Lambert's Azimuthal Equal-Area Projection

While Lambert's cylindrical equal-area projection has been used only moderately (although nineteenth- and twentieth-century derivatives have been offered and even strongly promoted; see chaps. 3 and 4), and its transverse aspect is very rare, Lambert's next new projection, the azimuthal equal-area, is now commonly seen in atlases. He discussed the polar aspect (fig. 2.16) as a "different method of [equal-area] representation in which the meridians are straight lines intersecting at the proper angle at the pole" (Lambert 1772; 1972 trans., 75). For this he again used the calculus, integrating a differential equation that required area equivalence of map and globe elements, to derive the relationship equivalent to

$$\rho = 2R \sin [(90° - \phi°)/2]$$

for "subdivision of the meridional circles" at the various latitudes ϕ (1972 trans., 75). He described the equatorial aspect (fig. 2.17) in similar detail, but he only implied the suitability of the oblique aspect, which is at least as popular (1972 trans., 75–78).

Like the azimuthal equidistant, the azimuthal equal-area is not perspective. Both projections are similar in appearance if limited to a hemisphere or less, but the parallels on the polar aspect of Lambert's projection gradually become closer together as the distance from the pole increases. The other aspects of the equal-area projection show a similar compression near the edges. The entire earth may be shown within a circle from one center, but the outer edge of the map is severely distorted, resembling the edge of one hemisphere of the orthographic projection.

The azimuthal projections are closely related, as modern mathematical relationships help confirm. The following equations provide rectangular coordinates for all azimuthal projections of the sphere, using symbols and reference systems described

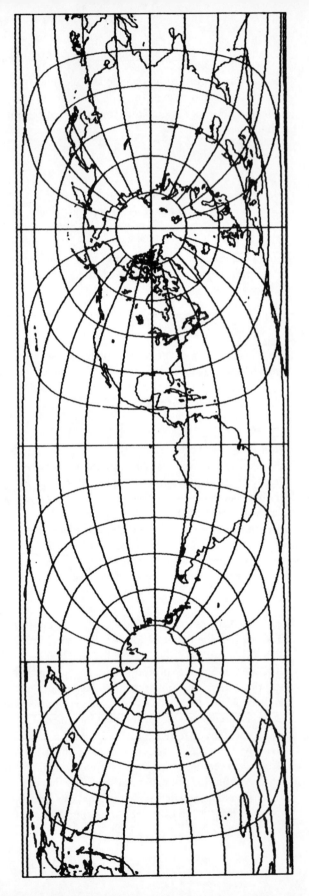

FIGURE 2.15. A modern map of the world based on the transverse cylindrical equal-area projection devised by Lambert. There is no shape or scale distortion along the central meridians 90° W and E. 15° graticule. Compare figs. 2.5 and 2.11; the only construction difference between them is in the spacings in a direction perpendicular to the central meridian.

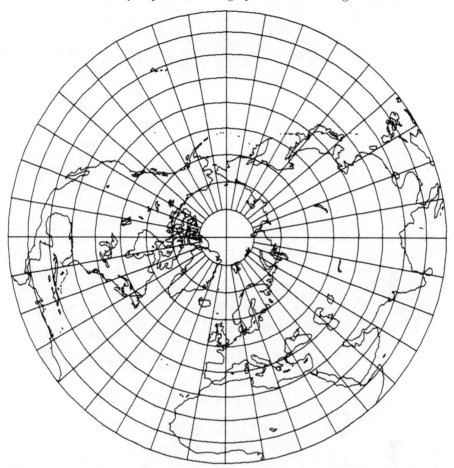

FIGURE 2.16. A polar Lambert azimuthal equal-area projection of the northern hemisphere. Parallels are closer together nearer the equator. 10° graticule.

earlier, except that ϕ_1 is the latitude of the center of the projection, which is the origin for the rectangular coordinates x and y, z is the great-circle distance from the center, and (ρ, θ) are the polar coordinates, radius and azimuth east of north, respectively:

$$x = \rho \sin \theta;$$
$$y = \rho \cos \theta;$$
$$\sin \theta = \cos \phi \sin \lambda / \sin z;$$
$$\cos \theta = (\cos \phi_1 \sin \phi - \sin \phi_1 \cos \phi \cos \lambda)/\sin z; \text{ and}$$
$$\cos z = \sin \phi_1 \sin \phi + \cos \phi_1 \cos \phi \cos \lambda.$$

The y axis increases northerly, the x axis easterly. The following distinguish the projections mathematically: for the orthographic, $\rho = R \sin z$; for the gnomonic, $\rho = R$

FIGURE 2.17. An equatorial Lambert azimuthal equal-area projection of the western hemisphere. Central meridian 100° W. 10° graticule.

tan z; for the stereographic, $\rho = 2\,R \tan (z/2)$; for the azimuthal equidistant, $\rho = R\,z$; and for the Lambert azimuthal equal-area, $\rho = 2\,R \sin (z/2)$. To further simplify the formulas for x and y, note that the term $\rho/\sin z$ is common to both. For the five projections, this common term reduces to R, $R/\cos z$, $2\,R/(1 + \cos z)$, $R\,z/\sin z$, and $2^{1/2}R/(1 + \cos z)^{1/2}$, respectively, thus eliminating several of the trigonometric calculations.

Lambert's Conical Equal-Area Projection

Last and perhaps least of Lambert's cornucopia for cartography is an equal-area conic projection that is very rarely used. Although an equal-area conic projection is commonly used now, it is Albers's, developed thirty years later and discussed in chapter 3. Lambert's projection is actually a special case of the Albers in which one of the two standard parallels of the latter is the North or South Pole.

For Lambert, his conical equal-area, called his isospherical stenoteric projection by Germain and O. S. Adams,[66] was an adaptation of the polar aspect of his azimuthal equal-area projection: "We have thus far assumed that the meridians intersect at the correct angles at the pole. This condition can be modified to take the angles to be *m*-times larger or smaller than the true ones" (Lambert 1772; 1972 trans., 78). To preserve true area, it is consequently necessary to multiply the radius of each parallel of latitude by an appropriate constant related to but not the same as *m*. By properly selecting *m*, the scale along a designated parallel ϕ_1 may be made correct. There is also no distortion of shape along this parallel; thus it is a standard parallel. Again Lambert utilized differential calculus (1972 trans., 78–80). The formulas he derived to produce polar coordinates for the projection centered on the North Pole may be expressed thus with only slight rearrangement:

$$\rho = 2R \sin (\pi/4 - \phi/2)/\sqrt{n},$$
$$\theta = n\lambda, \text{ and}$$
$$n = (1 + \sin \phi_1)/2.$$

The rectangular coordinates may then be found from the equations for *x* and *y* given with the simple conic projection. Symbols are also defined there, again using *n* for the cone constant instead of Lambert's *m*. If ϕ_1 is the equator, the earth is enclosed in a semicircle (fig. 2.18). Lambert chose $n = \frac{7}{8}$ for his example, an outline map of Europe.

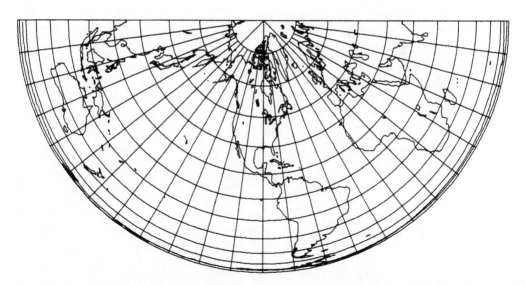

FIGURE 2.18. An outline map of the world based on the Lambert conical equal-area projection, with the equator as the standard parallel, giving the map a semicircular shape. 15° graticule.

CONCLUSIONS

The Renaissance had ended with the introduction of elementary mathematical analysis to the development of map projections. This analysis especially was applied to the projection set forth by Gerardus Mercator and refined by Edward Wright that shows rhumbs as straight lines, and to a pair of projections on which all parallels are cut by meridians at their true spacings whether the parallels are circular arcs (the Werner projection) or straight lines (the sinusoidal). Map projection had moved from easily plotted graticules of simple circular arcs and straight lines to those frequently plotted with the assistance of tables of trigonometric functions.

The fact that the Werner and sinusoidal projections are also correct in area scale appears to have been initially a fortunate coincidence unstated and apparently unknown by those who developed or used them. Lambert himself, who developed projections that he knew to be equal-area, at the same time discussed the sinusoidal with no mention of the fact that it too has no area distortion.

The Mercator projection was already known to be conformal by Mercator (although his understanding of the principle was rudimentary), and the stereographic was discovered by Edmond Halley to have this property. Murdoch consciously made the total area of his map correct with respect to the portion of the sphere being mapped, but he did not pursue constancy of local area scale. The first deliberate achievement of correct area scale at every point on a map was left to Lambert and his development of cylindrical, azimuthal, and conical equal-area projections. Halley applied only geometry to establish the conformality of the stereographic, but Lambert not only used the calculus to prove the validity of the properties but also developed three new conformal projections as well.

Lambert's new equal-area and conformal projections were major advances that would have been almost impossible without the calculus. The others of the approximately sixteen new projections of the eighteenth century (table 2.1 and fig. 2.19) were relatively more pedantic (except for the work of Lagrange and Parent), utilizing

TABLE 2.1. Map Projections Developed during 1670–1799

Modern Names	Figures	Inventor (Date)	Advantages Claimed or Presumed in Development
Bonne	2.1	Gradually improved from Ptolemy's 2d proj. (ca. A.D. 100) by Sylvano (1511), De l'Isle (ca. 1700), and Bonne (1752), with true scale along all parallels and central meridian, and proper curvature of central parallel.	
La Hire	2.2	La Hire (1701)	Perspective azimuthal with low error.
Parent	—	Parent (1702)	Three perspective azimuthals with low error.
(None)	—	Mead (1717)	Two modified globulars, perhaps not original.

(continued)

TABLE 2.1. Continued

Modern Names	Figures	Inventor (Date)	Advantages Claimed or Presumed in Development
Equidistant conic; simple conic	2.3	Gradually improved from Ptolemy's 1st proj. (ca. A.D. 100) by Contarini (1506), Ruysch (1507), and De l'Isle (1745), with true scale ultimately along one or two parallels and all meridians.	
Murdoch	—	Murdoch (1758)	Two equidistant conics with low error and total area correct; one perspective conic with total area correct.
Euler	—	Euler (1777)	Equidistant conic with equal-scale error for extreme and central parallels.
(None)	—	Colles (1794)	Perspective conic.
Cassini	2.5	C. F. Cassini (1745)	Central meridian and its perpendiculars all straight and at true scale.
Lambert-conformal conic	2.6, 2.7	Lambert (1772)	Conic proj. with local angles correct (conformal); proj. intermediate between Mercator and polar stereographic.
Lagrange	2.8, 2.10	Lambert (1772)	Modification of equatorial stereographic to show world conformally in circle with circular meridians and parallels; generalized by Lagrange (1779).
Transverse Mercator	2.11, 2.12	Lambert (1772)	Central meridian straight and at true scale; equator straight; conformal.
Cylindrical equal-area	2.13, 2.14	Lambert (1772)	Simple equal-area; equator at true scale; meridians and parallels all straight.
Transverse cylindrical equal-area	2.15	Lambert (1772)	Preceding proj. rotated so that central meridian is straight and at true scale; equator straight.
Lambert azimuthal equal-area	2.16, 2.17	Lambert (1772)	Azimuths correct from center; equal-area.
Lambert conical equal-area	2.18	Lambert (1772)	Conic adaptation of polar azimuthal equal-area to provide correct scale along chosen parallel.

Note: Map projections are listed in the order described in chapter 2. No contemporary names are known for the new projections.

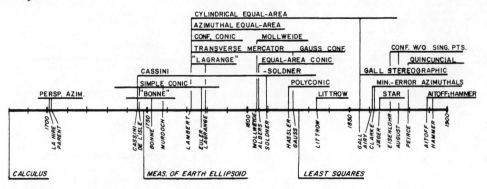

FIGURE 2.19. Time line, 1680–1900, including chapter 3, showing origin of important projections, inventors, and some related events. Vertical positions of projections are not significant. Extensions of horizontal lines to left or right of vertical lines imply indefinite extent.

geometry and trigonometry but essentially no calculus—the projections of La Hire, De l'Isle, Murdoch, Colles, Euler, and even Cassini. What made Cassini's projection so significant was its treatment of the earth as an ellipsoid, but this was accomplished by geodetic triangulation rather than by derivation using the geometry of the ellipse as achieved by Lambert.

Thus it was at the hand of Lambert, especially, that the development of map projections progressed from Renaissance adolescence to scientific maturity. Many more projections would be developed in the nineteenth and especially twentieth centuries, but many of the principles of analysis were well established by and particularly in 1772.

THREE

Map Projections of the Nineteenth Century

THE INCORPORATION OF differential and integral calculus into the numerous map projections developed by Lambert (1772) showed that the centuries-old subject was fertile for significant advancements. Lambert had not set the boundaries of the subject, however. The new mathematics of the nineteenth century brought forth such pertinent concepts as least squares, elliptic functions, conformal mapping of a circle into a polygon, and distortion ellipses. Such innovators as George Airy, Friedrich Eisenlohr, Charles Sanders Peirce, and Émile Guyou developed very complicated and academic projections utilizing one or more of these mathematical principles; others such as F. R. Hassler, James Gall, and G. Jäger presented projections that became popular partly because they could be more easily understood, even if they did not have a special property of conformality, area equivalence, or other mathematical features. Then there were the useful equal-area projections of Karl B. Mollweide and Heinrich Christian Albers, and Carl F. Gauss's ellipsoidal transverse Mercator.

The wealth of material available to us on the multifaceted approach to map projections during the nineteenth century (and the even larger amount available for the twentieth) requires condensation for this volume. In contrast, much more of the limited resource material was used to recount pre–nineteenth-century map projection history.

NINETEENTH-CENTURY USE OF EARLIER PROJECTIONS

By 1800, over thirty projections had been developed, many of them proposed or used as alternatives to earlier more rudimentary forms, such as the conic projections that much later replaced the coniclike forms developed by Claudius Ptolemy. It is almost impossible to speak accurately of a total number of projections, because they were not necessarily presented as discrete sets. In any case, by the end of the eighteenth century, about a dozen were in use for published maps. Of Lambert's seven projections presented in 1772, three received modest use in the nineteenth century and became very important in the twentieth. The rest have been relatively dormant.[1]

The commercial mapmakers of the nineteenth century included some, such as Hermann Berghaus (1828–1890), August Petermann (1822–1878), and Ernst

Debes (1840–1923), who seemed intrigued by projections and utilized a broad range or devised new ones, but most mapmakers were content to follow conventional practices in choosing projections for their atlases and wall maps. The perusal of some twenty nineteenth-century atlases, in addition to the observation of maps reproduced in modern studies of the history of cartography, has led to the analysis of contemporary commercial projection usage described below. In general, these nineteenth-century atlases do not identify the projections used for the maps, aside from the frequently named Mercator. The projections were determined by this writer from their appearance and any necessary measurements. A rare exception is the Debes (1895) atlas, in which all projections are identified at least by name.

Cylindrical and Rectilinear Projections

The Mercator Projection

The Mercator projection (fig. 1.37) conceived in the sixteenth century was emerging as a frequent basis for world maps, but circular hemispheres were just as common in providing the sole world map (in two parts) in atlases during the nineteenth century. The Mercator was the only world map projection in an 1814 geography by Morse, but in more complete atlases throughout the century it was supplemented by globular projections (or occasionally still the equatorial stereographic) portraying the eastern and western hemispheres.[2] The first American star chart was also based on the Mercator and extended between 66°30′ N and S declination; the work of William Croswell (1760–1834), it was published in Boston in 1810.[3]

Aside from its use for world maps, with its excessive areal exaggeration as the polar regions are approached, the Mercator was also the basis of most of the atlas maps of oceans, chiefly the Atlantic and Pacific, or of Australia and the islands of the South Pacific.[4] Since the Mercator was intended as a maritime aid because of its portrayal of rhumbs as straight lines, the mapping of oceans may seem an ideal use of the projection, but this is true only if an important purpose of a map is to lay down navigational routes. In geographic atlases this is rarely the case, where relative sizes and distances are just as important on sea as on land. Nevertheless, the practice has continued into the late twentieth century.[5]

A more appropriate use of the Mercator, ideal for a conformal map of a region straddling the equator, can be found in some atlases where it was used for maps of the East Indies.[6] The Mercator projection can also be justified because of its conformality for use on some earlier maps showing wind directions and ocean currents, but many later thematic or theme-based maps of the nineteenth (and twentieth) century have left the same sort of misconception in showing distributions and isolines that geographic use has caused.[7]

The Equirectangular and Trapezoidal Projections

By the 1800s the simply drawn equirectangular and trapezoidal map projections (figs. 1.4, 1.6) were nearing obsolescence as detailed geographical maps. Both projections contain equidistant, straight lines for parallels of latitude as well as for meridians; the latitude lines are parallel in each case, but the meridians are parallel for the equirectangular and converging for the trapezoidal. A 1791 map, *The United States of America* by Osgood Carleton, is based on the equirectangular with a standard parallel of about 40° N. (Ristow 1985, 68–69). The trapezoidal projection was used as late as 1859 for several maps of the United States and one of the Mediterranean Sea.[8]

Transverse Cylindrical Projections

By the beginning of the nineteenth century, there were three established transverse cylindrical projections—the transverse plate carrée, usually called the Cassini (fig. 2.5), the transverse Mercator (fig. 2.11), and the transverse cylindrical equal-area (fig. 2.15). They are transverse to the basic cylindrical projections that are "equidistant," conformal, and equal-area, respectively.

The first was developed in its initial form in 1745 by Cassini de Thury for the survey of France and was from the beginning applied to the ellipsoidal figure of the earth. The second and third were both presented by Lambert in 1772, and only for the earth as a sphere. Although the transverse cylindrical equal-area projection has rarely been used, the other two were frequently used during much of the nineteenth and twentieth centuries for large-scale topographic mapping. This required an increasingly precise understanding of the ellipsoidal form of each projection, as discussed below. In the spherical forms of the three, only the transverse Mercator projection has been found in nineteenth-century atlases, twice in Debes's atlas.[9]

The Cassini-Soldner Projection About 1810 Johann G. von Soldner developed a system of rectangular coordinates based on the earth as a sphere, for the large-scale mapping of Bavaria. He constructed the meridian of the origin or zone center at a constant scale. This served as the y axis, while the x coordinates were designed to maintain the same scale for the lengths of arcs of great circles perpendicular to this meridian.[10] The origin was placed along the meridian at an arbitrary central latitude. This followed (for the sphere) the same principles as the Cassini projection, thus leading to the name Cassini-Soldner for the projection as used for large-scale cadastral maps. During the nineteenth century, "this method of projection was used in Germany for the cadastral maps of most states; it constituted, up to the year 1927, also the basis of the cadastral maps in Prussia, where the ellipsoidal shape of the earth has been taken into consideration, however" (Jordan and Eggert 1939; 1962 trans., 191).

Beginning the last half of the nineteenth century, the same projection was used by the Ordnance Survey for topographic mapping of the United Kingdom, with separate origins first for individual counties and later, with some consolidation of the separate projections near the end of the century, for thirty-nine Cassini zones.[11] This projection, already replaced in France with the Bonne, was superseded in the United Kingdom by the 1920s, since it is neither conformal nor equal-area, but, according to Steers (1970, 229), "the reasons for its selection appear to have been primarily ease of calculation and plotting, coupled perhaps with the great prestige of" Cassini. During the twentieth century, it was replaced by the transverse Mercator projection (below) not only in England, but also in most other countries in which it had been adopted.

Using the spherical formulas for the ellipsoid involved a special radius of the sphere. This radius was an appropriate mean of the curvatures of the ellipsoid in various directions at a given point near the center of the region being mapped. Later formulas accommodated the varying curvature of the central meridian as well as the parallels and thus truly treated the earth as an ellipsoid rather than as a closely approximating sphere.

Ellipsoidal Transverse Mercator Projections Prompted by the studies of conformality published in 1779 by Lagrange, but apparently not by those of Lambert,[12] Carl Friedrich Gauss (1777–1855) (fig. 3.1), one of the leading mathematicians of all time, engaged in an extensive study of the transformation of one surface onto another. This included the conformal transformation of the ellipsoid onto the plane. While Lambert had developed the transverse Mercator projection for the sphere, in 1822 Gauss carried out the mathematical derivation for the ellipsoidal form with the central meridian at a constant scale. This projection, frequently referred to as the Gauss conformal, was used for the survey of Hannover during the same decade.[13] It was this form of the transverse Mercator that was reevaluated by Louis Krüger (1912), leading to the current name Gauss-Krüger.

Meanwhile, Gauss (1843), while originating the term *conformality*, used another important conformal transformation, first developed by Lagrange (1779), namely, the conformal projection of the ellipsoid onto a sphere. This sphere was in turn projected with the spherical transverse Mercator projection formulas onto a plane to provide a "double projection" adapted by Oskar Schreiber and used for the

FIGURE 3.1. Carl Friedrich Gauss (1777–1855), one of the greatest of all mathematicians. He developed the ellipsoidal version of the transverse Mercator projection, now the one most commonly used for large-scale mapping.

Prussian Land Survey of 1876–1923.[14] This version of the transverse Mercator is perfectly conformal, but the central meridian is not at a constant scale.

Gauss was born in the German town of Brunswick in 1777 to poor parents. At the age of ten his mathematical genius began to attract attention, and his education at Caroline College in Brunswick was paid by the duke of Brunswick. At eighteen Gauss invented the method of least squares, just before he began studies at the University of Göttingen, where his prolific output continued with the duke's financial support. Discovery of the first known asteroid, Ceres, by Giuseppe Piazzi in 1801 inspired Gauss to begin an intensive study of celestial mechanics with a calculation of the asteroid's orbit from only a few observations. This was followed by monumental studies including geodesy, conformal mapping, and mathematical physics. Gauss, director of the Göttingen Observatory from 1807, died in Göttingen in 1855.[15]

Azimuthal Projections
The Stereographic Projection

The stereographic projection (figs. 1.14, 1.16, 1.19), with its circular (or occasionally straight) meridians and parallels—whether the aspect is polar, equatorial, or oblique—combined with its well-known perspective nature and increasingly known conformality, remained popular. Its use was diminishing, however; it was employed almost exclusively for hemispheres, eastern/western, northern/southern, and occasionally oblique hemispheres. One exception was Debes's use of the oblique aspect for regional maps of southern and equatorial Africa.[16] A clear carryover from their own eighteenth-century atlases were equatorial stereographic maps of Africa (reconstructed with modern shorelines in fig. 3.2) and America in De l'Isle and Buache's (1833) world atlas. The stereographic projection was being outnumbered

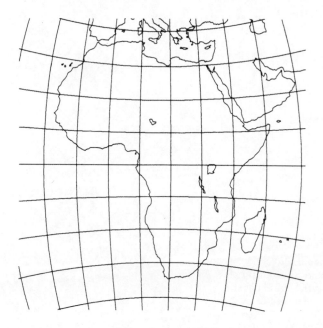

FIGURE 3.2. A modern outline map of Africa using the equatorial stereographic projection with the central meridian near 20° E as used by De l'Isle and Buache (1833) for "Carte de l'Afrique."

in all three aspects, however, by the globular, Lambert azimuthal equal-area, or azimuthal equidistant.

The Orthographic and Gnomonic Projections

Both the age-old orthographic and gnomonic projections (figs. 1.11, 1.12, 1.13) remained almost entirely in peripheral roles. The first map in Bartholomew's 1893 *Graphic Atlas* contained an oblique orthographic hemisphere centered on the North Atlantic Ocean, and Rand McNally's 1896 *Pictorial Atlas* included small oblique and polar orthographic hemisphere insets adjacent to the primary large globular eastern and western hemispheres, but the projection was not even used to that extent in other typical atlases.[17]

The gnomonic projection, with its great scale distortion usually offsetting its unique property of showing all great circles of the sphere as straight lines, was almost totally absent in geographic atlases, except for a six-sheet atlas of the world published at Weimar by C. G. Reichard (1758–1837). Reichard's (1803) work was equivalent to projecting the globe onto a circumscribed cube tangent at the poles (reconstructed in fig. 3.3); it may have been inspired by the series of celestial atlases

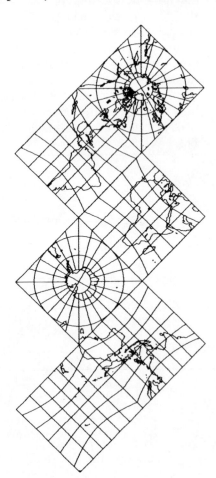

FIGURE 3.3. A modern gnomonic projection of the world onto a cube using the arrangement of Reichard (1803), although central meridians are uncertain. The poles are central on two faces. 15° graticule.

published by at least five mapmakers between 1673 and 1789, also using four equatorial and two polar gnomonic aspects.

The Azimuthal Equidistant Projection

The azimuthal equidistant projection was used, as before, almost entirely in its simplest aspect, for either hemispheres or polar regions centered on the North Pole and occasionally on the South Pole.[18] While the equatorial aspect has little practical value unless one needs distances and directions from a city or other point on the equator, Debes used it for a map of Africa in 1895, and the oblique aspect was used for an 1816 map of a hemisphere centered on London (reconstructed with modern shorelines in fig. 3.4). In 1895 Debes also used it for his map of Europe.[19] An ellipsoidal form was, in effect, devised by Philippe Hatt (1886), using series to calculate rect-

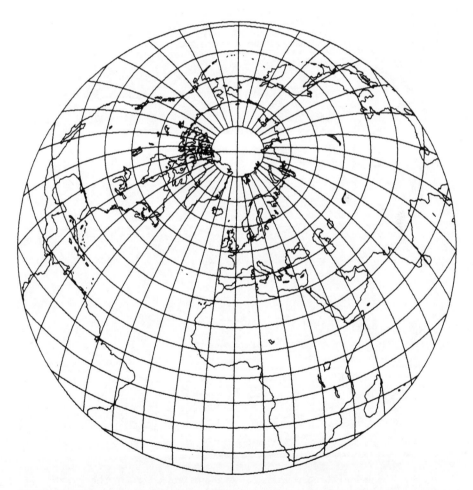

FIGURE 3.4. The oblique azimuthal equidistant projection of one hemisphere centered on London, an outline reconstruction of a map by George Buchanan in Thomson (1816). 10° graticule.

angular coordinates for azimuths and distances; this form was later adopted for some topographic mapping in Greece.

The Lambert Azimuthal Equal-Area Projection

Lambert's azimuthal equal-area projection of 1772 (figs. 2.16, 2.17) was slow to receive recognition. Apparently its first use for small-scale atlas maps did not occur until the 1890s, but this use included all three aspects—polar, equatorial, and oblique (fig. 3.5, a reconstruction of a hemisphere centered on Berlin).[20] The azimuthal equal-area projection was such a logical addition to the list of long-known azimuthal projections, once the calculus could be applied to calculations of areas on zones of the sphere, that the projection was apparently independently rederived in

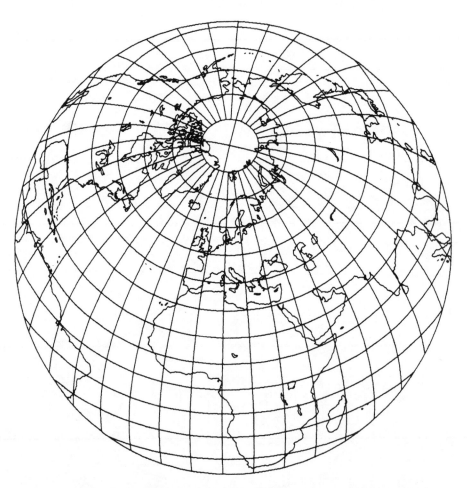

FIGURE 3.5. The oblique Lambert azimuthal equal-area projection of one hemisphere centered on Berlin, an outline reconstruction of map 16 in the Berghaus atlas by Justus Perthes (1892). 10° graticule.

its polar aspect by Anton-Mario Lorgna (1730–1796) in 1789 and called the Lorgna projection in treatises from 1799 to at least 1882.[21]

Conic and Sinusoidal Projections

Cylindrical and azimuthal projections were used infrequently in atlases for maps of continents, countries, and intermediate regions. The simple conic and the Bonne projections (figs. 2.1, 2.3), or the sinusoidal projection (fig. 1.39b), which is the equatorial form of the more general Bonne, were the much preferred forms. The azimuthal projections were often suitable for regional maps, but they were more complicated to construct unless centered on a pole.

The simple conic is constructed using straight equidistant meridians converging to a common point that also serves as the center of the equidistant circular arcs representing the parallels of latitude. It provided the base for maps of small- and medium-sized countries throughout the century, as well as for several maps of the United States, Canada, and Europe. The less satisfactory modification in which most meridians do not intersect the parallels at right angles still appeared into the early nineteenth century, but only one example was found after 1840.[22] The simple conic projection was appropriate for small countries and regions regardless of shape, or for large countries or continents of predominant east-west extent or at worst extending about equally in all directions. In one example, however, North America, a region more nearly north-south in extent, was mapped on the simple conic projection in an atlas of 1896 (Rand McNally 1896, 40–41).

The conformal conic appeared in maps of Europe, Australia, and portions of Asia and the Americas in the Debes atlas of 1895.[23] Debes also innovatively used the conformal conic in its oblique aspect for West Africa.[24] In several other cases, however, the conformal conic was described as if the projection originated during the nineteenth rather than eighteenth century.

For atlas maps of large countries and continents, the Bonne projection was clearly the favorite, giving the regions both a uniform area scale and the aesthetically pleasing combination of curved meridians and curved parallels befitting a representation of the globe. Well over half the maps of the five most populated continents were prepared on either the Bonne projection or, in the case of equatorially centered Africa and near-equatorial South America (fig. 1.39b), the sinusoidal. No maps of Africa were found on the Bonne, but it was occasionally used for maps of South America, with the center lying some distance south of the equator.[25] Australia was usually plotted according to the Bonne projection if shown separately, but it was just as often part of a map of Oceania on a projection such as the Mercator. Other regions occasionally mapped on the Bonne included the Arabian Peninsula, China, Russia, and the United States.[26]

In addition to its employment for small-scale maps in atlases and for sheet maps, the Bonne projection played a significant role in nineteenth-century larger-scale topographic mapping using ellipsoidal formulas. This began with its adoption in France after steps begun in 1802 by Bonne, finally leading after wartime delays to its acceptance in 1821 by a special commission of the Dépôt de la Guerre (war office). The Bonne projection thus replaced the Cassini used for the then obsolete maps of France prepared during the 1700s.[27] It received the name *projection du dépôt de la guerre* in other European countries, where it was used for similar purposes. These regions included Austria-Hungary (1:750,000 scale maps), Belgium (1:20,000 and reductions), Denmark (1:20,000), Italy (1:500,000), Netherlands (1:25,000), Russia (1:126,000), Spain (1:200,000), Switzerland (1:25,000 and 1:50,000), Scotland and Ireland (1:63,360 and smaller), as well as France (1:80,000 and 1:200,000) (Hinks 1912, 65–66).

The Bonne may have been used for some of the U.S. Coast Survey charts of the Delaware Bay in 1844–45 (Schott 1882, 293). One larger-scale American use of the Bonne was for a nine-sheet map of the state of Virginia (then including West Virginia) prepared in 1825 by Herman Böÿe, a Richmond engineer of German ancestry, at a scale of five miles to the inch (1:316,800).[28] As stated on the map, "Flamsteed's Projection, as modified by French geographers, was adopted for the projection of the map, the earth being considered a sphere." (The modified Flamsteed is another name for the Bonne projection, as applied to the topographic mapping of France [Hunt and Schott 1854, 98].)

The Globular Projection

Emerging as the chief projection used for maps of the eastern and western hemispheres was the globular (fig. 1.32), invented by al-Bīrūnī about 1000, then reinvented by Nicolosi in 1660 and independently promoted by Arrowsmith beginning in 1794. It replaced the longstanding equatorial stereographic, retaining the appeal of a graticule consisting entirely of circular arcs or straight lines. On the other hand, it replaced preservation of local shape and angles (conformality) on the stereographic with preservation of correct scale along the equator and along the central meridian on the globular. Although principally used for maps of these two hemispheres in atlases throughout the nineteenth century, the globular projection was occasionally used for some continents, specifically the Americas and Asia, and even for atlas sheets of India.[29]

One other old globular projection, Apian's second of 1524, with equidistant elliptical meridians and equidistant straight parallels, was used independently by Dominique François Jean Arago (1786–1853) in 1835 in his *Astronomie populaire*.[30]

THE NEW PROJECTIONS OF THE NINETEENTH CENTURY

The number of atlas and wall maps of the nineteenth century not utilizing one of the above projections was small (table 3.1). Occasionally, however, new projections by mapmakers or influential scholars found their way into a commercial atlas. The polyconic projection developed for United States coastal mapping in the 1820s is an important example of a new projection that became so well accepted because of its use by the United States government for large- and medium-scale maps that it began to appear in several atlases for small-scale maps as well.

New Cylindrical Projections

Perhaps because of the very simplicity of cylindrical projections in the normal aspect, they were generally ignored by mathematicians and the more scientific mapmakers especially attracted to the development of new projections. Cylindrical

TABLE 3.1 Map Projections in Nineteenth-Century World Atlases, Major
Regions

Atlas	Region											
	1	2	3	4	5	6	7	8	9	10	11	12
Warner and Carey (1820)	ME	GL	—	—	SI	GL	—	BO	BO	SI	BO	—
De l'Isle and Buache (1833)	—	ES	—	—	ES	BO	—	BO	ES	ES	—	—
Bradford (1835)	—	GL	ST	—	SI	BO	—	BO	BO	SI	SC	ME
Stieler (1855)	ME	ES	—	ST	SI	BO	BO	SC	BO	BO	SC	ME
Cornell (1859)	—	GL	AE	—	SI	BO	—	BO	BO	SI	TZ	ME
Mitchell (1866b)	ME	GL	—	—	SI	BO	—	BO	BO	SI	BO	ME
Appleton (1872)	—	GL	—	—	SI	PO	—	SC	BO	SI	SC	ME
J. G. Bartholomew (1890)	ME	GL	LA	LA	SI	BO	BO	BO	BO	SI	SC	SI
Justus Perthes (1892)	ME	EL	LA	—	SI	BO	—	BO	BO	BO	—	BO
J. G. Bartholomew (1893?)	ME	GL	—	LA	SI	BO	SC	SC	BO	BO	SC	ME
Debes (1895)	ME	EE	—	—	EE	OE	LC	LC	OB	OB	LC	ME
Rand McNally (1896)	ME	GL	—	AE	PO	BO	—	SC	SC	SI	SC	ME
B. E. Smith (1897)	ME	GL	—	AE	SI	BO	BO	BO	PO	SI	PO	ME

Notes. Stieler (1855) also used the Bonne for a map of Europe. Bartholomew (1890) also used the Gall stereographic for a world map. Justus Perthes (1892) also used the Mollweide for a world map. Inset maps are not included.

Region number		Projection symbol	
1	world	AE	polar azimuthal equidistant
2	eastern and western hemispheres	BO	Bonne
		EE	equatorial azimuthal equidistant
3	northern and southern hemispheres	EL	equatorial Lambert azimuthal equal-area
4	polar areas	ES	equatorial stereographic
5	Africa	GL	globular
6	Asia	LA	polar Lambert azimuthal equal-area
7	Australia	LC	Lambert conformal conic
8	Europe	ME	Mercator
9	North America	OB	oblique Breusing
10	South America	OE	oblique azimuthal equidistant
11	United States	PO	polyconic
12	Oceania.	SC	simple conic
		SI	sinusoidal
		ST	polar stereographic
		TZ	trapezoidal.

projections have equidistant parallel straight lines for meridians and parallel straight lines for latitudes, so the only option open to the innovator is the spacing of the latter. Furthermore, the only conformal cylindrical projection is Mercator's of 1569, and the only equal-area cylindrical projection is Lambert's of 1772, except for affine reshaping of the latter. Other than the equidistantly spaced parallels of the plate carrée and the related equirectangular, almost any other cylindrical projection has to be a compromise, with an arbitrary choice of parallel spacing.[31]

During the 1800s new regular cylindricals, all of them for world maps, were announced by only two individuals, both of them clergymen. The first made no reference in his paper to the fact that Lambert's work was closely related to one of his presentations, probably because he was not aware of it, and he included none of the rather simple mathematics involved in spacing the parallels. In addition, one obvious but almost useless cylindrical projection with no credited originator seems to have found its way into textbooks: the central cylindrical, produced by projecting the globe geometrically from the center onto a tangent (or secant) cylinder (fig. 3.6). The distortion of shape and area increases rapidly away from the equator, and, unlike the somewhat similarly derived gnomonic (projected geometrically from the center onto a tangent *plane*), it has no compensating favorable property. Its inclusion in textbooks is most useful in showing that it is not the same as the Mercator, even though both have gross area distortion and cannot show either pole.[32]

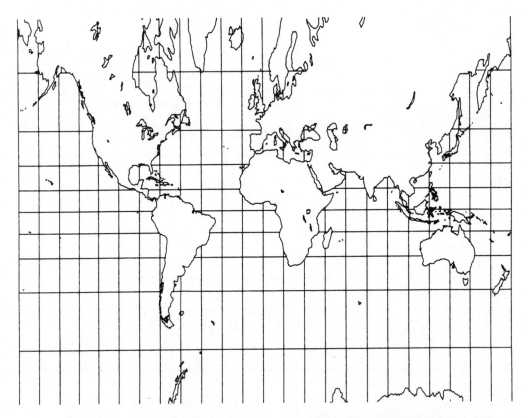

FIGURE 3.6. The central cylindrical projection of most of the world. 15° graticule. A perspective projection from the center of the globe onto a tangent cylinder, it has little value except as a textbook comparison with the Mercator projection, because of its extreme distortion.

The formulas are simply

$x = R \lambda$, and

$y = R \tan \phi$,

where variables are as described on p. 49.

Although the central cylindrical seems to have an unknown origin, sometime before the mid-nineteenth century J. Wetch proposed a transverse aspect, rotating the poles of the globe to coincide with the equator of the base projection.[33] This has the formulas

$x = R \cos \phi \sin \lambda/(1 - \cos^2\phi \sin^2\lambda)^{1/2}$, and

$y = R \arctan (\tan \phi/\cos \lambda)$.

(Variables are defined on pp. 49 and 76.)

A more useful new aspect of an existing cylindrical projection also seems to have casually appeared without treatment as a formal innovation—the oblique Mercator, or oblique conformal cylindrical projection (fig. 3.7). Charles Sanders Peirce (pronounced "purse") (1839–1914), whose development of the quincuncial projection is discussed below, recorded his concept of the oblique or "Skew Mercator" with the U.S. Patent Office in 1894 (Eisele 1963, 305–6). He applied it to a small portion of the ellipsoidal earth, but did not complete the development. The first use of the oblique Mercator is found (twice, in the spherical form) in Debes's 1895 atlas, and Alois Bludau refers to Debes's use and to the projection in 1899.[34] The two maps portray Southeast Asia and Central America; Debes names the projection and states the locations of the poles of the oblique aspects, each tilted 45° from the normal poles.

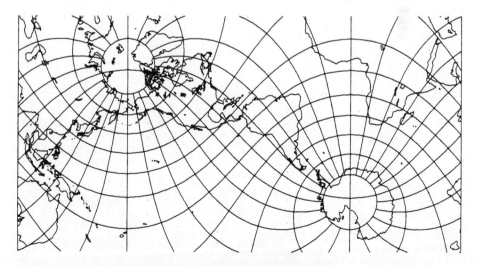

FIGURE 3.7. An example of the oblique Mercator projection with the North Pole falling where 45° N latitude and 90° W of center would fall on the regular Mercator. Central meridian 90° W (defined as meridian at map center in this case).

The first new regular cylindricals of the nineteenth century that were formally presented were those of James Gall (1855, 1871, and 1885; Freeman 1963), a clergyman of Edinburgh, in 1855. Born in 1808 to a publisher whose firm he joined in 1838, Gall left the business after ten years to become an evangelist. Ordained in 1861, he held a pastorate for ten years, but continued mission work, dying in 1895. Prompted, like many others, by the desire for a world map that avoided some of the scale exaggerations of the Mercator projection, Gall (1885, 119, 121) stated that

> I had published *An Easy Guide to the Constellations,* in which each constellation had a page for itself; and it became necessary for me to publish an *Atlas of the Stars* to show their connection. In planning the atlas, it occurred to me that a magnificent panorama of the stars, including three-fourths of the heavens, might be brought into one map by using a cylindrical projection, with the latitudes drawn stereographically and rectified at the 40th parallel. . . .
>
> It then occurred to me that the same, or a similar projection, would give a complete map of the world, which had never been done before; and, on drawing a projection with the latitudes rectified at the 45th parallel, I found that the geographical features and comparative areas were conserved to a degree that was very satisfactory.
>
> . . . I read a paper before the British Association in 1855, and exhibited the three new projections—the Stereographic, and Isographic, and the Orthographic.

His "Orthographic" is equal-area (fig. 3.8), actually a cylindrical equal-area modified to obtain two standard parallels at 45° N and S. Gall (1885, 121) said of this, "it is true that the geographical features are more distorted on this than on any of the others, but they are not distorted so as to be unrecognisable; and so long as that is the case, its advantages are not too dearly bought." His "Isographic" (fig. 3.9) is an equirectangular projection with the same two standard parallels. He preferred his

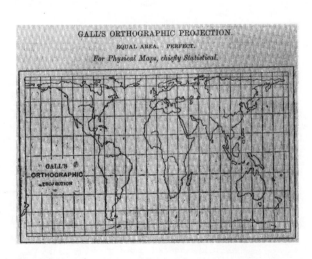

FIGURE 3.8. The Gall orthographic projection reproduced from Gall (1885, 121). The cylindrical equal-area projection with standard parallels at 45° N and S latitude. 15° graticule. Revived a century later as the "Peters" projection. *U.S. Geological Survey Library.*

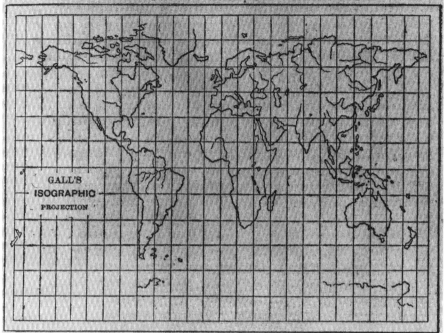

FIGURE 3.9. The Gall isographic projection reproduced from Gall (1885, 123). The equirectangular projection with standard parallels 45° N and S. 15° graticule. *U.S. Geological Survey Library.*

"Stereographic" (fig. 3.10) for general purposes. It is a perspective projection of the globe onto a cylinder from a point on the equator of the globe opposite the meridian being projected. The cylinder is not tangent to the globe at the equator, but is secant, cutting the globe at 45° N and S, and giving the projection the same standard parallels as those of his other two projections. It is neither equal-area nor conformal, but its balance of distortion has led to its use in several atlases published in Great Britain, including some by the Bartholomew family, beginning with John G.[35]
    For its coordinates,

$$x = R\lambda/\sqrt{2}$$
$$= 0.7071 \ R\lambda, \text{ and}$$
$$y = R(1 + \sqrt{2}/2) \tan (\phi/2)$$
$$= 1.7071 \ R \tan (\phi/2),$$

with symbols as given above.

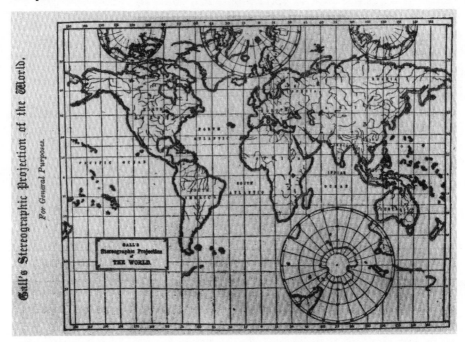

FIGURE 3.10. The Gall stereographic cylindrical projection, or just the Gall projection, reproduced from Gall (1885, 120). 15° graticule. The one of his three that he preferred, it is a perspective of the globe onto a cylinder secant at 45° N and S, the standard parallels. The point of perspective is on the equator opposite a given meridian. *U.S. Geological Survey Library.*

Gall (1885, 122) declared that his stereographic, "though inferior to Mercator's for navigation, is superior to it" in that is shows the entire world and does not have the "great waste of room" in the polar regions or the "grossly misrepresented" areas. He says, with exaggeration in his own statements, that in "the Stereographic [Spitzbergen and Borneo] appear in their true proportions," and polar regions "are as accurately represented as the equatorial."

Gall (1885, 123) wanted no copyright, but said "all that I would ask is that, when they are used, my name may be associated with them . . . as Gall's Stereographic, Isographic, and Orthographic Projections of the World." Nevertheless, a century later, his "Orthographic" projection was being heavily promoted as a new projection by Arno Peters, in spite of repeated statements in critical cartographic literature stressing the lack of novelty.

The other new cylindrical projection involved a cylinder tangent at the equator rather than secant elsewhere. It was proposed by Carl Braun (1831–1907), a Jesuit priest (Braun 1867; Germain 1868). On his version (fig. 3.11), the graticule is geometrically projected from the same moving point as that used by Gall for his stereographic, namely, a point on the equator of the globe opposite the meridian being projected. The tangency of Braun's cylinder leads to a standard parallel at the equa-

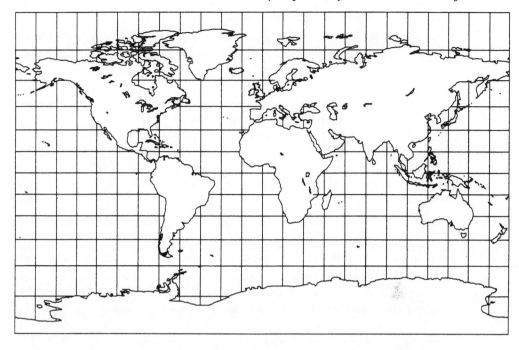

FIGURE 3.11. The Braun stereographic cylindrical projection. Developed like Gall's (fig. 3.10) but with the cylinder tangent at the equator, which is the standard parallel. 15° graticule.

tor, rather than at Gall's 45° N and S, thus closely resembling the Mercator near the equator (with spacing from the equator agreeing within 2% in a band about 25° to either side). In a footnote, Braun points out that to emulate approximately the Mercator with a perspective cylindrical projection, one can place the point of projection in the equatorial plane, but at ⅗ of the globe radius from its center in a direction opposite the meridian being projected; then "one obtains rather accurate spacing of parallels from the equator for the Mercator projection," except near the poles.[36] The resulting spacing from the equator is within 2% up to 80° N or S. For both of these cylindrical projections,

$$x = R \lambda,$$

but for the first, as stated by Braun (1867, 272 with different symbols),

$$y = 2R \tan (\phi/2).$$

And for the footnoted modification, formulas are not given by Braun, but

$$y = 1.4R \sin \phi/(0.4 + \cos \phi).$$

Braun's (first) projection, like Gall's stereographic, is an attempt to balance distortion and avoid the excesses of the Mercator projection (as stressed by both Gall and Braun), but Braun's approach has been used much less than Gall's.

New Pseudocylindrical Projections

Until 1805 the only pseudocylindrical projection with important properties was the well-known sinusoidal or Sanson-Flamsteed. Pseudocylindricals have in common straight parallel lines of latitude and curved meridians. The sinusoidal has, in addition, equally spaced parallels of latitude, true scale along parallels, and equivalency or equal-area. As a world map, it has the disadvantage of high shear at latitudes near the poles, especially those farthest from the central meridian.

In 1805, Karl Brandan Mollweide (1774–1825) announced an equal-area world map projection that is aesthetically more pleasing than the sinusoidal because the world is placed in an ellipse with axes in a 2:1 ratio, and all the meridians are equally spaced semiellipses, except for limiting forms—a straight line for the central meridian and a circle for the two meridians 90° east and west of the central meridian (fig. 3.12).[37] Mollweide presented his projection in response to a new globular projection of a hemisphere, described by Georg Gottlieb Schmidt (1768–1837) in 1803 and having the same arrangement of equidistant semiellipses for meridians. But Schmidt's curved parallels do not provide the equal-area property that Mollweide obtained. (Schmidt's projection is described below.)

Mollweide was born in Wolfenbüttel, Germany, and graduated from the University of Halle. In 1811 he began work in the astronomical observatory of Leipzig University, where he became professor of astronomy and in 1814 professor of mathematics.[38] While he was mainly a teacher, he published several mathematical papers, including six on map projections. He died in Leipzig.

Although the mathematics for coordinate calculation of the Mollweide projection may be derived relatively simply, the particular uneven spacing of the parallels

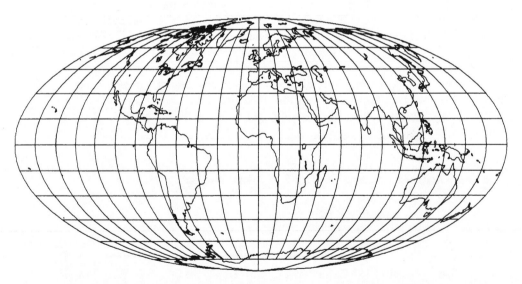

FIGURE 3.12. The Mollweide projection, an equal-area pseudocylindrical projection with all meridians shown as semiellipses and parallels as straight lines. Central meridian 0°. 15° graticule.

involves iteration or else inverse interpolation of a table computed without iteration for transformation from rectangular to geographic coordinates. A parametric angle (θ here) was used by Mollweide (1805b, 156–59, with different symbols), and similar formulas are still used:

$$x = (2\sqrt{2}/\pi)R\lambda \cos \theta, \text{ and}$$
$$y = \sqrt{2}R \sin \theta,$$

where $2\theta + \sin 2\theta = \pi \sin \phi$. Except for the intermediate angle θ, symbols are as given on p. 49. Iteration to find θ for a given ϕ may now be facilitated by using techniques such as the Newton-Raphson, but this was less practical before the advent of modern computers. Jules Bourdin, however, computed a seven-place table (five or six of the places are accurate) giving $\sin \theta$ and $\cos \theta$ for every half degree of latitude. This was reproduced by Germain and others, thus replacing iteration for most precomputer users.[39] These formulas contrast with the much simpler, noniterative pair for the sinusoidal projection:

$$x = R\lambda \cos \phi;$$
$$y = R\phi.$$

The Mollweide projection lay relatively dormant until Jacques Babinet reintroduced it in 1857 under the name *homalographic*.[40] As a result it is found in some atlases of the same century, including some editions of *Berghaus' Physikalischer Atlas*, where it appears both as a single world map for numerous thematic features and as a pair of eastern and western hemispheres, each consisting of the circle-bounded inner hemisphere of the Mollweide world map.[41]

The Mollweide projection, also called the Babinet, homalographic, homolographic, and elliptical projection for reasons indicated above, was the only new pseudocylindrical projection of the nineteenth century to receive much more than academic interest. De Prépetit Foucaut and Édouard Collignon (1831–1897) of France contributed novelties, and Adam Maximilian Nell (1824–1901) of Germany more of an innovation, later in the century.

Foucaut (1862, 5–10) presented what he called the stereographic equivalent projection (fig. 3.13) because the spacing of the parallels along the central meridian is the same as that of the equatorial stereographic azimuthal projection (or of Braun's stereographic cylindrical) (also see O. S. Adams 1945, 23). The meridians are then curved to make the projection equal-area, giving a projection with very sharp poles and much more extreme shearing in the polar regions than that of the sinusoidal. The formulas may be given as follows:

$$x = (2/\sqrt{\pi})R\lambda \cos \phi \cos^2(\phi/2);$$
$$y = R\sqrt{\pi} \tan (\phi/2).$$

In the same 1862 paper, Foucaut also weighted an average of the y coordinates of the Lambert cylindrical equal-area and the sinusoidal projections and then determined x to make the final projection equal-area:

$$x = R\lambda \cos \phi / (a + b \cos \phi), \text{ and}$$
$$y = R(a\phi + b \sin \phi),$$

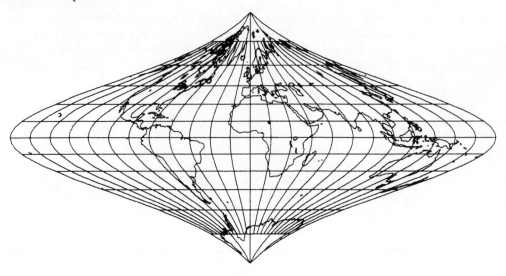

FIGURE 3.13. The Foucaut stereographic equivalent projection. Considerable shape distortion, but equal-area, and the parallels are spaced like those of the equatorial stereographic azimuthal projection along the central meridian. 15° graticule.

where *a* is the weight (0 to 1) given to the sinusoidal, and *b* (equal to 1 − *a*) is that of the cylindrical equal-area projection.

Collignon's (1865) equal-area projection (fig. 3.14) is still more of a novelty, consisting entirely of straight lines for the graticule.[42] In the form with unbroken meridians, the parallel latitude lines are spaced so that straight lines serving as meridians are equally spaced along the parallels of latitude and converge on the North Pole from a South Pole that is 1.414 or $\sqrt{2}$ times as long as the equator. It may also be constructed with meridians broken at the equator, symmetry about the equator, and the inner hemisphere in a square with the poles at opposite vertices. The northern

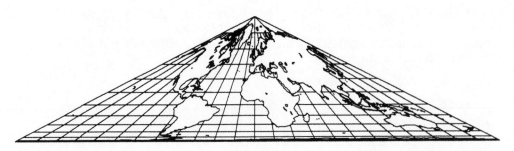

FIGURE 3.14. Collignon's rectilinear equal-area projection of the world. 15° graticule. A mathematical novelty.

hemisphere is the same in either case. The formulas for the first form may be written thus:

$$x = (2/\sqrt{\pi})R\lambda(1 - \sin \phi)^{1/2};$$
$$y = R\sqrt{\pi} \, [1 - (1 - \sin \phi)^{1/2}].$$

Nell (1890) described an equal-area pseudoconic projection (discussed later), of which the equatorial limiting form is a pseudocylindrical which resembles, like Foucaut's, an average of the cylindrical equal-area and sinusoidal projections, but with differences. Nell derived the formulas for the ellipsoid, but they can be simplified for the sphere as follows:

$$x = R\lambda \, (1 + \cos \theta)/2, \text{ and}$$
$$y = R\theta,$$

where $\theta + \sin \theta = 2 \sin \phi$. Iteration is involved in the last equation. The result is a projection (fig. 3.15) with poles as straight lines 0.724 times the length of the equator. This is apparently the first true pseudocylindrical projection with poles as lines rather than points.

New Conic Projections

The new cylindrical and pseudocylindrical projections were primarily intended for world maps, but conic projections were better suited to the extensive requirements for regional maps. Paralleling to some extent the limitations on new cylindrical projections, there is only one type of conformal conic projection, that developed by Lambert, one type of equal-area conic, that invented by Albers in 1805 (Lambert's conical equal-area is a limiting case), and one type of true equidistant conic, the one to which De l'Isle's name is often given.

Variations in conic projection can take four forms: (1) varying the manner in which standard parallels or other constants are selected, (2) spacing parallels to pro-

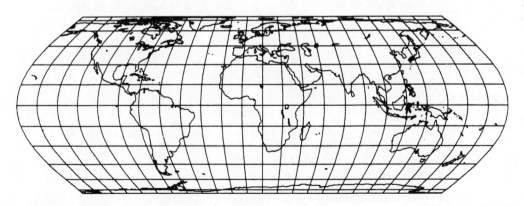

FIGURE 3.15. The Nell pseudocylindrical projection with sinusoidal meridians and a pole line 0.724 as long as the equator. The first flat-polar true pseudocylindrical. Central meridian 15° E; 15° graticule.

vide some arbitrary compromise of distortion, (3) adapting the conic projection of the sphere to that of the ellipsoid, and (4) developing pseudoconic projections, such as the Bonne or other modifications that are not true conics. Several of these approaches were taken during the nineteenth century.

Albers's Equal-Area Conic Projection

The last of the basic conic projections to be developed with one of the three major properties of conformality, equivalence, or equidistance along meridians was the equal-area or equivalent type presented by Heinrich Christian Albers (1805a), three months after Mollweide presented his elliptical world map in the same journal. Albers (1773–1833), a native and lifelong resident of Lüneburg, Germany, derived the formulas for the projection of the sphere using two standard parallels (fig. 3.16)

FIGURE 3.16. The Albers equal-area conic projection applied to an outline map of Europe with standard parallels of 20° and 70° N latitude. 10° graticule. No distortion occurs along the two standard parallels, and parallels gradually decrease in spacing away from the central parallel. Compare figs. 2.3 and 2.7.

(Bonacker and Anliker 1930). He itemized eight merits for his projection, including the fact that its area preservation applied not merely to the entire zone, but also to each "small individual part," unlike Murdoch, who had (in 1758) applied only the former property.[43]

Reichard used Albers's conic at Nuremberg in 1817 for a general map of Europe, and it was later used for maps of central Europe and Russia by Austrian military and Russian geographical organizations, respectively (Deetz and Adams 1934, 93). Scholarly attention before 1900 was uneven. Although Germain devoted three pages to the projection about 1866, the detailed 1863 history of map projections by d'Avezac-Macaya only mentions Albers in connection with the latter's 1805 critique of Murdoch's three conic projections. Thomas Craig's extensive 1882 treatise treats the Albers conic as a projection deserving only a few passing comments when describing Lambert's azimuthal equal-area projection.[44] Before long, however, Ernst Hammer (1889) was proposing the oblique equal-area conic (not naming Albers) for a map of Africa. In the twentieth century Oscar S. Adams (1927) of the U.S. Coast and Geodetic Survey derived the ellipsoidal form and made the projection standard for equal-area maps of the United States.

Relatively little is known about Albers himself. Although called "le docteur Albers" by d'Avezac-Macaya and by Germain, he apparently did not have this title and was listed elsewhere as a merchant and as one with a reputation for his knowledge of mathematics and natural sciences.[45]

In contrast with the Lambert conformal conic, the parallels of the Albers become closer together north and south of the central latitudes (those of the Lambert are increasingly spread apart), and both poles are shown as circular arcs (on the Lambert one pole is a point and the other is at infinity). The equally spaced straight meridians converging to the center of the concentric circular arcs representing parallels of latitude are common to both conics as well as to all other regular conics. The formulas for the spherical form may be written as follows:

$$x = \rho \sin \theta, \text{ and}$$
$$y = \rho_0 - \rho \cos \theta,$$

where $\rho = R (\cos^2\phi_1 + 2n \sin \phi_1 - 2n \sin \phi)^{1/2}/n$,

$$\theta = n\lambda, \text{ and}$$
$$n = (\sin \phi_1 + \sin \phi_2)/2,$$

while ϕ_1 and ϕ_2 are the two standard parallels (in either order), n is the cone constant, (ρ, θ) are polar coordinates, ρ_0 is the radius ρ calculated for the latitude of origin ϕ_0 instead of ϕ, and other symbols are as before.

Hassler's Polyconic Projection

Chronologically, the next new projection of the conic family, but the only other new conic-type projection of the nineteenth century to be used significantly, is the one variously called polyconic, American polyconic, or ordinary polyconic. The term *polyconic* has also been generically applied to any projection with circular arcs for parallels of latitude, whether or not they are concentric, but as a specific projection

FIGURE 3.17. Ferdinand Rudolph Hassler (1770–1843), Swiss-born and the first head of the U.S. Survey of the Coast. About 1820 he developed the polyconic projection, of great importance in U.S. large-scale mapping. *U.S. Geological Survey.*

it was applied in 1853 by Edward Bissell Hunt of the U.S. Coast Survey to one first proposed by Ferdinand Rudolph Hassler (1770–1843).[46]

Born in Switzerland, Hassler (fig. 3.17) arrived in the United States in 1805 and two years later was appointed head of the new Survey of the Coast, to implement a law passed that year authorizing the president to "cause a survey to be taken of coasts of the United States."[47] After receiving delayed funding, Hassler visited Europe to obtain the necessary equipment and returned to the United States in 1815 to begin a geodetic triangulation survey near New York City. An impatient Congress replaced him in 1818, so the work nearly ceased until he was made superintendent in 1832. This appointment ended in 1843, when he died from exposure following a fall while protecting instruments during a severe storm in Delaware. By then, the renamed Coast Survey had been well established under his stubborn and intolerant but technically excellent leadership. This agency was renamed the Coast and Geodetic Survey in 1878 and given several other names beginning in the 1970s, reverting to the 1878 name in 1991.

By 1820 Hassler (1825, 232, 406–8) had decided that a new projection was needed for the survey:

> The projection which I intended to use was the development of a part of the earth's surface upon a cone, either a tangent to a certain latitude, or cutting two given parallels and two meridians, equidistant from the middle meridian, and extended on both sides of the meridian, and in latitude, only so far, as to admit no deviation from the real magnitudes, sensible in the detail surveys. I

had just commenced some calculations relative to the question,—which radius of the earth was most advantageous to admit the greatest extent to the projection under the above condition, whether the geocentric radius of the latitude, the radius of curvature of the meridian at the tangent point, or the radius of the sphere tangent to the spheroid at the point. . . .

In each of these sheets, it was intended to bring the results of several parallels, so that the central meridian alone should become a straight line, and all the other meridians and parallels broken lines, nearest the curve, to which they belong. . . .

This distribution of the projection, in an assemblage of sections of surfaces of successive cones, tangents to or cutting a regular succession of parallels, and upon regularly changing central meridians, appeared to me the only one applicable to the coast of the United States. Its direction, nearly diagonal through meridian and parallel, would not admit any other mode founded upon a single meridian and parallel, without great deviations from the actual magnitudes and shape, which would have considerable disadvantages in use.[48]

It became used commonly, but not exclusively, for coastal charts of the United States over the next hundred years.[49] When the U.S. Geological Survey came into existence in 1879 and began issuing maps of land surveys, the polyconic was the only projection used for the agency's topographic quadrangles until the mid-twentieth century (see Snyder 1987b, 2, 126–28). This emphasis on usage by U.S. government agencies led to its use in several nineteenth-century commercial atlases as well, for some maps of the United States, Canada, North America, Asia, and Oceania.[50]

The polyconic projection (fig. 3.18) of Hassler is simultaneously universal for a given figure of the earth (sphere or ellipsoid), simply drawn, even for the ellipsoid, and employs useful scale characteristics. The projection is true to scale along the central meridian and along each parallel. It is neither conformal nor equal-area, and is only free of distortion along the central meridian. Therefore, it should only be used for regions of predominant north-south extent. For topographic quadrangles, the distortion is nearly negligible because of the small east-west extent.

The name *polyconic* results from the fact that each parallel of latitude is represented by a circular arc whose radius would be obtained by unrolling a cone tangent to the sphere or ellipsoid at that latitude, and marked with the latitude line. Thus many (poly-) cones are involved, rather than the single cone of a regular conic projection. After the various latitudes are laid with true spacing along the central meridians, each is marked at correct intervals for the meridians, which are then passed through the points. Although the ellipsoidal form was properly used for topographic maps, the spherical formulas are given here for simplicity and comparison with other projections:

$$x = R \cot \phi \sin (\lambda \sin \phi);$$
$$y = R \{\phi + \cot \phi [1 - \cos (\lambda \sin \phi)]\}.$$

If ϕ is zero, these formulas are indeterminate, but $x = R\lambda$ and $y = 0$.

In 1853 the first reference to the rectangular polyconic projection was made. This projection was derived and used at that time by the U.S. Coast Survey for portions of the United States exceeding about a square degree.[51] It has since been

FIGURE 3.18. An ordinary polyconic projection of North America. 10° graticule. A simply drawn compromise projection used for U.S. topographic and coastal maps for decades. There is no distortion along the central meridian (95° W here), and scale is true along all parallels.

used for topographic maps by the British War Office and thus has been called the War Office projection as well.[52] The parallels of latitude are drawn in the same manner as those of the American polyconic: equally spaced along the straight central meridian and as circular arcs each with a radius equal to its radius on a developed cone tangent at that parallel. The meridians are nonuniformly spaced along each parallel except one so that the resulting meridian curves intersect each parallel at right angles, hence the name *rectangular* (fig. 3.19). Any parallel other than the pole may be made true to scale. If this parallel is the equator, the resulting formulas for the sphere may be written

$$x = \rho \sin \theta \text{ and}$$
$$y = R\phi + \rho (1 - \cos \theta),$$

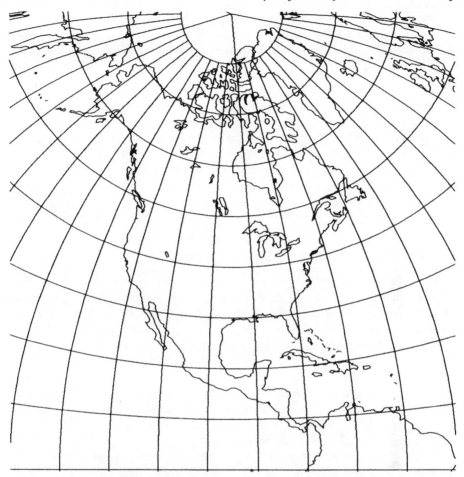

FIGURE 3.19. The rectangular polyconic projection with the central meridian (95° W) and equator at true scale. Meridians and parallels intersect at right angles, but the projection is neither equal-area nor conformal.

where $\rho = R \cot \phi$ and

$$\theta = 2 \arctan [\lambda (\sin \phi)/2],$$

but if $\phi = 0$, $x = R\lambda$ and $y = 0$. For large-scale mapping, ellipsoidal formulas and provision for other true-scale latitudes are available (O. S. Adams 1919a, 13–18).

A third kind of polyconic originating within the Coast Survey is called the equidistant polyconic projection in the same 1853 annual report; it was apparently used for field sheets and some charts of small areas until at least 1882.[53] The central meridian and central parallel are constructed as for the ordinary polyconic. Temporary limiting parallels are located from tabular values, and meridians are connected at their true-scale points on each parallel. Other parallels are then marked at true distances from the central parallel on the theoretically curved meridians, and other parallels are drawn, the temporary parallels then being removed. Thus the meridians are at true scale, although distances in most other directions are not precise. As Shalowitz (1964, 140) states,

> This should be regarded as a convenient graphic approximation, admissible within certain limits, rather than as a distinct projection, although it is capable of being extended to the largest areas with results quite peculiar to itself. If extended to include the entire earth, with the equator as the central parallel, all the parallels would become concave toward this line, for the distance between parallels measured along the curved meridians being constructed equal to that along the central straight meridian, it necessarily follows that the parallels must converge in receding from the central meridian. This is exactly the reverse situation that exists in the ordinary and the rectangular polyconic projections.

Reinvention of Lambert's Conformal Conic Projection

Charles Louis Harding, an astronomer who in 1804 discovered Juno, the third known asteroid, was an early user of the Lambert conformal conic projection, but apparently without knowing of Lambert's work. He used the projection for eight plates in his star atlas of 1808–22.[54] Thomas Craig (1882, 37–38) called it Harding's projection. Harding was followed by Gauss, who, in his general work on conformality (1822, but published 1825), developed the conformal conic in considerable detail, knowing of Harding's work but not Lambert's (Gauss 1825; Herz 1885, 254). A large map of Russia based on this "Gauss's projection" was prepared by Nicolas de Khanikov and published in twelve sheets at a scale of 1 : 1,680,000 by the Society of Geography of St. Petersburg.[55] To accommodate the range of latitudes 36° to 68° N, the standard parallels were made 46° and 58° N.

British scientist John F. W. Herschel (1792–1871) announced that he had developed a conic projection that provided the intermediate link between the equatorial Mercator and the polar stereographic projections (Herschel 1859). This is exactly the goal Lambert had stated in 1772 in developing the conformal conic projection. Lambert derived both spherical and ellipsoidal formulas, stressing their use for regional maps; Herschel concerned himself only with the sphere and world maps, for which the sphere is sufficient.

After presenting his derivations, Herschel considered world maps using "his" projection and recommended a cone constant (or ratio of the angles between meridians on the map to the true angles) of ⅓ (fig. 3.20), thus permitting the world map to be "comprised in a sector of 120°, . . . which is preferable to [other cone constants], and seems to me not unlikely to supersede all other projections for a general chart" (1859; 1860 paper, 106). Herschel, whose more famous father, William, discovered the planet Uranus, was an outstanding astronomer and physical scientist in his own right. He too was apparently unaware of Lambert's work on cartography.[56]

George Boole (1815–1864), a renowned British mathematician, was inspired by Herschel's work to develop the more general case of the conformal conic, including transformation of the ellipsoid onto the plane. He thus independently re-created

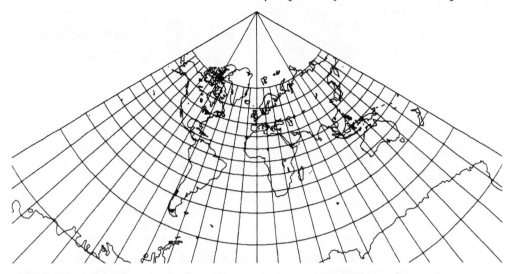

FIGURE 3.20. An outline reconstruction of J. F. W. Herschel's world map using his reinvention of the Lambert conformal conic projection with a cone constant of 1/3. Central meridian 0°; 15° graticule.

Lambert's ellipsoidal version of the same projection, which was given Boole's name by Craig in 1882.[57]

Perspective Conic Projections

Like the central cylindrical projection described earlier, a conic projection may be developed by geometrically projecting the globe from its center onto a cone either tangent at one parallel or secant at two. It is so elementary that there seems to be no formal origin of it, except for the obscure use by Colles in 1794, described in chapter 2, and it is useful only to illustrate its possibility (Fisher and Miller 1944, 51–52). For two standard parallels ϕ_1 and ϕ_2, the polar coordinates are as follows:

$$\rho = R \cos [(\phi_2 - \phi_1)/2][\cot \phi_0$$
$$- \tan (\phi - \phi_0)], \text{ and}$$
$$\theta = \lambda \sin \phi_0,$$

where $\phi_0 = (\phi_1 + \phi_2)/2$. Symbols are as before. For the tangent form with one standard parallel, ϕ_1 may be equated to ϕ_2. If the standard parallels favor the northern hemisphere, only part of the southern hemisphere can normally be shown.

In 1867, with his stereographic cylindrical projection, Braun proposed as his second projection a stereographic conic projection (fig. 3.21), projecting the earth sphere from the South Pole onto a cone with its apex as usual along the north polar axis, but tangent at latitude 30° N (see also Germain 1868). Braun (1867, 276–77) called the projection a combination of the principles of the Murdoch projections and the polar stereographic azimuthal projection, and claimed that its distortion is

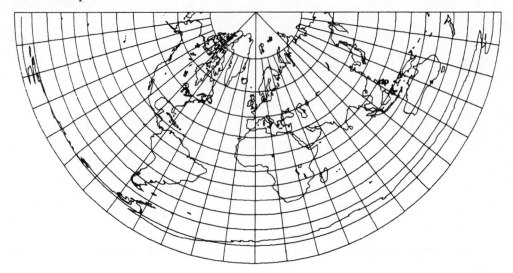

FIGURE 3.21. The Braun stereographic conic projection of the world, perspectively projecting the globe from the South Pole onto a cone tangent at 30° N latitude. Central meridian 0°; 15° graticule.

moderate. The cone constant is found to be 0.5, enclosing the world map in a semicircle, and the polar coordinates are as follows, using Braun's formula but changing symbols:

$$\rho = 3R \{\cos 30° - [\tan ((\phi + 30°)/2)]/2\};$$
$$\theta = \lambda/2.$$

In 1876, Johan Lidman, a Swedish navy lieutenant, took a further step by preparing a double transformation to an azimuthal projection via a perspective conic.[58] The globe is first projected perspectively from its center onto a cone tangent at latitude 45° N, and the positions on the cone are then plotted orthographically onto a plane coinciding with the equatorial plane of the globe.

Other Conic Variations

Nicolas Auguste Tissot (1824–1897), best known to cartographers for his indicatrices (discussed later), and often referred to as M. A. Tissot even in non-French works, although M. stands for Monsieur, used the related principles in 1881 to select parameters for a low-error Albers equal-area conic projection.[59] Rather than selecting standard parallels as such, he selected the cone constant and the radius of the equator so that for a given range of latitudes the scale factors along the limiting parallels are equal to each other and are the reciprocal of the least scale factor along any of the central parallels. (The scale factor is the ratio of scale on the map at a given point and in a given direction to the stated map scale. At the standard parallels, the scale factor is 1.0 in all directions.) Tissot, admissions examiner at the École Polytechnique (Paris), said this projection is the "périgonale" equal-area conic for a zone limited by two parallels, and V. V. Kavrayskiy later found that this approach gives the least maximum angular distortion for an equal-area conic.[60] In polar coordinates, the Tissot equal-area conic has formulas equivalent to the following:

$$\rho = R \ [2(C - \sin \phi)/n]^{1/2}, \text{ and}$$
$$\theta = n\lambda,$$

where $n = \sin \phi_0$,

$$C = (\sin \phi_0/\cos \delta + \cos \delta/\sin \phi_0)/2$$
$$\phi_0 = (\phi_A + \phi_B)/2, \text{ and}$$
$$\delta = (\phi_A - \phi_B)/2,$$

with (ϕ_A, ϕ_B) the northern and southern limiting parallels, respectively.

Nell (1890, 94–97) devised a combination of the Bonne and of the Lambert equal-area conic projection having the same standard parallel, and applied it to the ellipsoid. He used the arithmetic mean of the meridian spacing on corresponding parallels and then adjusted the radii of the still-circular parallels to retain the equal-area property, but the mathematics is iterative and complicated.

New Azimuthal Projections

With only two or three exceptions, the new nineteenth-century approaches to azimuthal projections involved varying the perspective projection of the globe onto tangent or more commonly secant planes. Like the works of La Hire and Parent in the early eighteenth century, the new developments resulted in the choice of an appropriate point of perspective from which to project the *far* side of the globe to produce a map with a minimum of distortion according to a chosen criterion.

John Lowry (1825), a London engraver, calculated that the point of perspective should be placed about 1.69 times the radius of the sphere, measuring from its center in the direction of the South Pole, in order to produce a low-distortion north polar projection.[61] This value (more accurately 1.6858) is the geometric mean of the eighteen values of this ratio which are calculated by requiring that each 5° of latitude from 0° to 85° be successively placed at its true proportional distance from the North Pole.

About 1850, Philipp Fischer (1818–1887) specified this constant at 1.752 (the reciprocal of $(\pi/2 - 1)$) so that the bounding circle of a hemispherical map would be placed at its true distance from the center for a perspective projection onto a tangent plane.[62] In 1857 Henry James (1803–1877), director general of the British Ordnance Survey (1854–75), brought forth his perspective projection onto a secant plane, with the point of projection placed at 1.50 times the radius of the globe from the center, but the range of the map is made 113°30' (90° plus 23°30', the obliquity of the ecliptic).[63] The value of 1.50 was intended to approximate the location providing minimal overall distortion of the "⅔ [actually 0.699] of the surface of the sphere" shown in the resulting map, but a better value for the point of perspective is 1.368, as determined by James's colleague Alexander Ross Clarke (1828–1914) in a joint paper (James and Clarke 1862, 309–11). James's published maps generally used the oblique aspect, centered at the Tropic of Cancer (23°30' N) and 15° E, thus showing all the continental masses except for Australia and much of Antarctica (fig. 3.22).

FIGURE 3.22. The James azimuthal projection, showing most landmasses, with a range of 113°30′, centered at 23°30′ N and 15° E. A perspective projection of the far side of the globe from a point 1.5 times the radius of the globe from the center. 15° graticule.

Heinrich Friedrich Gretschel (1830–1892), a professor in Freiberg, chose 1.6180 (half of $(1 + \sqrt{5})$) to produce on the polar aspect the same scale along any meridian at the equator as the scale at the pole (Gretschel 1873, 94).[64] A contemporary summary by Norbert Herz and Tissot of several existing approaches to the perspective projection made it clear that some had only slight variations in the location of the point of perspective.[65]

Attracting more attention academically was a series of azimuthal projections in which the relatively new mathematical concept of least squares was applied. Developed in the first two decades of the century by Gauss and by Adrien Marie Legendre (1752–1833), the principle states that the best or most probable value of a quantity is that value for which the sum of the squares of the errors is least.[66] George Biddell Airy (1801–1892), a leading British geodesist who in 1830 developed the dimensions of the Airy ellipsoid used in British geodetic surveying for a century, applied the principle of least squares to attempt a truly "minimum-error" azimuthal projection in 1861.

Airy was born in Northumberland, England, and died in Greenwich. He became Lucasian Professor of Mathematics at Cambridge in 1826 (like Newton in 1669) and director of the Cambridge Observatory in 1828, moving to Greenwich as astronomer royal in 1835. The latter post he held until his retirement in 1881 (Eggen 1970–80; Bell 1937, 106). He approached the development of his projection with a goal of having a minimum "total evil" determined by what he called "Balance of Errors" (Airy 1861, 410, 414). His "Projection by Balance of Errors" is neither perspective, conformal, nor equal-area, but it is a compromise appearing very much like the azimuthal equidistant projection, especially if limited to about one hemisphere (fig. 3.23) (see Snyder 1985a, 58–59, 63–65).

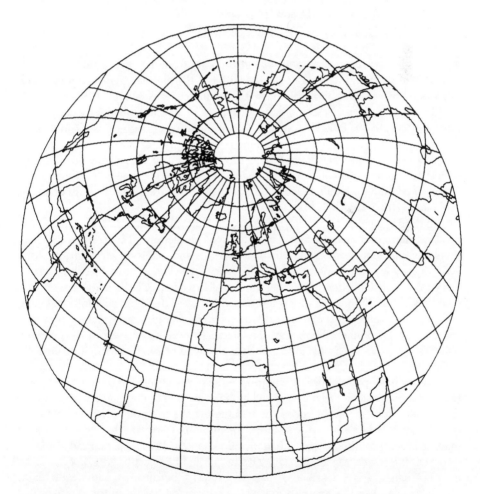

FIGURE 3.23. An oblique Airy minimum-error azimuthal projection of one hemisphere, centered on London. 10° graticule.

Airy's specific approach was to minimize the sum of the squares of the errors in scale both along and perpendicular to the radii from the center of projection. For the polar aspect, these measurements are along meridians and parallels, respectively. Airy had made an error in his derivations, but this was corrected by James and Clarke (1862, 306–12), the latter performing the mathematics. James's request for Clarke's assistance was no doubt prompted in part by the fact that Airy found his own projection superior to James's (described above) by a greater margin than was justified. In addition, James was notorious for self-promotion and credit-taking during his direction of the Ordnance Survey; his vanity was probably important in correcting the renowned Airy (Ordnance Survey 1991, 6–7).

The resulting formulas for the Airy projection are more complicated than those of the standard azimuthal projections, but they can be readily programmed. The radius from the projection center may be calculated thus:

$$\rho = 2R \, [\cot (z/2) \ln \sec (z/2)$$
$$+ \tan (z/2) \cot^2(\beta/2) \ln \sec (\beta/2)],$$

where z is the great-circle distance of a given point from the chosen center and β is the value of z corresponding to the outer limit of the circular map region to which minimum error is being applied. If $z = 0$, the equation fails, but $\rho = 0$. This formula may be used to develop rectangular coordinates in the same manner as other azimuthals (see pp. 89–90). The concept was stimulating enough to result in considerable scholarly attention in subsequent textbooks and articles, considering the minimal cartographic use made of the projection.[67] The Airy projection (as corrected) was used (1903–36) for an Ordnance Survey map of the United Kingdom at a scale of ten miles per inch (1:633,600), but was hardly used otherwise for atlas and wall maps (Hinks 1912, 37; Maling 1973, 173).

In the 1862 paper correcting Airy's work, Clarke also applied Airy's approach to produce minimum-error *perspective* azimuthal projections. This series of projections was thus more constrained than Airy's series and had somewhat greater "total evil," to use Airy's expression.[68] Clarke, like Airy, is well known as a British geodesist responsible for important ellipsoidal figures of the earth. Clarke made several calculations of the dimensions of a reference ellipsoid as he obtained additional data, and dates are therefore normally added to distinguish his various measurements. The two most prominent figures are the Clarke 1866 ellipsoid used for the geodetic network in the United States (and generally throughout North America) from 1880 to the 1980s, and the Clarke 1880 ellipsoid used particularly for Africa.

Locating the point of perspective on Clarke's projections and the intersection of the secant plane along the radius of the globe extending to the map center requires the calculus or iteration, but once the parameters are computed, they apply to the entire map. Clarke provided the values for minimum-error perspective azimuthal maps of several continents and of a hemisphere, and in 1879 he presented his counterpart of the James projection: calling it the "Twilight Projection" (fig. 3.24), Clarke gave it a range of 108° (instead of James's 113°30'), because astronomical twilight officially ends when the sun is 18° below the horizon or 108° from the zenith.[69] The

FIGURE 3.24. Clarke's "Twilight Projection," azimuthal with a range of 108° centered at 23°30′ N latitude, 0° longitude. Like the James projection (fig. 3.22), a far-side perspective, but from a point 1.4 times the radius from the center.

point of perspective for the "Twilight Projection" is 1.4 times the radius from the center, and the secant plane is 1.7572 times the radius from the point of perspective. Modern recomputations of Clarke's constants produce minor variations. Clarke's minimum-error perspectives were used for some British weather maps prior to 1955 (Maling 1973, 173).

A few more azimuthal projections surfaced during the nineteenth century. Lidman's azimuthal projection of 1876 via a double projection through a cone was described previously under conic projections. In the 1880s, F. A. Arthur Breusing (1818–1892) conceived a nonperspective azimuthal projection that balances the shape distortion of the azimuthal equal-area projection and the area distortion of the conformal stereographic projection by using for a given point a radius from the projection center equal to the geometric mean of the radii as calculated for these two projections.[70] Therefore, the radius is

$$\rho = 2R \left[\tan (z/2) \sin (z/2)\right]^{1/2}.$$

Like Airy's projection, Breusing's geometric projection closely resembles the azimuthal equidistant projection. Its first practical usage was for two maps of North and South America by Debes in 1895.[71] There is also a Breusing harmonic projection, but this was devised by A. E. Young (1920, 7–8), who so named it as a projection even closer to Airy's, provided that the globe radius R is multiplied by an appropriate constant dependent on the map range β:

$$\rho = 4R \tan (z/4) \sin^2(\beta/4)/[\ln \sec (\beta/2) - \tan^2(\beta/4)],$$

with symbols as given for the Airy projection.

In 1887, Ernst Hammer (1858–1925), a professor of surveying and related fields in the Technische Hochschule of Stuttgart for his last forty-one years, chose a perspective azimuthal projection in which the *total* area within the bounding circle is held true to scale for a tangent plane. Then

$$P = 4 \cos^2(\beta/4) - 1,$$

where P is the distance of the point of perspective in radii from the center of the sphere, in a direction away from the plane, or $(1 + \sqrt{2})$ for a hemisphere.[72]

Modified Azimuthal Projections

The first projection to fall into a classification called *pseudoazimuthal* was invented by H. Wiechel (1879). Like the pseudoconic projections (notably the Werner and Bonne), the pseudoazimuthals in the normal (polar) aspect have curved meridians and concentric circular arcs for parallels, but the circles are complete. The Wiechel projection, which is only of interest in the polar aspect (fig. 3.25), is interesting because all meridians are identical circular arcs with radii equal to that of the globe, the scale is correct along every meridian, the parallels are concentric circles spaced as they are on the polar Lambert azimuthal equal-area projection, and the projection is, like Lambert's, equal-area. In fact, it can be developed as a polar Lambert divided into an infinite number of concentric rings which are then all rotated clockwise (or all counterclockwise) by an angle equal to half the angular distance of the ring from the pole; thus the projection does not lose its fidelity of area. However, it has been generally ignored.[73]

Russian cartographer David A. Aitoff (1854–1933) in 1889 devised an elementary but very appealing modification of one hemisphere of the equatorial aspect of the azimuthal equidistant projection. Introduced in a French atlas published that year, it consists of stretching the latter projection horizontally (in a direction parallel to the equator) by a factor of 2:1 and doubling the value of each meridian (fig. 3.26) (Aitoff 1889, 1892). Thus the world is shown in an ellipse with axes in a ratio of 2:1, and both the equator and central meridian are at true scale. The projection is no longer azimuthal, nor can distances be measured directly along any other straight line. The formulas may be shown thus:

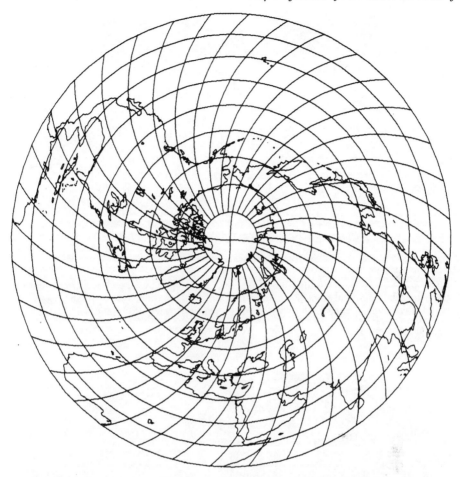

FIGURE 3.25. The northern hemisphere based on the Wiechel pseudoazimuthal projection. 10° graticule. An equal-area projection with circular arcs for meridians and parallels and true scale along all meridians.

$$x = 2Rz \cos \phi \sin (\lambda/2)/\sin z, \text{ and}$$
$$y = Rz \sin \phi/\sin z,$$

where $z = \arccos [\cos \phi \cos (\lambda/2)]$. If ϕ and λ are both zero, the equations for x and y are indeterminate, but $x = 0$ and $y = 0$.

The Aitoff projection soon inspired Hammer (1892) to invent a world map (fig. 3.27) looking much like Aitoff's, but maintaining equal area instead, with prominent credit to Aitoff in both the title and text of Hammer's paper. Instead of the azimuthal equidistant, Hammer expanded the equatorial aspect of the Lambert azimuthal equal-area projection by the same 2:1 factor in a direction parallel to the equator, then doubled the values of each meridian.

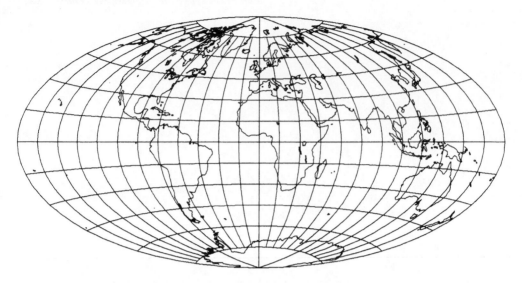

FIGURE 3.26. The world on the Aitoff projection. Developed by expanding a hemisphere of the equatorial azimuthal equidistant projection and doubling values of meridians, it maintains true scale along the equator and central meridian (0° here). 15° graticule.

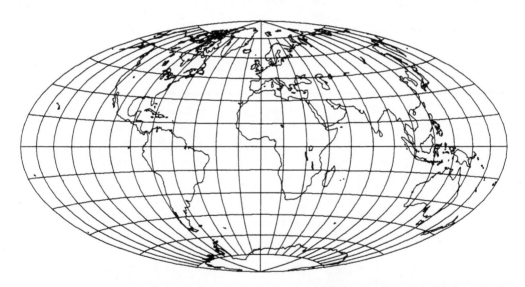

FIGURE 3.27. A Hammer projection of the world. This was inspired by and often mistakenly called the Aitoff projection, but involves expanding the Lambert azimuthal equal-area instead of the azimuthal equidistant projection and is therefore equal-area. Central meridian 0°; 15° graticule.

As with Aitoff's projection, Hammer's world is shown in an ellipse with axes in a 2:1 ratio, but the equator and central meridian are no longer true to scale, the scale gradually decreasing as the distance from the center increases. Unlike the other well-known elliptical equal-area projection, the Mollweide, the parallels are curved rather than straight, leading to less shearing of the polar regions away from the central meridian. The meridians of the Hammer are not the ellipses of the Mollweide except for the bounding 180th meridians. Its formulas may be written

$$x = 2R\sqrt{2} \cos \phi \sin (\lambda/2)/[1 + \cos \phi \cos (\lambda/2)]^{1/2};$$
$$y = R\sqrt{2} \sin \phi/[1 + \cos \phi \cos (\lambda/2)]^{1/2}.$$

The Hammer projection soon replaced the Aitoff and has been moderately used in the twentieth century because of its equal-area and elliptical features, but the name Aitoff was incorrectly applied to the new projection in English literature, possibly due to translation errors, to such an extent that H. J. Andrews (1952) and John B. Leighly (1955) on both sides of the Atlantic wrote papers trying to clarify the confusion.[74] Since then the name Hammer or Hammer-Aitoff has been used much more regularly.

A different sort of modification of an azimuthal projection was undertaken by Alexis E. Frye (1895, 5).[75] Used in several of his school geographies of the period, the modification consists of an oblique azimuthal equidistant projection centered on London for the inner hemisphere, with interrupted appendages to add South America and Australia. Frye did not provide details, but he called this a "Map Showing the World Ridge," with major mountain ranges appearing connected as a large arc.

Globular Modifications

Schmidt's 1803 globular projection of a hemisphere was the inspiration, as noted earlier, for Mollweide's elliptical equal-area pseudocylindrical projection of 1805. Schmidt (of Giessen) in turn used some of the precepts of Fournier's two globular projections of 1643, for both of which the meridians consist of equally spaced semi-ellipses joined at the poles, as do Schmidt's.[76] Instead of Fournier's simple straight or circular lines of latitude, Schmidt made his parallels complex curves intersecting each meridian at equal intervals along the meridian. Since the meridians vary widely in length, the result is no true scale anywhere except along the central meridian and equator. The formulas, furthermore, now involve the use of elliptic integrals, but these were just being developed in the early 1800s, and Schmidt apparently used less advanced techniques to calculate the positions of parallels (his work is not available to this writer). The complexity and lack of usable properties made Schmidt's projection academic, except as Mollweide's inspiration.

Nell (1852) modified Nicolosi's globular projection by producing a globular hemisphere which in a sense is an average of the Nicolosi and equatorial stereographic projections. In Nell's, all meridians and parallels remain circular arcs, except for the usual straight central meridian and equator, and the parallels continue to intersect equidistantly the outer circle representing the meridians 90° from center. The meridians and parallels, however, intersect the equator and central meridian, respectively, midway between the intersections for those of the Nicolosi and equatorial stereographic.[77]

Conformal Innovations

A more useful projection was presented by Joseph Johann von Littrow (1781–1840) (1833, 142). It falls into a special group of projections called retroazimuthal, to which there were several contributions during the next 140 years. While the ordinary azimuthal projection gives the true azimuth of all points from the center of the map, the retroazimuthal provides the true azimuth of the center from all other points, as an angle from vertical.

The Littrow projection (fig. 3.28) exceeds this requirement by also being conformal and by permitting the direct measurement of azimuth from any point on the map to *any* point along the central meridian, instead of just to one point. This latter feature may be made more universal by working with only the graticule, plotting the central point in question at its appropriate latitude along the central meridian, then plotting the other points at their proper latitudes and differences in longitude from

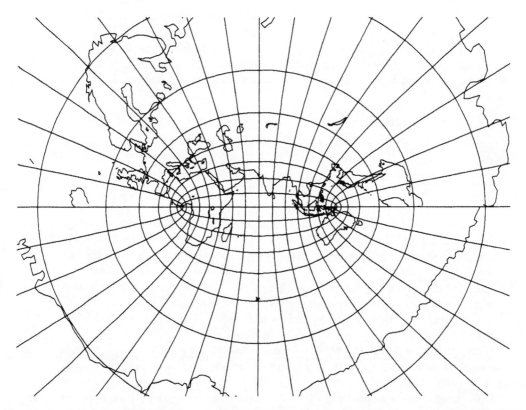

FIGURE 3.28. The Littrow projection of most of the eastern hemisphere. Conformal and with great area distortion, but having special retroazimuthal properties. Central meridian 70° E; 10° graticule.

that of the central point. The azimuth from any of these other points to the central point is the angle between vertical and the direction of the straight line connecting the two points, even though this straight line is not the great-circle path. This feature was hardly recognized until Patrick Weir of the British Merchant Navy developed it independently as part of a navigational diagram published in 1890, thus leading to a common description as the Weir azimuth diagram.[78]

Visually the Littrow projection is also of special interest because, except for the straight equator and central meridian, the parallels of latitude are ellipses, the meridians are hyperbolas, the meridians 90° from center are shown as horizontal straight lines in which the northern half coincides with the southern half, and less than one hemisphere can be shown on a given map. The elliptical parallels and hyperbolic meridians are all confocal, that is, their foci are all at the same two points, the intersections of the 90° meridians with the equator. The relatively simple formulas, essentially as presented by Littrow, are as follows:

$$x = R \sin \lambda / \cos \phi;$$
$$y = R \cos \lambda \tan \phi.$$

The Littrow is also a transverse aspect of one of the more general forms of the Lagrange projection, and vice versa.[79]

The heavily mathematical era relating conformal mapping, complex algebra, elliptic integrals, and simple geometric figures, initiated in part by Lambert and Lagrange with respect to the circle in the 1770s, continued with three independent but related developments in the latter half of the nineteenth century. Hermann Amandus Schwarz (1843–1921) demonstrated how to use complex (real and imaginary) integrals to transform conformally the interior of a circle to fill the interior of any regular polygon (Schwarz 1869). Charles Sanders Peirce (1879) announced the first use of elliptic functions for a map projection, and Émile Guyou (1843–1915) developed a projection transverse to that of Peirce (Guyou 1886, 1887).[80]

Schwarz showed that if the radius of the circle is one unit, its interior can be conformally represented by the interior of a regular polygon with the following computation:

$$z' = \int_0^z (1 - z^n)^{-2/n} dz,$$

where $z = y + ix$, $z' = y' + ix'$, (x, y) are the rectangular coordinates for the circle, and (x', y') are those of the polygon of n sides. While Schwarz did not use this principle cartographically, it was extensively used in the next century, especially by O. S. Adams (1925) and L. P. Lee (1976, 3, 38–74) for polygonal maps, which are intriguing but still at best novelties. Most directly, the conformal stereographic hemisphere in a circle and the "Lagrange" world in a circle developed by Lambert can thus be placed conformally into an equilateral triangle, square, pentagon, hexagon, and so on. The novelty loses its impact if the number of sides is so large that the polygon resembles a circle.

Schwarz was born in Silesia and died in Berlin. After receiving his doctorate in 1864, he became assistant professor at Halle in 1867, but moved as full professor to Zürich in 1869, Göttingen in 1875, and the University of Berlin in 1892. He lectured at Berlin until 1917 and was the leading mathematician there, succeeding and much influenced by the work of the prominent mathematician Karl Wilhelm Theodor Weierstrass (1815–1897). His solution to conformal mapping of the circle into regular polygons was prompted by Riemann's important general theory of conformal mapping in mathematics, presented in his 1851 doctoral dissertation.[81]

Peirce called his projection the quincuncial because of the five portions of his arrangement (fig. 3.29): the North Pole is at the center of an inner square bounded by the equator, and the South Pole is at each corner of the outer square, the southern hemisphere consisting of four triangular sections. Thus each of eight isosceles triangles has an identical graticule of meridians and parallels, symmetrical about its own central meridian and bounded by the equator and two meridians 90° apart.

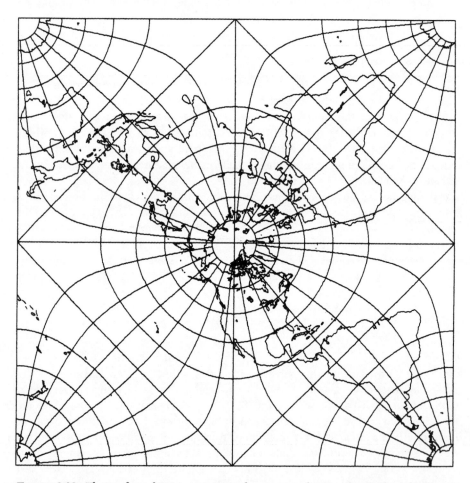

FIGURE 3.29. The conformal Peirce quincuncial projection showing the northern hemisphere in a square and the world in a larger square. The projection involved the first use of elliptic functions in mapping. This map employs the same dividing meridians as those used by Peirce, but with a 15° graticule instead of his 5°.

There are "singular" (nonconformal) points at each bend in the equator, but the projection is otherwise completely conformal. The quincuncial projection has attracted few uses, one being a U.S. Coast and Geodetic Survey world map of air routes about 1946, but the mathematical ingenuity prompted several papers and comments in subsequent years.[82]

Peirce, born in Cambridge, Massachusetts, was the son of Harvard mathematician and scientist Benjamin Peirce. After graduating from Harvard, the younger Peirce began work for the U.S. Coast Survey in 1859. He held the title of assistant there from 1867, when his father not coincidentally became superintendent of the agency (1867–74), until he retired in 1891.[83] His mathematical skills resulted in many papers on physics, philosophy, astronomy, and geodesy, but he died in Milford, Pennsylvania, in near poverty.

Guyou, a French lieutenant and professor, showed the world map in a rectangle with a length twice its width, placing the eastern and western hemispheres conformally in squares side by side (fig. 3.30). For his projection, the singular points occur at the bends (at 45° N and S) of the meridians bounding the hemispheres.

In his own illustration Peirce repeated his basic projection so that it also showed northern and southern hemispheres side by side in squares, like Guyou's eastern and western hemispheres (Peirce 1879, following 396). The two projections are transverse to each other, that is, the graticule of Peirce may be rotated to place the poles at the centers of Guyou's squares and vice versa. In addition, both projections may be tesselated, or mosaicked, indefinitely to cover a wall with many

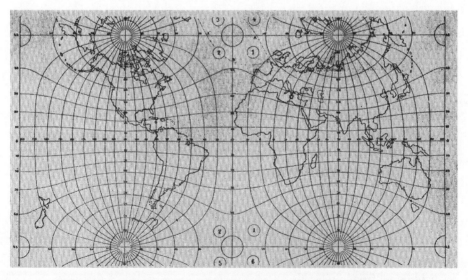

FIGURE 3.30. The conformal Guyou projection of the eastern and western hemispheres in squares. This and the Peirce quincuncial (fig. 3.29) are transverse to each other, and either may be indefinitely mosaicked. 10° graticule. Reproduced from Deetz and Adams (1934, 159).

small connected world maps. The formulas are omitted here due to their relative complexity.[84]

About the time of Schwarz's work on conformal projections, Karl VonderMühll (1868) of Leipzig enlarged on Lagrange's important work of 1779. Whereas Lagrange developed the general case of conformal projections with all-circular meridians and parallels, VonderMühll generalized further to develop the case of conformal projections with conic sections (circles, ellipses, parabolas, and hyperbolas) and straight lines (the limiting forms) for all meridians and parallels. This seems to have escaped the interest of subsequent cartographers until mathematician B. H. Brown (1935) referred to VonderMühll's work in developing his corresponding series of equal-area projections, but both studies tended to produce bizarre novelties except for the previously developed projections that fell into the series.

Star Projections

Star projections are generally centered on the North Pole, with the land-intensive northern hemisphere shown as an azimuthal projection and the southern hemisphere divided into several pointed appendages attached to the equator, interrupted primarily in the oceans, and terminating at the South Pole. The emphasis is on general appearance, and cartographic qualities of conformality, equivalence, or uniform scale are usually missing, especially in the southern hemisphere.

Apparently the first star projection, a "Polygonal North-Polar Star Projection," was announced in 1865 by G. Jäger, director of a Vienna zoo. His projection has eight points unequally spaced to minimize cutting through southern landmasses.[85] The interruptions occur at longitudes 10°, 60°, 100°, and 155° E and 35°, 85°, 120°, and 155° W. The northern hemisphere is contained within an irregular octagon, equivalent to eight unequal triangles with vertices at the North Pole. The points are triangles fitting the edges of the octagon and extending to the South Pole. Within each of the sixteen triangles, all parallels are equidistant straight-line segments spaced at true distances along the central meridians of the southern triangles, and the meridians are straight lines radiating from the North Pole at their true angles. The southern points are thus unequal in distance from the North Pole.

The next version of this type of projection was by August Petermann, a German geographer and cartographer mentioned earlier; his name continues with the prominent geographical journal *Petermanns Geographische Mitteilungen*, which he founded in 1854 with Justus Perthes at Gotha (Robinson 1982, 123–25). Petermann essentially reprinted Jäger's paper from its original source into his own journal, where the only accompanying illustration is a modification by Petermann, using equidistantly spaced circular arcs for all parallels and retaining Jäger's eight points but spacing them uniformly (Jäger 1865b, 67–68, illus. following 70). Petermann's interruptions occur at longitudes 10° E and every 45° thereafter, with central meridians midway. All meridians remain straight, with breaks at the equator; the northern hemisphere is an azimuthal equidistant projection, and the eight southern points are all equidistant from the North Pole.

The most commonly seen star projection was originated by another prominent German geographer-cartographer also mentioned earlier, Hermann Berghaus (1828–1890). Born in Herford, Westphalia, he spent most of his life at Gotha pro-

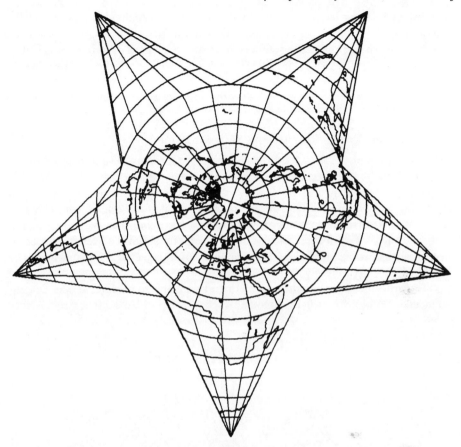

FIGURE 3.31. The Berghaus star projection of the world. The northern hemisphere is on a polar azimuthal equidistant projection, with the arrangement chosen by Berghaus. 15° graticule.

ducing maps and continuing the work of his uncle Heinrich Berghaus, producing physical atlases (J. G. Bartholomew 1891). In 1879 he emulated Petermann's projection with the azimuthal equidistant northern hemisphere, circular arcs for parallels, and straight lines for meridians broken at the equator, but he used five identical points (fig. 3.31) instead of Petermann's eight. The interruptions occur every 72° (360°/5) of longitude, beginning with 16° W (or 16°, 88°, and 160° W; 56° and 128° E).[86] Australia is split once, and Antarctica is in five parts, but the projection's similarity to a common symbol for a star may have influenced its use in Adolf Stieler's *Hand-Atlas* of 1880 and Russell Hinman's *Eclectic Physical Geography* of 1888, where it is the principal world projection (Pattison 1960). In 1911 the Association of American Geographers adopted the Berghaus star projection as the heart of its insignia.

A fourth and more distinct star projection was devised by Anton Steinhauser (1883) about the same time and called a conoalactic projection.[87] The northern hemisphere is shown as an equidistant conic projection contained in a sector of 240° split at longitude 180°, with the pole as a point and latitude 4°18′ N as the other effective standard parallel. The southern hemisphere consists of four identical lobes with straight meridians and concentric equidistant parallels attached to the equator.

Like other projections, star projections have been plotted recently by computer, but the simplicity of their designs belies the complexity of the necessary plotting equations. These are therefore omitted here.[88]

Conformal Projections without Singular Points

Friedrich Eisenlohr (1831–1904) of Heidelberg, Germany, invented a conformal projection (fig. 3.32) showing the entire world in two cusps (Eisenlohr 1870). The projection is significant for two reasons: it shows the world with the least overall distortion for a conformal map, and it contains no "singular" points where conformality fails. Most conformal maps, when extended to world maps, have singular points that are not conformal, such as the two poles on either the Mercator or Lambert conformal conic projections, where meridians intersect at the wrong angles or else the points are infinitely distant.

The criterion for minimum overall distortion is not so obvious. It was first announced by Pafnutiy L'vovich Chebyshev (1821–1894) of Russia that a region may

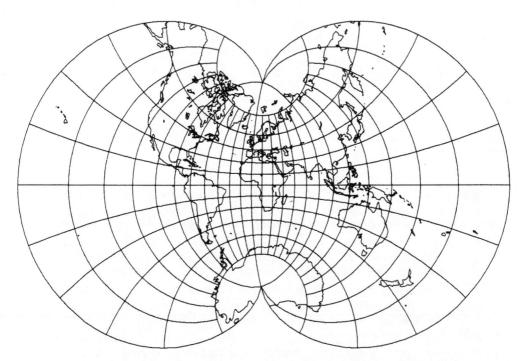

FIGURE 3.32. The minimum-error conformal Eisenlohr projection of the world. Scale around the outer meridian is constant, and there are no "singular" points of nonconformality. Central meridian 15° E; 15° graticule.

be best shown conformally (Chebyshev 1856) if the range of scale errors (scale factors minus 1) over the region is a minimum, and that this results if the region is bounded by a line of constant scale.[89] This was proven by the Russian mathematician Dimitry Aleksandrovich Grave (1863–1939) (1896), and this principle has led several Russian and Polish writers of the twentieth century to refer to projections derivable from this criterion as Chebyshev projections.[90]

The scale factor along the entire two-cusped boundary of Eisenlohr's projection (he does not refer to Chebyshev) is $(3 + 2\sqrt{2})$ or 5.828 times the scale factor at the projection center. As with most conformal projections, the equations, omitted here, may be expressed in complex algebra.[91] Largely because of the difficulty of the formulas, the projection has been little used, since the August projection (below) is simpler and only slightly more distorted.

Crediting Eisenlohr with his inspiration and G. Bellermann with codevelopment, Friedrich W. O. August (1840–1900) described a conformal projection (fig. 3.33) looking much like Eisenlohr's, but with a wider range of scale factors (8:1 instead of 5.83:1) and a varying scale around the map boundary, which is a two-cusped epicycloid (August 1874). Like the Eisenlohr, the August (or August epi-

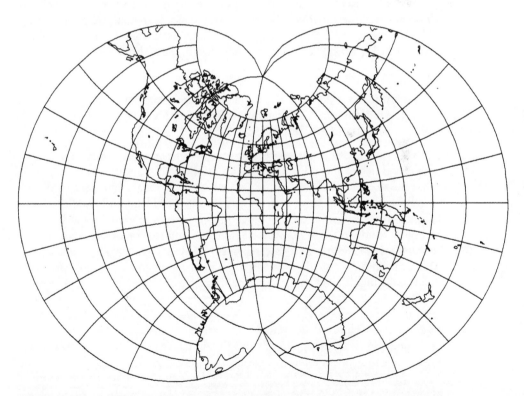

FIGURE 3.33. The conformal August epicycloidal projection of the world. The outer meridian is a two-cusped epicycloid, and there are no singular points. Central meridian 15° E; 15° graticule.

cycloidal) projection has no singular points. This epicycloid is the curve traced by a point on the circumference of a circle rolling around the circumference of another circle with twice its radius. August claimed that Eisenlohr's projection is too complicated to be practical, and August included a geometrical construction to offset the fact that his formulas are also complicated. Eisenlohr (1875) responded a year later in considerable detail, complete with tables of scale factors, but August's was the projection discussed much more extensively in the next century.[92]

Polyhedric and Polyhedral Projections

At some time during the nineteenth century, although it is not clear when, the polyhedric projection was developed in various forms and used for topographic mapping in Europe. None of the slight variations is precisely equal-area or conformal. By the early twentieth century, varieties had been used in a half dozen European countries and Japan.

François Reignier (1957, 271–81) describes in some detail five versions: (1) the ordinary polyhedric, used for a map of France at 1:50,000 between 1900 and 1914 until its replacement by the Tissot projection described just below; (2) the Prussian polyhedric or natural projection, used for Austria-Hungary, Japan, and several states or provinces of Germany and Russia in the nineteenth and early twentieth centuries; (3) the projection for a map of France at 1:100,000; (4) the natural projection derived from the sinusoidal projection and used for maps of Italy at scales of 1:25,000, 1:50,000, and 1:100,000; and (5) the polyhedric or natural projection used for the national map of Spain at 1:50,000.[93] There has apparently been no subsequent use for new maps.

These projections do not directly relate to more common projections, except for possible similarities to the gnomonic and orthographic. The basic principle was described thus by Arthur R. Hinks (1912, 122):

> This [polyhedric projection] is of no scientific interest, except that it is much used in European maps. Take on the spheroid the four points that are to be the corners of the sheet, and pass a plane through them; this will be strictly possible if the sheet is to be bounded by meridians and parallels. Let fall perpendiculars from each point of the enclosed spheroidal trapezium to this plane, and we have the projection. The formulae for the calculation of coordinates are complicated, but in practice they are scarcely required, since within the limits of a single sheet not more than one degree square the projection is indistinguishable from the polyconic and from many other projections. Adjacent sheets fit along the edges, and the whole series of sheets representing a zone of latitude can be fitted together and laid out flat, but will not fit the adjacent zone. Thus it is not possible to combine a number of small sheets to make one large one, though in practice the difficulty would be felt only when it was a question of combining the original engraved plates. The deformations of the printed sheets would be much larger than those due to the projection alone.

The name *polyhedric* results from the fact that if the individual trapezoid-shaped maps are joined together for a large region, a portion of a polyhedron with a great many sides would result (Zöppritz and Bludau 1899, 127).

The concept of the polyhedral globe itself was initiated by Albrecht Dürer in 1538 and resurrected during the nineteenth century in flat maps by Reichard in 1803. In 1833, J. W. Woolgar introduced a stereographic projection of the entire globe onto the four triangular faces of a regular tetrahedron. The North Pole is one vertex, the South Pole is at the center of the opposite face, and the centers of all other faces are consequently at latitude 19°28′ N (arcsin (⅓)), but spaced 120° in longitude.[94] The edges of the faces do not match (there is overlapping), but the range of scale for each face is 1.5 : 1 for the 70°32′ maximum extent from the projection center. Later, in 1882, Charles A. Schott displayed a globe projected obliquely onto a circumscribed cube.[95]

The polyhedral concept also appeared in three-dimensional form at least twice in the form of patents.[96] In 1851 a British patent was awarded to Juan Nepomuceno Adorno, and it contained a casual reference to an icosahedral map (an icosahedron is a regular polyhedron with twenty equilateral triangles constituting its faces).[97] J. Marcus Boorman (1877) of New York received a U.S. patent for an "Improvement in geometric blocks for mapping." He proposed "a system of new and useful geometrical solids, other and distinct from those commonly termed 'the five platonic solids' and their compounds" (Boorman 1877, 1), with solids of 15, 22, 23, 24, 32, and 37 faces of various regular and irregular polygons of from four to seven sides each. The projection of the globe onto these polygons was left rather vague, especially with the proviso that, if desired, the "portions to be mapped . . . may be accurately transferred to a flat surface or map by any suitable process" (Boorman 1877, 4).

Tissot's Optimal Projection

Another projection presented by Tissot in 1860 and repeated in his landmark treatise of 1881 on map projections was included (in 1881) under the topic "Recherche du système de projection le mieux approprié à la représentation d'une contrée particuliére"—to find the best projection for a particular region.[98] His approach was to write the equations for rectangular coordinates as power series in terms of functions of the differences of latitude and longitude from the central parallel and central meridian, respectively. The terms of the series were developed using the derivatives required for formulas of distortion (also developed by Tissot), minimizing the distortion and truncating the series to include only terms up to the third order in the final equations for coordinates.

The resulting minimum-error projection is not quite equal-area or conformal, but, like the polyconic, has negligible distortion for a very small region. It was adopted by the French Service Géographique de l'Armée about 1920, but was called Système Lambert, even though it is not the same as the Lambert conformal conic.[99]

Fiorini's Projections

Matteo Fiorini (1827–1901), an engineer and professor of geodesy at the University of Bologna, was the leading nineteenth-century Italian authority on map projections, studying them from both analytical and historical standpoints. In his general work of 1881, he is said to have presented two projections, one duplicating part of A. R. Clarke's earlier work and the other strictly a novelty, if it exists at all.[100]

Fiorini's first projection is the minimum-error perspective azimuthal projection (or *scenographic* projection, as Fiorini called it) of a hemisphere, one of the series of perspective projections already developed by Clarke in 1862, whom Fiorini references.[101] The second Fiorini projection is said to be an adaptation of Littrow's conformal retroazimuthal projection of 1833, described above and conspicuous for its elliptical parallels of latitude and hyperbolic meridians. Hammer says that Fiorini obtained a conformal projection with elliptical meridians and hyperbolic parallels of latitude, and Hans Maurer repeated this, indicating specifically that Fiorini's change involved interchanging latitude with longitude and x with y.[102] Unlike the Littrow, however, such a projection is not conformal, even though the meridians and parallels still intersect at right angles, and a careful reading of the pages Maurer references in Fiorini does not provide any such adaptation.

The Jervis Cycloidal Projection

Developed presumably near the middle of the century, but apparently first published posthumously in Turin, Italy, by his son, was Thomas Best Jervis's (1796–1857) cycloidal projection (Jervis 1895). According to his son, Jervis, who was an officer of the Bombay Engineers and director of the Topographical and Statistical Depot of the British War Department,

> long studied a method for the most faithful possible delineation of the sphere on a plane surface, and was led to devise his beautiful cycloidal projection, in which the distortion is minimized over large surfaces of the globe. Within 40° on either side of the central meridian the curves are very gentle; extending the latitude further distortion commences. . . . This projection is not suited for maps embracing latitudes stretching beyond 15° across the Equator [that is, the range may be from the North Pole to 15° S].
>
> To construct Jervis's cycloidal projection graphically draw a vertical line AB [no diagram given] for the central meridian, dividing it into degrees B′, B″, B‴, and so on. Then draw the horizontal line C A D, passing through the pole A. The revolution of a circle along the line C A D, and whose diameter is represented by the distance of the given parallel from the pole, will give the curves of latitude. The longitude is determined by the corresponding arc of the revolving circle. (W. P. Jervis 1898, 254)

If the last statement is taken to mean that the longitude from center equals the angle by which the circle has rotated, the formulas can be written

$$x = r (\lambda + \sin \lambda), \text{ and}$$
$$y = -r (1 + \cos \lambda),$$

where $r = (R/2)(\pi/2 - \phi)$, with the y axis along the central meridian, positive north, and the x axis perpendicular through the North Pole. As Fiorini (1900, 186) concluded, Jervis's son was unclear about the construction of the meridians; the cycloidal projection seems to have been ignored in any later literature. Figure 3.34 shows this projection using these formulas.

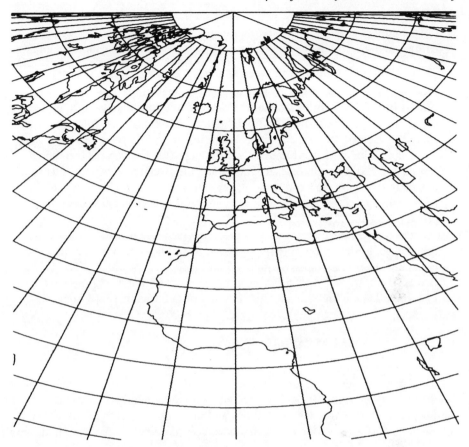

FIGURE 3.34. A presumed construction of the Jervis cycloidal projection, centered on the Greenwich meridian. 10° graticule.

Projections to Promote Commerce

Discussions of new nineteenth-century map projections are incomplete without reference to representations of all or portions of the United States in ways that cannot be attributed to any mathematical or artistic concept as such, but only to a desire to favor a particular commercial enterprise. If these were only sketches favorably showing the locations of a few industrial or commercial sites, they would be considered artistic renderings, but the detailed geographic information included gives a deceptively cartographic flavor that places them in the category of map projections. More specifically, they can be called cartograms, which are further discussed in the next chapter (Raisz 1962, 216).

These maps delineate the routes of particular railroads as smoothly curved channels passing through the hearts of major portions of America and affording easy access to the remaining regions. To accomplish this favorable image, many states are mapped with significant reshaping and changes in areas, and the locations of towns and connecting railroads are moved accordingly.

Rand McNally and Company emerged in the late nineteenth century as a leading commercial publisher of railroad maps as well as world atlases. In a booklet distributed by the mapmaker about 1879, the following candid comments were made:

> Map "Designing," to other than a railroad official, might seem a peculiar phrase, but the majority of railroad maps have some "peculiar designs" hidden under the careful pencil of the draughtsman. It requires a faculty only acquired by experience and a perfect knowledge of the railroad system of the country, to "design" a good railroad advertising map. The various friendly interests must be shown to best advantage, and the rival interests disposed of in a manner that "no fellow can find out." The drawing of a good map is a matter of considerable difficulty, but the "designing" of a good map involves the exercise of tact and ingenuity.
>
> Probably more *original* map projections have been made by our map drawing room than have ever been produced in the United States. It is not generally known that our large railroad and county map, which is 58 × 100 inches, is the second *original* projection of a United States map ever made. Our United States and Canada Atlas is made from the same projection.[103]

General Treatises and Journals

The increasing number of map projections and the abundance of sophisticated mathematics led logically to several detailed treatises that not only describe nearly all the projections available, but also present their mathematical derivations in detail and occasionally propose new projections and new approaches to analysis. In most of them, the history of earlier developments is considered incidental; emphasis is on the principles and features.

By far the most important historical treatise is that of d'Avezac-Macaya (1863); it is without illustrations and entirely nonmathematical, except for passing reference to some formulas in a few of the voluminous footnotes. With only a little exaggeration, Thomas Craig (1882, xii) stated that d'Avezac-Macaya's "complete historical account . . . leaves absolutely nothing to be said on the subject" of history of map projections.

For the most important analytical treatises "comparatively easy of access," Craig (1882, xii), author of the only significant general treatise of the century in English, listed those by Littrow (1833), Germain (1866), and Gretschel (1873). Five more treatises, published during the last two decades of the century, could be added to Craig's and his list: ones by Fiorini (1881), Tissot (1881, plus an 1887 translation and revision by Hammer), Karl J. Zöppritz (1884; the 1899 2d ed. with Bludau), Herz (1885), and Trutat (1897). The twentieth century saw several treatises and textbooks written in English and Russian as well as other languages, but of the ten listed here, counting d'Avezac-Macaya, the dominant languages are German (four) and French (four), with one each in Italian and English.

A handful of technical journals contained several papers each on map projections during the 1800s. First was the shortlived *Zach's Monatliche Correspondenz zur Beförderung der Erd- und Himmels-Kunde,* with nine such papers, all published during 1805–8, but including all six of Mollweide's and both of Albers's. In 1854 Petermann launched his *Mittheilungen,* which became *Petermann's Mittheilungen* in 1879 (and it received its current name, *Petermanns Geographische Mitteilungen,* in 1938) and included a similar number of projection papers by the end of the century. In France, the Société de Géographie, founded in 1821, promptly began its *Bulletin,* which contained over a dozen such papers, all by 1879, while the periodicals of the Royal Geographical Society in Britain led English-language journals in the subject, with only a half dozen projection-oriented papers in its *Proceedings* and *Journal,* all in 1857–71.[104]

THE TISSOT INDICATRIX

Although some of the detailed treatises presented one or two new projections (generally described above), they basically discussed those existing previously, albeit with very thorough analysis. One scholar, however, proposed an analysis of distortion that has had a major impact on the work of many twentieth-century writers on map projections. This was Tissot, who introduced the concept of what he called the "ellipse indicatrice" or an ellipse which, translating, "at each point [on the map] establishes a kind of indicatrix of the system of projection."[105] This has become known as the Tissot indicatrix, or ellipse of distortion. His classic 1881 exposition contains pertinent extracts from his paper of 1878, although concepts appeared in an article as early as 1859.[106]

The concept of the indicatrix may be summarized as follows: Any very small (theoretically infinitesimal) circle on the sphere or ellipsoid representing the earth is plotted on a flat map as a circle or ellipse centered about the same point. If these very small circles are plotted at various locations on the maps, using the same radius for each circle on the globe, the distortion at any given point on the map is indicated by the shape and size of the circle or ellipse.

On a conformal map projection, where there is no local shape distortion, all the very small circles on the globe will be plotted as circles on the map, but their radii will vary in direct proportion to the scale at that point (fig. 3.35). On an equal-area map projection, all these small circles on the globe will be plotted as circles or ellipses with the same areas, the shape of the ellipse being an indication of the local shape distortion in that its major, minor, and intermediate axes are all directly proportional to the scale on the map in the same direction at the same point (fig. 3.36). On projections that are neither equal-area nor conformal, the Tissot indicatrices vary in both size and shape, depending on the local characteristics of the projection (fig. 3.37).

Tissot developed formulas from which the relative axes of the indicatrices may be calculated for a given projection and location. With modern computers, indicatrices at each intersection of a plotted meridian and parallel can be automatically plotted as the representation of the corresponding circles on the globe with a constant radius of say 0.1° in great-circle distance from the successive centers, the resulting circles and ellipses uniformly enlarged by a factor of say 50 to 1 on the map so that they are readily visible.

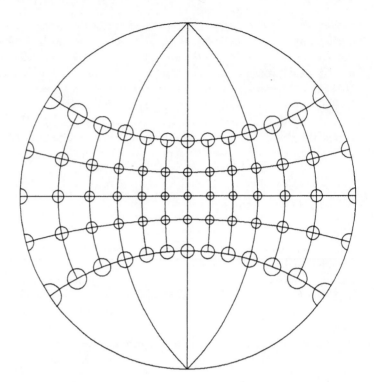

FIGURE 3.35. The "Lagrange" projection of the world (cf. fig. 2.8) with Tissot indicatrices shown at every 30° graticule intersection. Equal small circles on the sphere, these circles, shown equally enlarged, are of varying size on the map, indicating conformality but linear and areal scale variation.

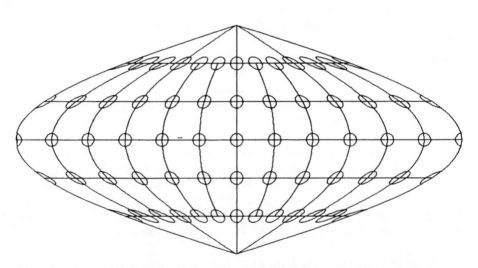

FIGURE 3.36. The sinusoidal projection of the world (cf. fig. 1.39) with Tissot indicatrices shown at every 30° graticule intersection. Unlike the varying circles of fig. 3.35, the indicatrices are circles or ellipses of the same area for this equal-area projection, but linear scale and local shape vary as the dimensions and shape of the ellipse.

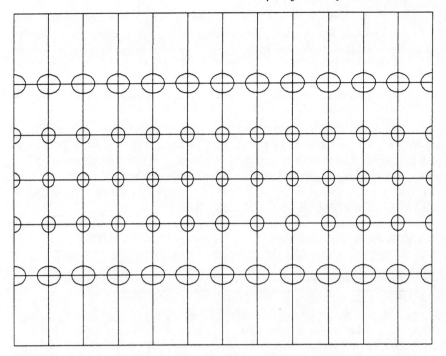

FIGURE 3.37. The Gall stereographic projection of the world (cf. fig. 3.10) with Tissot indicatrices every 30°. Since this projection is neither equal-area nor conformal, the ellipses vary in both area and shape.

In computing this distortion, Tissot also used other concepts. For any given point, the scale factor along the meridian, called h by Tissot and most later writers, the scale factor k along the parallel of latitude, and the angle of intersection Θ between the meridian and parallel, once calculated, are sufficient to determine the scale factor in any other direction and therefore the distortion as well as the shape and size of the indicatrix.[107] Furthermore, any pair of lines intersecting on the globe will intersect on the map at the same or a different angle. At any point, there is one pair of lines intersecting at a right angle on both the globe and the map. Unless the map is conformal, all other pairs of lines intersect at incorrect angles on the map at this same point. The greatest deviation from the correct angle of intersection, or maximum angular deformation, as it is usually now called ("maximum de l'alteration d'angle" by Tissot), may also be calculated from h, k, and Θ, and was designated 2ω by Tissot.[108]

CONCLUSIONS

The most significant single nineteenth-century contribution by cartographers, geodesists, and mathematicians to map projection science was the establishment of sound mathematical principles. Lambert and Lagrange had made important beginnings in the 1770s, but Gauss and Tissot, especially, added broad-based contributions by the 1880s. Others such as Airy, Clarke, Schwarz, and Peirce concentrated on more specific but complicated tasks, while treatise writers like Germain, Gretschel, and Craig collected the diverse works of others into single interpretive volumes. The new projections of Mollweide, Albers, Gall, and others involved simpler mathematics and were more easily rederived, like some of Lambert's, but they well deserve their originators' names.

The accelerated development of the subject during this period (see fig. 2.19) is indicated by both the number of new projections and the number of books and papers published. The growth of new projections may be seen by comparing tables 1.1 and 2.1 with table 3.2 in this chapter. The number of projections listed in the tables as separate items are as follows (some items embrace a series of projections): pre-1670, 16; 1670–1799, 16; and 1800–1899, 53.

A tabulation of books and papers in a recent bibliography (Snyder and Steward 1988, vi–vii) shows the following:

Language	German	French	English	Italian	Latin	Others	Total
pre–1700	1	2	2	—	9	2	16
1700–1799	6	6	8	2	10	3	35
1800–1899	125	71	60	25	—	11	292

These numbers do not necessarily indicate quality, but the trends are clear. It remained for scientists and others of the twentieth century, with their greater ease of computation and publication, plus their greater heritage, to dwarf the quantity of nineteenth-century innovation, but also to emphasize the importance of much of the effort so laboriously carried out earlier.

TABLE 3.2.   Map Projections Developed during 1800–1899

Modern Names (Contemporary Names)	Figures	Inventor (Date)	Design
Cylindrical			
Cassini-Soldner	—	Cassini (1745) Soldner (1809)	Transverse equidistant cylindrical. Scale correct along and perpendicular to central meridian.
Gauss conformal	—	Gauss (1822)	Transverse Mercator. Scale correct along central meridian for ellipsoid; conformal.
Central cylindrical	3.6	?	Perspective projection from center onto tangent cylinder.

(continued)

TABLE 3.2.   *Continued*

Modern Names (Contemporary Names)	Figures	Inventor (Date)	Design
Cylindrical			
Wetch	—	Wetch (18?)	Transverse aspect of preceding.
Oblique Mercator	3.7	?	Oblique Mercator.
Gall orthographic	3.8	Gall (1855)	Cylindrical equal-area with std. parallels 45° N and S.
Gall isographic	3.9	Gall (1855)	Equirectangular with standard parallels 45° N and S.
Gall stereographic; Gall	3.10	Gall (1855)	Stereographic cylindrical with std. parallels 45° N and S.
Braun stereographic cylindrical	3.11	Braun (1867)	Stereographic cylindrical with std. parallel at equator; compromise.
2nd Braun cylindrical	—	Braun (1867)	Perspective cylindrical resembling Mercator to about 80° N and S; not promoted by Braun.
Pseudocylindrical			
Mollweide; homolographic (Babinet; homalographic; elliptical)	3.12	Mollweide (1805) reinvented by Babinet (1857)	Equal-area pseudocylindrical world map in ellipse.
Foucaut stereographic equal-area (stereographic equivalent)	3.13	Foucaut (1862)	Equal-area pseudocylindrical world map with parallels spaced as on stereographic.
Foucaut averaged equal-area[a]	—	Foucaut (1862)	Equal-area combination of cylindrical equal-area and sinusoidal.
Collignon	3.14	Collignon (1865)	Equal-area pseudocylindrical world map with straight meridians and parallels, pointed pole(s).
Nell pseudo-cylindrical	3.15	Nell (1890)	Equal-area combination of cylindrical equal-area and sinusoidal.
Conic			
Albers equal-area conic	3.16	Albers(1805)	Equal-area conic with two standard parallels, for sphere.

(continued)

Table 3.2. *Continued*

Modern Names (Contemporary Names)	Figures	Inventor (Date)	Design
Conic			
Polyconic; American polyconic; ordinary polyconic	3.18	Hassler (1820)	Universal, simply drawn, with circular arcs at true scale and curvature for parallels; central meridian at true scale.
Rectangular polyconic; War Office	3.19	U.S. Coast Survey (by 1853)	Circular arcs at true curvature for parallels; meridians intersect at right angles.
Equidistant polyconic	—	U.S. Coast Survey (by 1853)	Modification of ordinary polyconic.
Lambert conformal conic (Harding; Gauss; Herschel; Boole)	3.20	Lambert (1772) reinvented by Harding (1808), Gauss (1822), Herschel (1859), Boole (ca. 1860)	Conformal conic with one or two standard parallels for sphere and/or ellipsoid.
Perspective conic	—	?	Perspective projection from center onto tangent or secant cone.
Stereographic conic	3.21	Braun (1867)	Perspective projection from South Pole onto cone tangent at 30° N.
Tissot equal-area conic	—	Tissot (1881)	Equal-area conic with special selection of standard parallels.
Nell modified conic	—	Nell (1890)	Equal-area combination of Bonne and Lambert equal-area conic.
Azimuthal			
Lowry	—	Lowry (1825)	Perspective azimuthal with low error.
Fischer	—	Fischer (ca. 1850)	Perspective azimuthal with low error.
James	3.22	James (1857)	Perspective azimuthal with low error.
Gretschel	—	Gretschel (1873)	Perspective azimuthal with low error.
Airy (Proj. by balance of errors)	3.23	Airy (1861) corr. by James and Clarke (1862)	Minimum-error azimuthal using least squares.

(continued)

TABLE 3.2. *Continued*

Modern Names (Contemporary Names)	Figures	Inventor (Date)	Design
Azimuthal			
Clarke (incl. "Twilight")	3.24	Clarke (1862, 1879); part reinvented by Fiorini (1881)	Minimum-error perspective azimuthal using least squares.
Lidman	—	Lidman (1876)	Double perspective projection through tangent cone to plane.
Breusing geometric (Breusing)	—	Breusing (1880s)	Azimuthal mean of stereographic and Lambert azimuthal equal-area.
Modified azimuthal			
Wiechel	3.25	Wiechel (1879)	Pseudoazimuthal equal-area.
Aitoff	3.26	Aitoff (1889)	World map modified from azimuthal equidistant.
Hammer; Hammer-Aitoff	3.27	Hammer (1892)	World map modified from Lambert azimuthal equal-area.
Frye	—	Frye (1895)	Interrupted world map using azimuthal equidistant.
Miscellaneous			
Schmidt globular	—	Schmidt (1803)	Globular modification.
Littrow (Weir azimuth diagram)	3.28	Littrow (1833)	Retroazimuthal.
Nell globular	—	Nell (1852)	Globular modification.
Peirce; quincuncial	3.29	Peirce (1879)	Conformal polar hemisphere in square using elliptic functions.
Guyou	3.30	Guyou (1886)	Conformal equatorial hemisphere in square using elliptic functions.
VonderMühll	—	VonderMühll (1868)	Generalized conformal projections using conic sections.
Star			
Jäger star (polygonal north-polar star)	—	Jäger (1865)	Eight-pointed star with broken straight meridians and parallels.
Petermann star	—	Petermann (1865)	Eight-pointed star with broken straight meridians and curved parallels.

(continued)

TABLE 3.2.   *Continued*

Modern Names (Contemporary Names)	Figures	Inventor (Date)	Design
Star			
Berghaus star	3.31	Berghaus (1879)	Five-pointed star.
Steinhauser conoalactic	—	Steinhauser (1883?)	Four-pointed star with conical northern hemisphere.
Other			
Eisenlohr	3.32	Eisenlohr (1870)	Minimum-error conformal world map without singular points.
August; August epicycloidal	3.33	August and Bellermann (1874)	Low-error conformal world map without singular points.
Polyhedric	—	?	Compromise for large-scale maps.
Woolgar	—	Woolgar (1833)	Stereographic projection onto tetrahedron.
Polyhedral globe variations	—	Adorno (1851) Boorman (1877)	Globe mapped onto various polyhedra.
Cycloidal	3.34	Jervis (1850?)	
Tissot minimum-error (système Lambert)	—	Tissot (1860)	Compromise for large-scale maps.
Fiorini	—	Fiorini (1881)	Alleged modification of Littrow.
"Railroad maps"	—	Rand McNally and others (various dates)	Distortions to favor certain railroad companies.

Note: Map projections are listed in the order discussed in chapter 3. The contemporary name is the same as the modern name unless a parenthetical name is added.

[a] Contemporary name not known.

FOUR

Map Projections of the Twentieth Century

THE ACCELERATION OF map projection development in the nineteenth century was catalyzed by the development and application of mathematical principles. The calculus, least squares, and complex algebra opened up challenging approaches to mapping the increasingly explored and measured face of the earth. Reuse of old concepts also flourished, and the most-used new and old projections generally continued to be the simplest to construct.

The twentieth century continued this mixture of complicated theory and simple, frequently naïve approaches to "new" projections that were claimed in some cases to be not merely good but the best possible projection of the world without restriction. Inevitably, many of the new map projections were intended for world maps since this could result in products visibly different to the viewer. Improvements in projections for medium- and large-scale maps, however, produced changes that could be detected only by careful measurement. In some cases, advocacy of new projections appeared motivated by desire for self-promotion, but this is a subjective charge occasionally confused with zeal.[1]

Concurrent with the actual development of new projections during the twentieth century has been a proliferation of grid systems, used in almost every country as a means of locating a point by its rectangular coordinates according to an official ellipsoidal map projection, rather than by just its latitude and longitude. This method was formally used for systems beginning in central Europe in the late nineteenth century, although square or rectangular grids had originated with early Chinese maps and portolan charts.[2] At the same time, the number of projections used for most large-scale mapping dwindled to two or three, in particular the transverse Mercator and the Lambert conformal conic, although in a few countries the problem of map revision was not considered worth the effort of changing from such nonconformal projections as the Bonne or the azimuthal equidistant projection.

The more complicated projection developments were in some cases carried out like the proverbial reason for mountain climbing—because the challenge was there. Beyond the achievement of placing the world map conformally in a triangle, for example, there is little to recommend such a projection. The inventors of many of these mathematical novelties generally did not promote their work beyond modest scientific publication. Others applied older spherical projections, like the Albers **155** equal-area conic, to the ellipsoid, so that they could better meet the needs of twentieth-century precision.

The fact that, in spite of the many journals and diverse book publishers, very few projections have been independently duplicated is testimony to the great number of possibilities. The percentage of projections reinvented is about 5%, both before and after 1900.

The greatest new noncartographic boon to twentieth-century map projection development has been the modern computer. Older projections that lay dormant due to complexity could now be plotted quickly by persons following elementary instructions after programming requiring only moderate proficiency in mathematics. Projections can now be developed using iterative solutions to complicated mathematical requirements, because the computer is so fast that an expert mathematician's analytical solution may be unnecessarily efficient, if indeed it is even possible in some cases.

The modern computer, however, is quite recent, and desktop crank calculators and logarithm tables were the fastest forms of calculation to more than two or three significant digits for innovators like Ernst Hammer (1858–1925) and Hans Maurer (1868–1945) of Germany, and Oscar S. Adams (1874–1962) of the U.S. Coast and Geodetic Survey, who flourished during the early years of a century that saw an almost incomprehensible scope of changes.

Twentieth-Century Use of Earlier Projections

Because they are more easily constructed, because the mapmaker is familiar with them, and, in many cases, because they have very useful properties, the projections predating 1900 or even 1800 are still the ones most commonly seen in atlases and on sheet maps. Just as only a few of the nineteenth-century commercial mapmakers (like Berghaus, Petermann, and Debes) used projections innovative for the time, the same is true of this century. John Bartholomew, whose family began to publish maps and atlases early in the nineteenth century, used several novel combination projections by mid-twentieth century.[3] Major English-language atlases like *The Times Atlas* and those of the National Geographic Society and Rand McNally and Company include projections prepared by or for them (such as The Times, the Chamberlin trimetric, and the Robinson projections, respectively), but most of their maps are based on standard old projections.[4] The older projections are discussed by category.

Cylindrical Projections

The Mercator Projection

As much as the Mercator projection (fig. 1.37), designed in 1569 as an aid to navigation, is criticized as a basis for general world maps, it continues in that highly visible role. A few examples of the extensive criticism of the Mercator by leading cartographers follow: "The great distortion in the north and south makes Mercator's projection altogether unsuitable for a land map" (Hinks 1912, 29; 2d ed. 1921, 29).

"Its common use for world maps is very misleading, since the polar regions are represented upon a very enlarged scale" (Deetz and Adams 1934; 1st ed. 1921, 32; 5th ed. 1944, 31). "It is largely responsible for many geographical misconceptions, e.g. the misleading appearance of the polar areas" (Steers 1970, 139; 1st ed. 1927, 99). "The Mercator world map enjoys an unmerited popularity. . . . its use should be restricted" (Raisz 1938, 87; 2d ed. 1948, 68). "It is of little use for purposes other than navigation" (Robinson 1953a; 2d ed. 1960, 82; 5th ed. 1984, 93). The *New York Times* added its editorial voice in 1943, saying, "the time has come to discard [the Mercator projection] for something that represents continents and directions less deceptively. . . . We cannot forever mislead children and even college students with grossly inaccurate pictures of the world."[5] Although its usage as the world map in atlases has diminished in the latter part of the twentieth century, it is still highly popular as a wall map apparently in part because, as a rectangular map, it fills a rectangular wall space with more map, and clearly because its familiarity breeds more popularity.[6]

The objections have thus had only a modest effect on public consumption, and several scholars and mapmakers consequently have been led to advance map projections, professionally or flamboyantly, as *a* or *the* solution to the problems posed by the geographical distortion of the Mercator. James Gall and Carl Braun made an attempt in the nineteenth century, and Alphons J. van der Grinten, Walter Behrmann, O. M. Miller, Trystan Edwards, and Arno Peters were among those most clearly invoking the name of Mercator in justifying their own solutions after 1900. The work of each of the latter group is described later.

The Mercator projection also continued to be used as recommended, especially for navigational purposes and occasionally for conformal maps of equatorial regions. For example, in the U.S. Coast and Geodetic Survey during the early twentieth century, the Mercator superseded the polyconic formerly used on coastal charts for scales from 1:10,000 to 1:1,200,000, because of "the desirability of meeting the special requirements of the navigator" (Deetz and Adams 1934, 104). A given chart was constructed so that a latitude near its center was made true to scale.

In world atlases, regions such as Indonesia and Indochina, close to or on the equator, were occasionally shown on separate maps based on the Mercator.[7] The first detailed map of an entire planet other than the earth was issued in 1972 on the Mercator projection: a map of Mars at a scale of 1:25,000,000 (Wilford 1972). It was followed by Mercator maps of portions of Mercury, Venus, and numerous natural satellites (Snyder 1987b, 42–43). Similarly, a handful of large-scale topographic mapping grids have been based on the Mercator (e.g., for six zones of Indonesia), but most zones elsewhere along the equator in Africa and South America have used the transverse Mercator or other projections (Mugnier 1985).

As the simplest conformal projection, the Mercator has been one of the projections most likely to be modified to apply to conformal mapping of the triaxial ellipsoid. This figure occasionally is attempted for the earth, but more seriously applied to Mars and some of the more irregularly shaped natural satellites in the solar system. The mathematics is very complicated, but there have been studies of such conformal mapping, especially during the space-conscious 1980s (Serapinas 1984; Snyder 1985b).

The Equirectangular Projection

By the twentieth century, use of the plate carrée (fig. 1.3) and the more general equirectangular or equidistant cylindrical projections (fig. 1.4) was almost nonexistent for detailed geographic maps. The simplicity of construction and some elements of scale preservation (along two parallels or else the equator, and along all meridians) gave this type of projection a role as an outline map, appearing on some computer screens of the 1980s as a quickly drawn base for insertion of other data. The U.S. Geological Survey has used it for index maps to show the status of mapping of topographic quadrangles and the like. Since quadrangle maps are bounded by constant intervals of longitude and latitude for a given series, such as those 7½ minutes of latitude by 7½ minutes of longitude, all appear the same size on a map of the United States based on the equirectangular projection, whether in the northern or southern portion of the country, and identical symbols can be used for current status.

John Senex had prepared a star map on the oblique equirectangular projection in 1718, but the first use of the oblique aspect for geographic maps was proposed by Charles Arden-Close (1941) for a map of the Middle East and the Indian Ocean, with the central line reaching a maximum latitude of 54°30′ N at 28° W. He reduced the scale of the central line by 10%, in effect using transformed standard parallels of arccos 0.9, or ±25°51′. A second use of the oblique equirectangular was proposed by F. V. Botley (1951) (fig. 4.1): by choosing a given point, for example, London, as

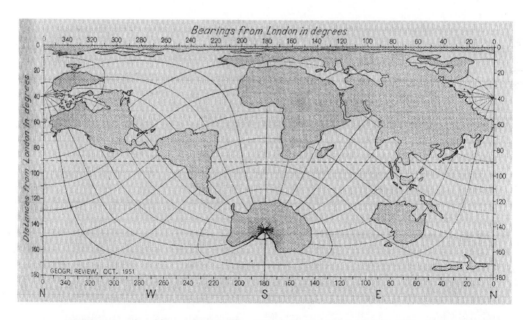

FIGURE 4.1. The oblique plate carrée with London at the north pole of the base projection (across the north edge of the map). Reproduced from Botley (1951, 641), who proposed it to determine the great-circle distance (vertically) and the direction clockwise from north (horizontally) in degrees from London to any other location on earth. 20° graticule. *U.S. Geological Survey Library, by permission of the American Geographical Society.*

the north pole on the base plate carrée and projecting the geographic meridians, parallels, and other features accordingly, the bearing and distance of any other point from the base point, London in this case, can be read directly as two rectangular coordinates. The map thus serves as an azimuth-distance diagram with rectangular instead of the polar coordinates of the azimuthal equidistant projection.

The Cassini Projection

The Cassini (fig. 2.5) and related Cassini-Soldner and Soldner projections, in principle transverse plate carrée or transverse equidistant cylindrical projections, still required attention, but only because of continuations of official mapping programs of the past. In France, where the projection began in the eighteenth century, new official use had disappeared by the early nineteenth century. In England, which began using it about then, the Cassini was replaced by the transverse Mercator beginning in 1919, but old maps remain to the present conforming to some of the thirty-nine Cassini grids for counties.[8] In numerous central European and other scattered areas, Cassini grids also survive (Mugnier 1985). All are large-scale, ellipsoid-based applications. For smaller-scale maps, the Cassini projection (using the sphere) is almost entirely a cartographic exercise for textbooks.

The Transverse Mercator Projection

The emergence of the transverse Mercator projection of the ellipsoid as the major projection for topographic mapping began in the nineteenth century, not long after Gauss developed the mathematics in one of his many landmark studies. Because of the constant scale along its central meridian in the ellipsoidal form normally used (as is the case with the Cassini as well), the transverse Mercator is especially suited for mapping north-south regions. There are many other possible versions of the ellipsoidal transverse Mercator, theoretically an infinite number, without constant scale along the central meridian, but they are rarely used (see Lee 1976, 92, 100–101).

In the first half of the twentieth century, the transverse Mercator received major boosts in the English-speaking world from a number of adoptions for grid systems. First, in 1919, the British adopted it in place of the thirty-nine county zone Cassini as a standard for the United Kingdom (see n. 8). Using a single transverse Mercator projection and zone, the British Grid System was then replaced with the Modified British System in 1927, both for military grids. For general topographic mapping, the National Grid was adopted in 1945. All three employed the same central meridian (long. 2° W), a constant scale factor of 0.9996 along it, and an origin at lat. 49° N. Only the system of annotating the grids changed.

Second, in the early 1930s the U.S. Coast and Geodetic Survey made the transverse Mercator the standard projection for the State Plane Coordinate System (SPCS) zones in states that are predominantly north-south in extent, or that were so specified.[9] This led to its use for fifty-six zones in twenty-two states (counting the later states of Alaska and Hawaii). All zones are bounded by county (or state) lines, except for meridian boundaries in Alaska, which has no counties; the zones were sized so that the scale does not vary by more than about one part in ten thousand from the stated map scale.

Rectangular coordinates for the SPCS are in feet when based on the 1927 *datum* or geodetic reference surface and in meters for the new datum of 1983. To minimize the overall variation, the scale along the central meridian of the zone is depressed by a ratio varying with the zone, so that the lines of true scale are roughly two-thirds of the distance to the zone limits. The 0.0001 limit was chosen to correspond approximately with the level of accuracy of surveying instruments of the time; thus the surveyor could satisfactorily use the nominal scale of maps prepared on this grid system without adjusting for the exact scale at the point in question. The zones of U.S. states predominantly east and west in extent are based on the Lambert conformal conic projection with two standard parallels.

The third and ultimately most widespread usage of the transverse Mercator for a single application resulted from the establishment of what was immodestly named the Universal Transverse Mercator (UTM) grid and projection system by the U.S. Army Map Service in the late 1940s.[10] Replacing the World Polyconic Grid (discussed later), the UTM system consists of a simple zone arrangement: Numbering eastward from the 180th meridian from Greenwich, the sixty equal zones (with rare exceptions in the North Sea) are bounded by meridians successively 6° apart, with latitudes extending from 84° N to 80° S. Rectangular coordinates are given in meters.

The central meridian for each UTM zone is midway between the bounding meridians, and it is given a scale factor of 0.9996, as in the British National Grid, so that the lines of true scale are approximately 180 km on either side. This results in a scale varying no more than about one part in one thousand from the nominal scale. The UTM formulas can be the same as ellipsoidal transverse Mercator formulas used for the U.S. State Plane Coordinate System, or the Gauss-Krüger projection used for zones elsewhere; the constants entered as parameters are changed to suit. The formulas may occasionally appear different, due to different means of derivation, but numerical differences in coordinates should result only from the effects of truncation of the series used.

The ellipsoidal transverse Mercator projection is used for over 80% (by area) of the large-scale topographic grid systems throughout the world.[11] There are hundreds of such grid zones, on every continent except Antarctica and on numerous islands. The U.S. Geological Survey has applied the projection to maps of Mars, treated as an ellipsoid, at scales from 1:1,000,000 to 1:250,000 (Snyder 1987b, 42–43, 57).

In developing the series used to adapt Lambert's spherical transverse Mercator to the ellipsoidal versions in practice, four names predominate. Gauss laid the foundations and provided the equations used for the Gauss conformal projection of the nineteenth and twentieth centuries (Lee 1976, 100–101). Louis Krüger (1912) derived refinements leading to the commonly used Gauss-Krüger projection. The Gauss-Boaga projection, also using truncated series, was adopted about 1950 by the Istituto Geografico Militare Italiano.[12] Giovanni Boaga, a professor of engineering at the University of Rome, was also the general director of Catasto e dei Servizi Tecnici Erariali (Cadastral Survey and Technical Services of the Department of Finance) of Italy.

The Gauss-Schreiber projection is so-called because Oskar Schreiber about

1880 developed series to project the ellipsoid onto a conformal sphere, with latitudes positioned to retain the conformality of the ellipsoid, and with this sphere then projected onto the plane map using the spherical transverse Mercator.[13] In this form of double projection, unlike the others above, the central meridian has a varying scale. Schreiber's series were used in the Prussian land survey.

A recent transverse Mercator design with varying scale along the central meridian was presented by Guozao Li (1981) of the People's Republic of China. He used series to establish two standard meridians symmetrical about the central meridian, thus reducing distortion for zones bounded by meridians. The concept of the conformal sphere was originated by Lagrange in 1779 and given further emphasis by Mollweide (1807), Gauss (1825), O. S. Adams (1921, 7–10), and J. H. Cole (1943).[14]

In addition to the practical series derived by Gauss, Krüger, Boaga, and Schreiber, exact formulas providing a constant scale along the central meridian were developed in different forms by several investigators, beginning with unpublished work by E. H. Thompson in 1945, and involving elliptic and other functions.[15] These formulas can provide any desired accuracy at any distance from the central meridian by standard iterative techniques. Because of computation time, these formulas are suitable only for modern computers, and are normally useful only in extending the transverse Mercator to the entire ellipsoid (for one "zone"), an exercise having almost no practical significance because of the great distortion due to flattening of the round earth. One exception is found in use of the ellipsoidal transverse Mercator for large-scale maps of transpolar flights, for which the desire for conformality with a constant scale along a meridian over a pole rules out the usual polar azimuthal projections.[16] The series equations for the ellipsoidal transverse Mercator projection are limited by longitude range and are therefore unsuitable for this application.

Use of the (spherical) transverse Mercator in atlases is moderate. It appears in recent National Geographic Society atlases for the "Western Soviet Union," India, and sectional maps of Africa, for the Orient in *The Times Atlas*, and for southeast Asia, eastern Australia, and eastern Africa in *The American Oxford Atlas*.[17]

The Oblique Mercator Projection

The oblique aspect (fig. 3.7) of the Mercator projection was just emerging in the late nineteenth century, and then only in the spherical form. During the twentieth century, the relatively complicated nature of the formulas and perhaps the lack of simple scale relationships along any meridian or parallel even for the spherical form discouraged extensive use. Oblique Mercator atlas maps include Hawaii, the West Indies, Southeast Asia, and New Zealand by the National Geographic Society, the Americas in *The American Oxford Atlas*, but none in *The Times Atlas*.[18] H. J. Andrews (1935, 1938b) suggested oblique Mercator maps of the United States and Eurasia. Hinks (1940, 1941a) encouraged the projection's use for world maps by grouping continents with seven different sample central lines, resulting in less distortion of landmasses than if he had used the regular or transverse Mercator projections, but these proposals remained academic.

Ellipsoidal forms of the oblique Mercator, although much more complicated, were not academic, however. At least four distinct ellipsoidal forms emerged, all retaining perfect conformality, but none retaining constant scale along the central line (an apparently impossible combination for the ellipsoid, except when the limiting case of the sphere is treated). Chronologically, Max Rosenmund's (1903) form, used for official topographic mapping in Switzerland, appeared first (see also Bolliger 1967). In effect the ellipsoid for the Swiss form is first transformed to a conformal sphere, using the conformal latitude (mentioned above). This conformal sphere is then transformed to another conformal sphere in the manner of the general "Lagrange" projection, with the 360° of longitude slightly overlapping one revolution around the globe, and with the equator of the first sphere slightly south of the equator of the second sphere. The second sphere is then transformed to a flat plane using the spherical oblique Mercator formulas, a central line with its northmost latitude at Bern, and a radius equal to the mean radius of curvature at Bern of the Bessel ellipsoid, used in central Europe after Friedrich Wilhelm Bessel (1784–1846) presented dimensions in 1841.

A second ellipsoidal version is called the Laborde projection; in a general sense it is an oblique Mercator, but it is mathematically more related to the transverse Mercator. Jean Laborde, chief of the Service Géographique of Madagascar, presented this version in 1926 specifically for the mapping of this obliquely oriented island, to replace the Bonne projection used since 1896.[19] After projecting the ellipsoid onto a conformal sphere in a manner favoring the mapped region, he then prescribed a transverse Mercator projection of this sphere (the Gauss-Schreiber approach above) and finally applied a third-order complex-algebra transformation to that projection. The last step retains the conformality, but has the effect of rotating the central line so that instead of its coinciding with a meridian near Tananarive, it falls approximately along the oblique lengthwise axis of the island. This last (complex-algebra) transformation was applied by several others to develop projections described later.

A third form of the oblique Mercator was applied to obliquely oriented Italy by Cole (1943, 16–30) of the Survey of Egypt. It consisted of using the conformal sphere once for the ellipsoid and then applying the spherical oblique Mercator formulas. The fourth version was called the rectified skew orthomorphic by its originator Martin Hotine, and the Hotine oblique Mercator when later used by the United States. Hotine (1898–1968), head of the British Directorate of Overseas Survey, achieved much less of a scale change than Cole along the central line of his oblique Mercator in 1946 by projecting the ellipsoid conformally onto an aposphere, a surface with constant curvature based on a combination of the radii of curvature along and perpendicular to the meridian of the ellipsoid at a chosen center point.[20] This surface could be a sphere, but Hotine used a nonspherical aposphere. His closed equations involve hyperbolic functions.

Hotine applied the projection specifically to the topographic mapping of Borneo and Malaya, but it was later used for mapping in Liberia, and was adapted about 1960 to a single zone of the U.S. State Plane Coordinate System for the southeast panhandle of Alaska, by Erwin Schmid of the U.S. Coast and Geodetic Survey.[21] In addition, it was adopted about 1970 by the U.S. Lake Survey to map the five Great Lakes and related waters and by the U.S. Geological Survey for mapping of Landsat satellite imagery in 1978, just before the space oblique Mercator projection became available.[22]

The Gall Stereographic Projection

Of the three projections Gall developed in 1855, his preferred compromise, the Gall stereographic (fig. 3.10), was moderately used in British atlases of the late nineteenth and the twentieth centuries, largely because of the interest of the Bartholomews.[23] It was also modified into a twentieth-century pseudocylindrical projection by Guy Bomford to decrease the lengths of near-polar parallels of latitude. As pointed out earlier, the parallels and meridians on the Gall stereographic are geometrically projected onto a cylinder secant at latitudes 45° N and S, from a point at the crossing of the equator of the globe with the opposite meridian.

Two more modifications, in which the only change from Gall's is the pair of latitudes of secancy for the cylinder, appeared in the Soviet Union. The first was used by V. A. Kamenetskiy for a 1929 population-density map of the Soviet Union, with standard parallels or a cylinder secant at latitudes 55° N and S, although a conic projection would seem more appropriate.[24] The second, labeled a "Cylindrical stereographic projection of Gall with parallels secant at 30°" (fig. 4.2), was used for about half the world-distribution maps (nearly all the rest were interrupted Eckert VI) in the first volume of *Bol'shoy sovetskiy atlas mira* (*B.S.A.M.*, "Great soviet world atlas"), published in 1937.[25] This version was later called the BSAM cylindrical projection; it was also independently proposed (but enlarged to give correct scale

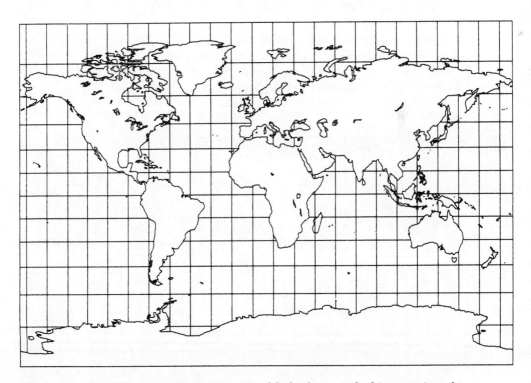

FIGURE 4.2. The Gall stereographic projection modified to have standard (or secant) parallels at 30° N and S latitude. Central meridian 10° E as used in *Bol'shoy sovetskiy atlas mira* (1937). 15° graticule. Cf. figs. 3.10, 3.11.

along the equator) as a new "modified Gall's" projection by O. M. Miller when he listed four generally new cylindrical projections in 1942. (Of the four, he recommended a different projection which came to be called the Miller cylindrical projection, described below.)[26]

All four of these stereographic cylindrical variations may be plotted using the formulas below, varying only the standard parallel ϕ_0 (with either sign):

$$x = R\ \lambda\ \cos\ \phi_0,\ \text{and}$$

$$y = R\ (1\ +\ \cos\ \phi_0)\ \tan\ (\phi/2),$$

where x and y are rectangular coordinates, R is the radius of the sphere at the scale of the map, ϕ is the latitude of the point, and λ is its longitude east $(+)$ or west $(-)$ of the central meridian, using radians.

Cylindrical Equal-Area Projections

The attractive concept of showing area relationships correctly on a map, combined with the simplicity of the cylindrical equal-area projection with its many, theoretically infinite, possible choices for standard parallels (always symmetrical about the equator), has led to several proposals for "new" projections in this category. The first, by Lambert in 1772, made the equator free of distortion (figs. 2.13, 2.14). In 1855, there was the Gall orthographic (fig. 3.8), with the parallels standard at 45° N and S.

Next, Walter Behrmann (1910) of Berlin, following mathematical analysis, concluded that the "best equal-area projection known for the entire world" is the cylindrical equal-area projection with standard parallels at 30° N and S (fig. 4.3). Based on his analysis, these parallels give the least mean maximum angular deformation for the entire map, combining the principles of both Gauss's least squares and Tis-

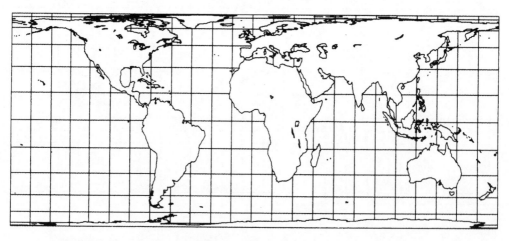

FIGURE 4.3. The Behrmann cylindrical equal-area projection. A modification with standard parallels 30° N and S latitude of the cylindrical equal-area projections of Lambert (fig. 2.13) and Gall (fig. 3.8). Central meridian 0°; 15° graticule.

sot's distortion analysis. (The Behrmann projection was used in both oblique and transverse aspects, without mentioning Behrmann, in the Swedish *Nordisk världs-atlas* beginning in 1926.)[27]

Another choice of standard parallels was made by Trystan Edwards (1953) of England when he tried to limit the "coefficient" of distortion (which he defined as the ratio of the scale factor along the meridian to that along the parallel at a given point) in a zone between the Tropic of Cancer (23°27′ N latitude) and some more-northern latitude (as well as the corresponding southern zone) to between ⅔ and ¾. This required that the more-northern latitude be arccos (0.75 cos 23°27′) or 46°31′, with the standard parallel equal to arccos (($\sqrt{3}$/2) cos 23°27′) or 37°24′, to retain the equal-area feature. Edwards, however, chose a standard parallel of 50°52′, or arccos (0.75 cos² 23°27′), and thus did not achieve the desired distortion characteristics. His very confident but confused tract made his projection sound much more complex than the slight modification of Lambert's, Gall's, and Behrmann's approaches that it is. Though Edwards predicted that his patented map "is likely to be generally acceptable," it was generally rejected.[28]

It was the quiescent Gall orthographic, however, that sprang forth in a reincarnation as the "Peters" projection a century after Gall; it had such an impact that it became a focus of vehement criticism by most cartographers who commented at all. German historian Arno Peters (1916–) apparently had not deliberately copied Gall when in 1967 he devised a projection essentially identical to the Gall orthographic and in 1973 presented it at a press conference; the design is too simple, and Gall's work, other than his cylindrical stereographic projection, was little known. Peters, however, continued to insist that he "created" the projection, in spite of numerous professional and newspaper articles to the contrary beginning in 1974.[29]

Of greater concern to cartographers were Peters's repeated claims for the panacea nature of "his" projection to replace the pervasive Mercator projection as a world map. These claims generally fell into two categories. One was the strong implication that this was the first equal-area projection developed, although Peters's more technical work *The New Cartography* mentioned several of them, including Lambert's original cylindrical equal-area projection.[30] The other controversial claim was that the less developed countries of the world were finally presented in a fair manner on a world map, rather than on a "Euro-centered" Mercator projection with severe distortion.

On the "Gall-Peters" projection, as one cartographic committee chose to call it, all the equatorial regions, such as Africa and northern South America, are stretched to appear about twice as long in a north-south direction as they should appear relative to east-west dimensions.[31] Thus the less-developed regions have far more shape distortion on the Gall-Peters projection than do the industrialized regions near latitude 45° N, which have no shape distortion at that latitude. The polar areas are even more distorted in shape.

In spite of this shape distortion, the map was claimed to have "no extreme distortions of form," and further to be "totally distance-factual," impossible for any flat map (as proven by Euler 1777) and especially fallacious for this projection.[32] If it were not for the intense promotional activity on its behalf, the "Peters" projection would have been dismissed in somewhat the same manner as the Trystan Edwards version. As a result of the promotion of this projection in Europe and the United States, its use became widespread, especially among socially concerned agencies. Supporter Ward L. Kaiser (1987, 23, back cover), of the (U.S.) National Council of Churches, stated that

> the General Board of Global Ministries of The United Methodist Church has a [six-foot-high] Peters world map etched in glass at the entrance to its New York office. . . . This map is in use in the Vatican and the offices of the World Council of Churches. NATO [North Atlantic Treaty Organization] forces have used it. United Nations agencies are among the strongest supporters. . . . It is now available in six languages [and] . . . has reached a worldwide distribution of more than sixteen million copies.

Atlas versions began to appear, including two in English (Arno Peters 1989–90), showing each region of the world (except for most of the oceans) at a constant area scale, but basing these regional maps on some form of the cylindrical equal-area projection as far from the equator as 80° N. The polar regions are based on the uncredited Lambert azimuthal equal-area projection, but it was claimed on the book jacket that *all* maps in the atlas use the Peters projection, and in the introduction that they are given a local standard parallel to reduce shape distortion, resulting in many non-"Peters" regions. Erroneous claims aside, failure to use equal-area conic or oblique azimuthal equal-area projections for the midlatitudes, in the name of "fidelity of axis," made the atlas still less appropriate than the world map for geographic use.

The cylindrical equal-area projection was appropriately used for central Africa in one twentieth-century series of atlases, with the equator as the standard parallel (Hammond 1966, 114–15).

The ubiquitous and inappropriate use of both the Mercator and the "Peters" projections, as well as the existence of many of the same problems in any cylindrical projection (all parallels equal in length, etc.), led the American Cartographic Association to issue a resolution in 1989 urging publishers and agencies "to cease using rectangular [cylindrical] world maps for general purposes or artistic displays." It was endorsed by seven other major geographic and cartographic organizations.[33]

Pseudocylindrical Projections: The Sinusoidal and Mollweide

Distinguished from cylindrical projections by curved instead of straight meridians parallel to each other, but sharing the pattern of straight parallels of latitude, the pseudocylindrical type of projection was about to become a favorite design concept for new projections as the twentieth century began. Nevertheless, the time-tested sinusoidal projection of 1570 and the Mollweide of 1805 were hardly ignored.

A choice projection for atlas and wall maps of South America and Africa continued to be the sinusoidal (fig. 1.39b), with its combined simplicity of construction, equivalence of area, and useful linear scale characteristics.[34] This projection is also used for large-scale topographic maps in Ecuador, as the Ecuador Ellipsoidal Flamsteed Grid.[35]

On an uninterrupted world map, there is extensive shearing at the polar portions of the outer meridians with either projection. In 1916 this inspired John Paul

FIGURE 4.4. John Paul Goode
(1862–1932), a leading educator in geography at the University of Chicago who developed several interrupted projections, especially the homolosine. *Photo courtesy of Robert B. McMaster.*

Goode (1862–1932) (fig. 4.4) of the University of Chicago to apply the long-known principle of interruption of a projection along meridians to the sinusoidal projection.

While he was not original in applying the basic concepts, Goode was more successful in promoting the use of interrupted projections. The goal was to reduce scale and shape distortion by choosing several central meridians to coincide with large land (or ocean) masses, and constructing the projection in the regular manner to the east and west of each of these meridians until a particular landmass had been included. Reduction in the distortion of shape and scale within the various lobes is offset by the increased discontinuities between parts of the world, where large gaps replace short distances. The central meridians north of the equator do not have to be the same as those south of it. Another term for this type of interruption is *recentered*.

Goode was dissatisfied with the results of interrupting the sinusoidal, although the projection has subsequently been used in this manner a number of times.[36] That same year (1916) he interrupted the Mollweide (fig. 4.5) and found it more satisfactory because of rounding at the poles.[37] This arrangement was used in *Goode's School Atlas* from the first edition of 1923 until 1948.[38] It also served as a further step in Goode's work, which led to the often-used interrupted homolosine projec-

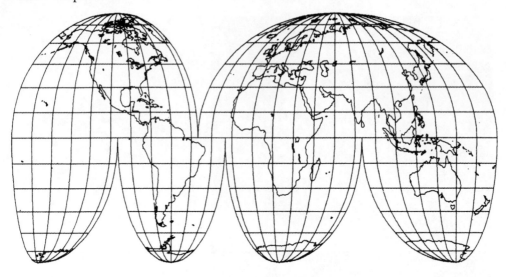

FIGURE 4.5. The interrupted Mollweide (homolographic) projection, with interruptions used by Goode for preserving landmasses, except for his extensions near Greenland and eastern Siberia.

tion, combining the sinusoidal and Mollweide (or homolographic) projections, and discussed later.

A native of Minnesota, Goode graduated from the University of Minnesota and, in 1903, received his doctorate from the University of Pennsylvania (Haas and Ward 1933). He taught in both states until assuming a permanent position with the University of Chicago in 1903. He lectured in the Philippines in 1911, was active in many geographical associations, and sang professionally in choirs and quartettes. A staunch opponent of use of the Mercator for school maps, he developed his interrupted projections and published *Goode's School Atlas* through Rand McNally as steps in presenting realistic maps.

While the uninterrupted Mollweide projection has seldom been used for formal maps, it has regularly been discussed as a classic textbook projection.[39] For example, Robert H. Bromley (1965) pointed out that by compressing the vertical dimensions of the Mollweide and expanding the horizontal dimensions, all by a factor of $\pi/(2\sqrt{2})$ or 1.11072, the equator becomes the standard parallel, without affecting preservation of area scale.

Much more often, the Mollweide led to new projections as it was combined with other projections or concepts. Such developments came from J. Fairgrieve, S. Whittemore Boggs, John Bartholomew, and Allen K. Philbrick, as well as Goode, and are described later.

Azimuthal Projections

Conformality has kept the stereographic projection popular for several applications. Our increasingly global communication network, the use of orbiting satellites, and

war (especially the Second World War) have helped boost the use of the orthographic, general perspective, and azimuthal equidistant projections. The Lambert
azimuthal equal-area increasingly has replaced the Bonne projection as the choice
in atlases for continent-sized regions. Of the common members of the azimuthal
group, only the gnomonic projection continues to have minimal use, as an auxiliary
navigational tool (with its straight great-circle routes) and as a textbook topic because
of its central perspective, its conic-section latitude lines, and its rapid changes in
distortion.

The Stereographic Projection

The polar stereographic projection in ellipsoidal form became standard for large-
and medium-scale maps of the north and south polar areas. The Universal Polar
Stereographic (UPS) projection and grid system was adopted by the Army Map Service in 1947 when the Universal Transverse Mercator (UTM) system was adopted
for nonpolar areas. The UPS projection consists of two coordinate systems, one extending from the North Pole to 84° N, and the other from the South Pole to 80° S.
The scale factor at each pole is reduced to 0.994, making the latitude of true scale
about 81°07′ N or S (Snyder 1987b, 157). After 1962, the same polar portions of
new maps in the International Map of the World series (at a scale of 1:1,000,000)
were to be based on a polar stereographic "with scale matching that of the Modified
Polyconic Projection or the Lambert Conformal Conic Projection at latitudes 84° N
and 80° S."[40]

The U.S. Geological Survey has used the polar stereographic, also in the ellipsoidal form, for maps of Antarctica, with a standard parallel, depending on scale, of
71° S or 80°14′ S, and the USGS Astrogeology Center, Flagstaff, Arizona, used the
projection in ellipsoidal (for Mars) or spherical form for polar areas of every planet
and satellite for which there was sufficient information in these regions (Snyder
1987b, 42–43, 157). In atlases, polar regions are generally shown on the azimuthal
equidistant or on the Lambert azimuthal equal-area projections.[41]

The oblique and equatorial aspects of the stereographic projection (figs. 1.16,
1.19) have been officially used in the spherical form by the Astrogeology Center for
circular maps of portions of the Moon, Mars, and Mercury that are centered on a
basin, crater, or other round feature (Snyder 1987b, 156). For decades these aspects
have been used annually in *The Astronomical Almanac* and its U.S. predecessor before 1981, *The American Ephemeris and Nautical Almanac*,[42] to show the coverage
of total, annular, and partial solar eclipses.

The ellipsoidal form of the stereographic projection has been developed with
several approaches, all of which maintain conformality but lose the strict perspective
feature and in some cases precise azimuthality. The simplest form results from the
substitution of the conformal latitude for the geodetic latitude, an approach applied
to topographic mapping using this projection in Hungary and the Netherlands.[43]
The latter form is equivalent to projecting the ellipsoid onto a conformal sphere and
then using spherical formulas for the stereographic, but applied to the conformal
sphere.[44]

Henri Roussilhe took a more complex approach to the ellipsoidal oblique stereographic in 1922, with a truncated approximate series, and the "Roussilhe" pro-

jection received some attention in the Soviet Union.[45] Projection through an intermediate aposphere was later described by Hotine (1946–47, 165–66).

The Orthographic and General Perspective Projections

Use of the orthographic (fig. 1.11) and general perspective projections rightly remained pictorial rather than providing bases for measurements of distance and area. The increased global consciousness mentioned earlier led to whole atlases featuring this theme, as well as emphasis on these projections in other atlases. A widely known publication was Richard Edes Harrison's (1944) *Look at the World: The Fortune Atlas for World Strategy*, published during the Second World War, with carefully constructed relief maps using both the orthographic and perspective from finite points near the earth, usually showing just the region near the horizon for the particular view.[46]

Following the launching of artificial satellites, the perspective projection has been given considerable scholarly attention in papers such as those by Victor Dumitrescu of Romania, Deakin of Australia, and Schmid and this writer of the United States.[47] Weather patterns regularly displayed on television and in newspapers are based on a perspective projection as seen by the Geostationary Operational Environmental Satellite (GOES) 36,000 km above the earth at the equator (fig. 4.6) (Monmonier and Schnell 1988, 128).

The orthographic (or at least near-orthographic perspective) projection is featured in color photographs of large physical relief globes, with numerous full-page and smaller oblique views in Frank Debenham's (1958) *The Global Atlas* and in various atlases by Rand McNally (1968, 1971).

Even the far-side perspective projection of so much interest to La Hire, Parent, and Clarke prior to 1900 was put to practical use to show more than half of the Moon's disk at a time. Because the Moon, over a period of time, shows slightly more than one hemisphere to the earth, Albert L. Nowicki (1962) placed the point of his perspective 1.537 radii from the center of the earth in a direction opposite the projection center for the Army Map Service (AMS) Lunar Projection (see also Maling 1965). This "modified stereographic" projection, as Nowicki described it, was given a 100° range from center; the "horizon" would extend about 30° farther.

The Azimuthal Equidistant Projection

The global emphasis that attracted mapmakers to the orthographic and perspective projections led to world maps based on the polar or oblique azimuthal equidistant projection. These could comprise a page of an atlas or a wall map.[48] The U.S. Coast and Geodetic Survey (1951) published a "1:47,423,730"-scale world map centered on New York (reconstructed with simpler graticule in fig. 4.7), with small (ca. 1: 300,000,000) insets centered on London and Tokyo. About 1930 a world map centered on Schenectady, New York, "was prepared for the convenience of radio engineers of the General Electric Co. interpreting transmission tests" (Deetz and Adams 1934, facing 163, 164). Use of the projection for regional maps, less related to "global awareness," continued especially for polar regions and occasionally for continents and even Alaska.[49]

Near the beginning of the twentieth century (1892, 1912), the projection was stressed as a means of making the world appear round on a flat map: the Canadian

FIGURE 4.6. A perspective view of most of the western hemisphere as seen by a geosynchronous weather satellite (GOES) 36,000 km above the equator at 90° W longitude. 10° graticule.

Pacific Railroad published a polar azimuthal equidistant map showing the links between its routes across Canada and shipping routes to Europe and the Orient, saying,

> It is not the intention to turn the old world upside down, but it was necessary to publish the map in the manner it has been to properly show the course as well as the advantages of the Canadian Pacific Route around the world. And also the route to and from the far East. We are so accustomed to view the picture of the world on Mercator's projection that we are apt to forget that the world is round. (Quoted in Modelski 1984, 158–59)

In the ellipsoidal form of the azimuthal equidistant projection, three variations have been used for topographic mapping. The oldest is Hatt's (1886) third-order-series version, used by Greece for some of its twentieth-century mapping (see also

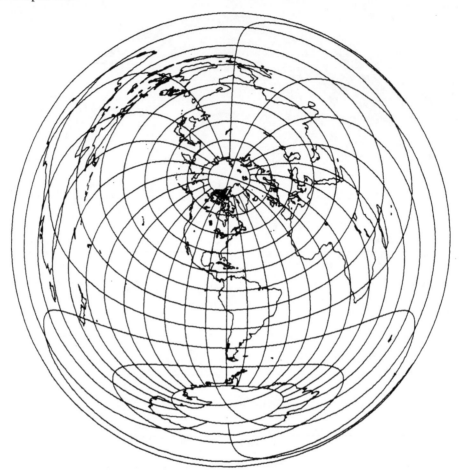

FIGURE 4.7. A simplified reconstruction of a U.S. Coast and Geodetic Survey map of the world using an azimuthal equidistant projection centered on New York City. 15° graticule.

Anatupon 1963). Two were devised after 1950 by the U.S. Coast and Geodetic Survey for mapping Guam and by the U.S. Geological Survey for mapping islands of Micronesia, respectively.[50] The Guam projection is a simple approximation, but the Micronesia version involves a more rigorous use of geodesic distances and azimuths on the ellipsoid.

The Lambert Azimuthal Equal-Area Projection

The Lambert azimuthal equal-area projection (figs. 2.16, 2.17, 3.5), with its concentric circles of constant distortion, is an ideal equal-area projection for circular regions and has been increasingly used in twentieth-century atlases. It is the basis for all polar and continental maps, except those of Europe, in Rand McNally's *Goode's World Atlas* and *International Atlas,* as well as for maps of Eurasia, the Americas, sectional Africa and South America, and three oceans in *The Times Atlas,* and for maps of five continents and an ocean in Hammond atlases.[51] It is generally the basis for the few remaining atlas maps of the eastern and western hemispheres.[52] It is used by the National Geographic Society (1988) for small polar insets on their new sheet map of the world. The formulas for the ellipsoidal form, adapting authalic latitude, are available, but only the spherical form has apparently been used for a published map (Snyder 1987b, 187–90). Authalic latitude, the equal-area analog of conformal latitude, was apparently introduced by O. S. Adams (1921, 10–12, 60–85). He also delineated rectifying latitude to maintain ellipsoidal meridian distances on a sphere.

Conic Projections

Regular conic projections as well as the Bonne and polyconic have been important to twentieth-century mapping. One of them, the Lambert conformal conic, became the second most prominent projection (after the transverse Mercator) for large-scale mapping. While the Bonne and polyconic have been used for a dwindling number of maps throughout the century, the Lambert, the Albers equal-area, and especially the simple or equidistant conic projections are common for maps of midlatitude countries and other regions smaller than continents.

The Bonne and Polyconic Projections

The Bonne projection (fig. 2.1) was the basis for some maps of North America, Asia, and Australia in atlases preceding 1950, and it is still used in *The Times Atlas* for a few regional maps.[53] Generally it has been replaced in its nineteenth-century prominence by the Lambert azimuthal equal-area or, for some maps of Europe, by the simple conic. For large-scale topographic mapping, the ellipsoidal form of the Bonne remains the projection used for parts of official grid systems of over a dozen countries, primarily in western Europe or bordering the Mediterranean Sea (Mugnier 1985).

The (ordinary or American) polyconic (fig. 3.18) can usually be distinguished from the Bonne by the fact that the parallels of latitude are nonconcentric on the former and concentric on the latter. The polyconic does not retain equality of area; the Bonne does. Until the 1950s, the polyconic remained the only projection used for U.S. Geological Survey (USGS) large-scale topographic quadrangles. For new quadrangles, the local projection of the State Plane Coordinate System is generally used, but the map legend may or may not reflect this.

The USGS (1916) used the projection for a map of the United States issued at a scale of 1:7,000,000, but, as a USGS publication stated in 1928, "the polyconic projection is not at all suitable for a single-sheet map of the United States or of a large State, although it has been so employed."[54] This base map of the United States may have encouraged regular use of the polyconic for atlas maps of the United States in *Goode's Atlas* and in Hammond atlases.[55] The polyconic also has been used for other regions such as Africa, the Caribbean, and much of Asia.[56]

In 1918 the U.S. Coast and Geodetic Survey and the Army Corps of Engineers established the progressive military grid system, based on the polyconic projection.[57] First applied to the conterminous United States, it consisted of seven zones each 9° of longitude wide with 1° of overlap, extending from 28° to 49°10′ N latitude, except for extensions to 24° in two zones to include southern Florida and Texas. Central meridians began at 73° W for zone A and continued every 8° to 121° W for zone G. Unlike the units in feet for the later State Plane Coordinate System and in meters for the later UTM system, the rectangular coordinates of the progressive military grid were in yards. Separate zones were established for other U.S. possessions, and the system was later extended as the World Polyconic Grid to cover all areas not covered by the British grid, namely, the Americas, eastern Asia, and the oceans.[58] This grid was made up of five 73°-wide bands, each with nine zones of the width used for the U.S. portion, but extending to 72° N and S. The UTM superseded all this in 1947.

The Lambert Conformal Conic Projection

After introduction of the conformal conic projection by Lambert (1772) (fig. 2.7), the projection was almost unknown as a Lambert development for over a century. In fact Harding, Herschel, and Boole had redeveloped it independently in both spherical and ellipsoidal forms during the nineteenth century. The First World War suddenly gave this projection new life, making it the standard projection for inter-mediate- and large-scale maps of regions in midlatitudes for which the transverse Mercator is not used. The Lambert became third to the latter and the polyconic in frequency of such use throughout the world. Desiring a conformal projection to facilitate angular measurements for surveys and artillery fire, the French changed their military mapping system by 1918 from the Bonne to a projection called Système Lambert, but actually using Tissot's projection of minimum deformation, developed in 1860.[59]

In 1918 the U.S. Coast and Geodetic Survey, inspired by the French, published tables based on the true Lambert conformal conic projection of the ellipsoid, with one minor stated simplification (Deetz 1918; O. S. Adams 1918): the geocentric latitude would be used in place of the rigorously required conformal latitude; the error is almost negligible, and calculations were considerably simplified.[60] The tables, calculated for standard parallels at latitudes 33° and 45° N, were used for almost all subsequent conformal maps of the conterminous forty-eight states, whether as a single map or as individual state maps prepared by the USGS at a scale of 1:500,000. The choice for state maps permitted edge-matching of any adjacent states, even though the standard parallels often did not touch the states under consideration.

Fifteen years after the USC&GS tables were published, the projection was officially adopted for the State Plane Coordinate System for states of predominantly east-west expanse. First was the North Carolina System established in 1933. Within a decade, systems were established for all then forty-eight states, the transverse Mercator being used for those not assigned the Lambert conformal conic. There were ultimately seventy-five Lambert zones in thirty-four states and territories. (Alaska, Florida, and New York use both the Lambert and the transverse Mercator for different zones, due to shape, and Michigan was changed from the latter to the former projection.) Unlike those for the 1:500,000-scale series, the standard parallels are different for almost every zone.

The Lambert projection is also standard for aeronautical charts, was recommended for the post–1962 quadrangle maps of the International Map of the World series at a scale of 1:1,000,000, and was adopted for topographic mapping in zones of scores of nations (Mugnier 1985; Snyder 1987b, 106). Midlatitude maps of several planets or their satellites are based on the Lambert (Snyder 1987b, 42–43, 106). For small-scale atlas maps, the Lambert conformal conic is common in some atlases and rare in others. It is identified (sometimes as conical orthomorphic) for most smaller-than-continental maps in *The American Oxford Atlas,* Rand McNally's *New International Atlas,* and the national atlases of countries such as Canada and Japan, but rarely in *Goode's Atlas* and *The Times Atlas,* and just once in the National Geographic *Atlas.*[61]

There is only one possible conformal conic projection for the sphere or for a given ellipsoid, once the standard parallel(s) have been selected. By establishing appropriate constraints to the selection of these parameters, the overall variation of scale may be reduced or minimized. Three Russians published procedures for minimizing scale variation according to specific criteria: For V. V. Vitkovskiy's (1856–1924) third projection (Vitkovskiy 1907) the maximum scale error between the standard parallels equals (with opposite sign) the scale error at each of the two limiting parallels.[62] V. V. Kavrayskiy's (1884–1954) third projection (1934, 92–94) is almost the same, but has the scale factors along the two limiting parallels equal and the reciprocal of the minimum scale factor. N. Ya. Tsinger (1842–1918), sometimes transliterated N. J. Zinger, devised as his first projection (Tsinger 1916) a conformal conic with minimum error, using least squares, for the actual territory represented on the map, regardless of whether it completely filled the rectangle or quadrangle bounding the map.[63]

There are much simpler rules of thumb provided for the selection of standard parallels for any of the conic projections, although these do not quite minimize variation of scale. Hinks (1912, 87) advised that "in general it will be found sufficient to take the standard parallels about one-seventh of the whole extent in latitude from the bounding parallels." In 1921 Deetz and Adams recommended, immediately after referring to Hinks's book on another matter,

> In general, for equal distribution of scale error, the standard parallels are chosen at one-sixth and five-sixths of the total length of that portion of the central meridian to be represented. It may be advisable in some localities, or for special reasons, to bring them closer together in order to have greater accuracy in the center of the map at the expense of the upper and lower border areas. (Deetz and Adams 1934, 79; also 1921, 1st ed., 79)

The Albers Equal-Area Conic Projection

After H. C. Albers introduced the equal-area conic projection (fig. 3.16) in 1805, it was used only minimally until, during the twentieth century, it became the dominant projection for maps of the United States and some other regions. The U.S. Coast and Geodetic Survey's booklet by O. S. Adams (1927, 1) contained for the first time ellipsoidal formulas and tables for the Albers, calling the projection "an equal-area representation [of the United States] that is as good as any other and in many respects superior to all others." Adams chose as standard parallels for maps of the conterminous United States 29½° and 45½° N latitude, with other values for outlying territories.

Because of these tables, since then practically every map of the United States using the Albers has been based on the same standard parallels and the ellipsoidal form, regardless of scale. This includes numerous maps prepared by the U.S. Geological Survey, such as all the sectional maps (at a scale of 1:2,000,000) in the USGS (1970) *National Atlas,* and two-page maps of the conterminous United States or sectional maps in several atlases.[64] The Albers has also been used with other standard parallels for other regions, including sections of Africa, North America, and Europe in several atlases, especially those of the National Geographic Society, and for maps of India smaller than a 1:1,000,000 scale in India's (1979) *National Atlas.*[65]

Like the conformal conic, there is only one equal-area conic for a given ellipsoid (or sphere) and given standard parallels. Also likewise, constraints for various low-error applications can be applied to the equal-area conics; they were developed by Tissot (1881) in France and by Russians Vitkovskiy (1907), Tsinger (1916), and F. N. Krasovskiy (1922).[66] Krasovskiy (1878–1948) is best known for the 1940 dimensions of the ellipsoid bearing his name and used in the Soviet Union. Vitkovskiy's second conic projection is equal-area, with the same scale error at each of the two limiting parallels as that of the central parallel (not quite the same as the constraints for his conformal third); for his equal-area (second) projection Tsinger repeated his principle of minimizing error for the territory of interest. Krasovskiy's projection, his first, was claimed to be a slight correction to Vitkovskiy's second; actually he produced one identical, but with rearranged formulas.

The Simple or Equidistant Conic Projection

The simple conic projection (fig. 2.3) is almost nonexistent as a basis of official mapping except in the former Soviet Union. As a compromise projection that is neither equal-area nor conformal, it has the special feature of maintaining correct scale along all meridians. The formulas have been given in detail in the Soviet Union and the United States for the ellipsoidal version, but it has been chiefly used in the twentieth century for regional atlas maps and as an inspiration for more scholarly derivations (Graur 1956, 151–55; Snyder 1978a). Rand McNally has used it for a substantial number of the less-than-continental regional maps in *Goode's* and other atlases, and it is common in *The Times Atlas.*[67] In older atlases this projection was not identified, but in the newer ones it is usually called simply a conic projection.

Equidistant conic projections have attracted more investigators trying to minimize distortion than have the other conics: In addition to Murdoch's (1758) and Euler's (1777) earlier versions, several were developed just before and during the twentieth century. In 1903 J. D. Everett (1831–1904) of the British Isles independently duplicated Murdoch's third projection, as corrected by A. E. Young, and the Ordnance Survey produced a map of the British Isles at a scale of 1:1,000,000 on another equidistant conic using the same scale restraints as those for Vitkovskiy's third (conformal conic) projection.[68]

The other five approaches were developed by Russians. Vitkovskiy's (1907) first and equidistant conic projection has the same scale constraint as his second (an equal-area conic), and A. A. Mikhaylov (1911–12) equalized the scale factors of the two limiting parallels and made them reciprocal to the minimum scale factor. Kavrayskiy (1934, 122–25, 133–34) presented two equidistant conics—his second projection applied a constraint similar to that of Mikhaylov, and his fourth applied the least-squares territorial constraint used by Tsinger (1916) for his conformal and equal-area conics.[69]

In addition, Krasovskiy (1922, 1925) described two "approximately equidistant" conics, his second and third. In these the scale along the meridians is constant, but at a scale factor slightly different from unity so that total area is preserved between limiting parallels. Scale variation is minimized using limiting constraints and least squares, respectively (Maling 1960, 261–62, 266).

A much simpler approach to the equidistant conic projection was suggested by Dmitri I. Mendeleev (1834–1907), a chemist renowned for the development of the periodic table of elements in 1869 (Mendeleev 1907; Maling 1960, 260–61). His conic is the same as the standard equidistant conic, except that one of the two standard parallels is the North Pole. It is thus an equidistant conic equivalent of the Lambert equal-area conic of 1772. A 1:15,000,000-scale map of Russia was constructed using Mendeleev's projection, with the other standard parallel at latitude 55° N, thereby having a cone constant of 0.939, almost a polar azimuthal equidistant projection.

Kavrayskiy's fourth projection was applied to the design of numerous maps of all or portions of the Soviet Union, with standard parallels of 47° and 62° (fig. 4.8),

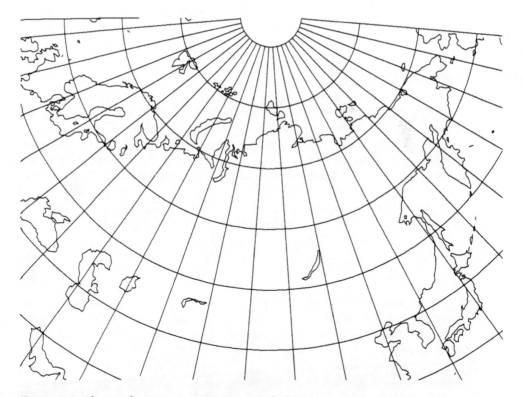

FIGURE 4.8. The equidistant conic projection as applied to numerous maps of Russia, using the standard parallels 47° and 62° N latitude selected by Kavrayskiy based on least squares for the Russian land shown. Central meridian 95° E; 10° graticule.

in the *B.S.A.M.* and elsewhere. The U.S. Central Intelligence Agency, with its intensive Soviet-watching during the cold war, applied the name *Kavraisky IV projection* to the general case of the equidistant conic with two standard parallels, even though this is unwarranted. (The agency uses *Ptolemy equal-interval conic projection* as the title if there is only one standard parallel.)[70]

Other Earlier Projections

The durable (Nicolosi) globular projection (fig. 1.32) reached the zenith of its usage for maps of the eastern and western hemispheres during the nineteenth century; during the twentieth century it was rarely seen except as a historically famous and simply constructed projection in textbooks.[71]

The Hammer projection (fig. 3.27), as an equal-area modification of the equatorial Lambert azimuthal equal-area projection, was devised in 1892 and has been occasionally used since. In a leading geography book by Finch and Trewartha, it was used just once (interrupted and condensed) in the first edition of 1936, but a dozen or more times in each of the second and third editions in the 1940s; in the later editions it was largely replaced by the McBryde-Thomas flat-polar quartic projection.[72] Aside from an occasional appearance in other geography texts, the Hammer has also been used for maps of the galactic universe as viewed from the earth.

NEW TWENTIETH-CENTURY PROJECTIONS

It was normal that many older projections would continue as staples of commercial maps and atlases (table 4.1), but a surprising number of the newer projections were incorporated as well. The development of new projections and of variations and different aspects of old projections proceeded in many directions. Several factors, including the greater foundation of map projection design on which to build, the greater number of technology centers in the world, increased communication, and improved techniques of computation culminating in high-speed computers, contributed to the exponential growth of map projections both in numbers and depth of science during the twentieth century. As previously stated, some of the proposals were quite simple, occasionally naïve or unoriginal, but often more popular and more usable than profound studies that produced projections that were slightly or significantly freer of distortion or that showed mathematical prowess.

Cylindrical Projections

Normal Aspects

Elementary modifications of older cylindrical projections such as a change of standard parallels on the Gall stereographic or on the cylindrical equal-area projection have been described above. There have also been several new regular cylindrical projections.

Russian Cylindrical Projections Several of the new cylindrical projections originated in the Soviet Union (Maling 1960, 205, 210–15, 303). Kavrayskiy's first projection uses the Mercator (with a chosen standard parallel) between latitudes 70° N and S. Beyond this is fused an equirectangular having the same meridian spacing and with parallels at uniform intervals equal to the spacing at 70° latitude on the Mercator, that is, the secant of 70° times the meridian spacing.[73]

Of special significance is a series of cylindricals initiated by N. A. Urmayev (b. 1895) in 1947.[74] These have preselected scale factors at each of three different parallels. The spacing of the parallels is then determined by series approximations fitting the selected requirements. Three specific graticules have been proposed, by Urmayev, A. A. Pavlov, and A. S. Kharchenko as revised by A. I. Shabanova. That of Urmayev (his third projection), used in the 1947 *Atlas ofitsera* (Officer's atlas) for the political world map, has formulas equivalent to

$$x = R\ \lambda,\ \text{and}$$
$$y = R\ (0.92813\ \phi\ +\ 0.37142\ \phi^3),$$

again using radians. Pavlov's version was presented in 1956 (Graur 1956, 119):

$$x = R\ \lambda,\ \text{and}$$
$$y = R\ (\phi\ -\ 0.0510214\ \phi^3\ -\ 0.0053457\ \phi^5).$$

The Kharchenko-Shabanova cylindrical of 1951–52 may be calculated thus:[75]

$$x = R\ \lambda\ \cos 10^0,\ \text{and}$$
$$y = R\ (0.99\ \phi\ +\ 0.0026263\ \phi^3\ +\ 0.10734\ \phi^5).$$

On the first and third of these projections, parallels gradually spread away from the equator, reminiscent of the Gall stereographic; on the second, they become closer together.

The Miller Cylindrical Projection Just as internally developed compromise cylindrical projections were commonly used in the Soviet Union, as above, and in the United Kingdom, with the Gall stereographic, so the United States, both federally and commercially, made considerable use of a compromise cylindrical projection (fig. 4.9) presented by Osborn Maitland Miller (1897–1979; fig. 4.10) of the American Geographical Society. Miller had been asked by the geographer of the U.S. Department of State, S. Whittemore Boggs, to study further alternatives to the Mercator, the Gall, and other cylindrical world maps. As Miller (1942, 424) stated,

> the practical problem considered here is to find a system of spacing the parallels of latitude such that an acceptable balance is reached between shape and area distortion. By an "acceptable" balance is meant one which to the uncritical eye does not obviously depart from the familiar shapes of the land areas as depicted by the Mercator projection but which reduces areal distortion as far as possible under these conditions.

Miller made four proposals, all of which he gave the same meridian spacing $x = R\ \lambda$. He called them, respectively, (1) perspective compromise, with $y = R\ [\sin (\phi/2)\ +\ \tan (\phi/2)]$; (2) modified Mercator (a), with $y = 1.5\ R\ \ln \tan (\pi/4\ +\ \phi/3)$; (3) modified Mercator (b), with $y = (R/0.8)\ \ln \tan (\pi/4\ +\ 0.8\phi/2)$; and (4) modified Gall's (30°), with $y = (1\ +\ 2/\sqrt{3})\ R\ \tan (\phi/2)$. "After some experimentation, the [modified Mercator (b)] was judged to be the most suitable for Mr. Boggs's pur-

TABLE 4.1. Map Projections in Twentieth-Century World Atlases, Major Regions

Atlas	Region																	
	1	2	3	4	5	6	7	8	9	10	11	12	13	14	15	16	17	18
Rand McNally (1916)	ME	—	—	—	PO	BO	BO	SC	BO	SI	SC	SC	SC	SC	—	—	—	ME
Zetterstrand and Rosén (1926)	SP	—	—	LA	TC	OL	—	AL	TC	TC	—	AL	—	—	RS	RS	RS	BE
Rand McNally (1932)	ME	EL	—	LA	SI	OL	BO	SC	BO	SI	SC	SC	SC	SC	—	—	—	ME
Bol'shoy sovetskiy (1937)	MK	EL	—	AE	—	—	—	—	—	—	—	—	SC	SC	MO	EC	EC	—
Cram (1941)	ME	GL	—	AE	SI	BO	BO	BO	BO	SI	SC	AL	SC	SC	ME	—	ME	ME
J. Bartholomew (1942)	MI	EL	—	AA	SI	BO	BO	BO	OL	OL	SC	AL	SC	BO	ME	—	PS	—
Lewis and Campbell (1951)	OH	EL	—	AE	EE	OL	LC	OE	OE	TM	OE	LC	LC	LC	MA	MA	MA	OE
J. Bartholomew (1958, etc.)	BA	—	—	AE	—	OL	—	BO	OL	OL	CT	CT	SC	—	EL	EL	—	BO
Hammond (1966)	AM	—	—	AE	OL	OL	BO	OL	OL	OL	SC	PO	SC	SC	—	—	EL	—
Union of Soviet (1954, 1967 ed.)	MP	EL	AE	AE	EL	OL	OL	LC	OL	OL	LC	SC	PC	LC	OT	UP	UP	—
Rand McNally (1971)	AC	—	—	LA	SI	OL	OL	SC	OL	SI	LP	LP	OL	PO	MC	—	MS	—
Natl. Geographic (1990)	RO	—	—	AE	CT	TE	CT	CT	CT	CT	CT	AL	TP	AL	ME	ME	ME	—
J. C. Bartholomew (1982)	MJ	—	—	LA	EL	OL	BO	BO	OL	OL	SC	SC	SC	—	EL	EL	EL	—
Rand McNally (1983)	RO	—	—	LA	OS	LC	LC	OS	—	OC	LC	AL	LC	LC	PS	PS	PS	—
Times Books (1985)	TI	—	—	AS	OS	OL	OL	BO	CT	OL	CT	CT	SC	SC	EL	OL	EL	BO
Rand McNally (1986)	GH	—	—	LA	OL	OL	OL	SC	OL	OL	SC	PO	SC	PO	PS	PS	GP	—

Notes. In Cram (1941), the line work is sometimes inconsistent; conic projections are uncertain. In Lewis and Campbell (1951), the modified Gall and Oxford projections are used for world thematic maps; North America and the United States are only shown sectionally except as part of larger regions, so the projection is listed for sectional parts. In Rand McNally (1983), the same applies to the USSR, Asia, and Africa. In Times Books (1985), the same applies to Australia, China, and North America.

Region number

1 world
2 eastern and western hemispheres
3 northern and southern hemispheres
4 polar areas
5 Africa
6 Asia
7 Australia
8 Europe
9 North America
10 South America
11 Canada
12 United States
13 USSR (European or all)
14 China
15 Atlantic Ocean
16 Indian Ocean
17 Pacific Ocean
18 Oceania.

Projection symbol (sometimes indicating several projections as noted below):

AA northern—AE, southern—LA
AC AE, Miller cylindrical, GH
AE polar azimuthal equidistant
AL Albers equal-area conic

AM AE, ME
AS northern—AE, southern—polar stereographic
BA "Regional," Lotus, Nordic, Tripel, ME
BE Behrmann
BO Bonne
CT Chamberlin trimetric
EC Eckert VI
EE equatorial azimuthal equidistant
EL equatorial Lambert azimuthal equal-area
GH Goode homolosine
GL globular
GP GH, PS
LA polar Lambert azimuthal equal-area
LC Lambert conformal conic
LP LC, PO
MA Mod. azimuthal equidistant (Bomford)
MC modified polyconic
ME Mercator
MI Kite, Gall stereographic, SI, ME
MJ "Regional," Gall stereographic, Nordic, TI
MK Gall stereographic, EC
MO Mollweide

MP Ginzburg modified polyconic
MS modified secant conic
OC (bipolar) oblique conic conformal
OE oblique azimuthal equidistant
OH oblique Hammer
OL oblique Lambert azimuthal equal-area
OS oblated stereographic
OT oblique TsNIIGAiK with oval isocols
PC oblique perspective cylindrical
PO polyconic
PS pseudocylindrical (unspecified)
RO Robinson
RS Rosén mod. Lambert azimuthal equal-area
SC simple (equidistant) conic
SI sinusoidal
SP interrupted and flat-polar sinusoidal
TC transverse cylindrical equal-area
TE two-point equidistant
TI The Times
TM transverse Mercator
TP transverse polyconic
UP Urmayev pseudocylindrical.

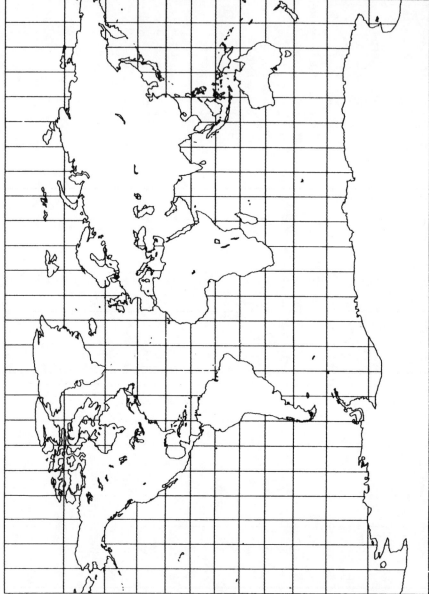

FIGURE 4.9. The compromise Miller cylindrical projection of the world, a modified Mercator. Resembling the Gall stereographic (fig. 3.10), it has been used considerably in the United States, as was the Gall in England. Central meridian 0°; 15° graticule.

FIGURE 4.10. Osborn Maitland Miller (1897–1979), director of the American Geographical Society and inventor of several commonly used map projections. *Courtesy of the American Geographical Society.*

pose," and that is the one called the Miller cylindrical projection (Miller 1942, 428). His spacing of parallels from the equator, then, is the same as if the Mercator spacings were calculated for 0.8 of the respective latitudes, with the result divided by 0.8. Consequently, the equator is free of all distortion.

Born in Perth, Scotland, and educated in Scotland and England, Miller came to the American Geographical Society in 1922, where he eventually became its director. There he developed several improved surveying and mapping techniques. An expert in aerial photography, he developed techniques for converting high-altitude photographs into maps. He led or joined several expeditions of explorers and advised leaders of others. He retired in 1968, having also developed the bipolar oblique conic conformal and the oblated stereographic projections.[76]

The *Esso War Map*, issued by the oil company in 1942, was the first general publication of the Miller cylindrical projection. Government use followed in 1943 in the Army Map Service and subsequently in the U.S. Geological Survey.[77] The Miller appeared in commercial world atlases beginning in 1949.[78]

Other New Cylindrical Projections Two British cylindrical-type projections remain novelties: One, a true cylindrical, was proposed by Charles F. Arden-Close (1864–1952), who was Charles F. Close until 1938 when he changed his name to comply with a bequest, director-general of the Ordnance Survey (1911–22), and the leading British map-projection innovator of the early twentieth century.[79] His projection is an arithmetic mean between the Mercator and the cylindrical equal-area (Arden-Close 1947). The poles are still at infinity, but parallels are spread less rapidly than they are on the Mercator.

H. Poole (1935) devised a sort of cylindrical projection with variably spaced straight meridians. His is a modified Mercator that is approximately true to scale as well as conformal along the true loxodrome route shown straight from Mildenhall, England, to Melbourne, Australia. This route roughly approximated the corresponding air routes. The normal Mercator spacing between each 10° of latitude along this route is compressed so that all parallels become equidistant straight lines. To retain conformality along the loxodromic path, meridian spacing is compressed by the same ratio as is the parallel spacing at each point along the path. Scale, shape, and area become increasingly distorted away from this line.

Although interest was shifting to noncylindrical projections in new research, one other was developed, like Poole's, with a special purpose. In 1977 this writer devised a cylindrical (and also conic) series of projections showing Landsat mapping-satellite groundtracks as straight lines (fig. 4.11) (Snyder 1981a; 1987b, 230–38). A given map applies to one set of orbit parameters and has two standard parallels. The series resembles the central cylindrical projection (fig. 3.6), with great polar-region distortion. These "satellite-tracking" projections have lain relatively dormant, the world orbit maps usually being plotted on equirectangular maps with curved groundtracks.

While the above is a strictly cylindrical projection, Jack Breckman (1962) of the Radio Corporation of America designed B-charts, which consisted of an initial square graticule with longitude as the horizontal coordinate and position along the artificial satellite orbit as the vertical. Groundtracks were plotted as sinuous curves; they were then made vertical and straight by sliding longitudes horizontally to be sinuous themselves, leading to a more pseudocylindrical projection.

Oblique and Transverse Cylindrical Projections

As calculating became easier with the advent of desktop mechanical and electro-mechanical calculators, electronic calculators, and finally computers with extensive electronic memory, oblique and transverse aspects of previously existing projections were increasingly proposed in scholarly papers and used for published maps, to rotate the lines of low distortion to suit the region being mapped.

For azimuthal projections, oblique and transverse (equatorial) aspects had been in use for centuries; some oblique azimuthal projections were as familiar as the normal (polar) aspects. For conic projections, other aspects are still rare, but for cylindrical and pseudocylindrical projections, the possibilities were often studied. The transverse and oblique Mercator and plate carrée or equirectangular projections had already appeared by 1900, as had the transverse cylindrical equal-area and even a transverse central cylindrical of almost no value.

Suggestions for new oblique and transverse cylindrical projections were chiefly related to the cylindrical equal-area and to some perspective cylindricals. Oswald Winkel (1873–1953) of Germany suggested using a steeply tilted oblique cylindrical equal-area projection for a map of the Americas (Winkel 1909). In 1926 Karl D. P.

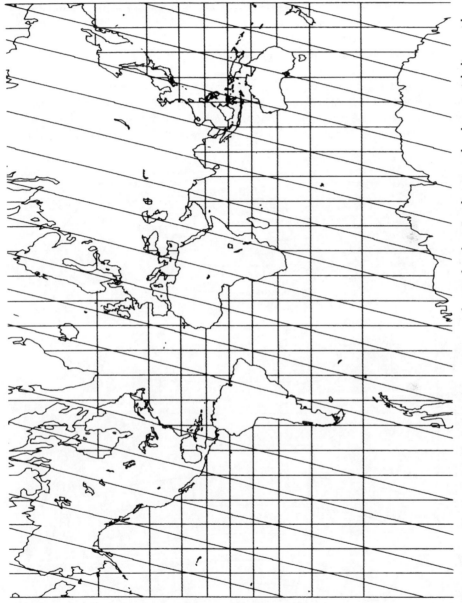

FIGURE 4.11. A cylindrical "satellite-tracking" projection designed to show Landsat groundtracks as straight lines, with standard parallels 30° N and S and other parameters used in Snyder (1981a). Groundtracks are shown for every fifteen paths of Landsat 1, 2, or 3, beginning with path 15 through Washington, D.C. Central meridian 0°; 15° graticule.

Rosén based maps of North America, South America, and Africa in the Swedish *Nordisk världsatlas* on the transverse aspect, and C. Warren Thornthwaite (1899–1963), an American climatologist, proposed the oblique aspect for a map of Eurasia and Africa (Thornthwaite 1927; Zetterstrand and Rosén 1926, pls. 29, 31, and 35). Poole (1934) used the oblique cylindrical equal-area to show some of the air routes of the British Commonwealth. This writer (Snyder, 1985c) derived formulas for ellipsoidal versions of all aspects of the cylindrical equal-area projection, including oblique, using authalic latitude.

Arden-Close (1943) derived a related projection by averaging arithmetically both the *x* and *y* coordinates of one hemisphere of the cylindrical equal-area projection and its transverse aspect, with central meridians and equators coinciding. The resulting hemisphere is bounded by a square with rounded corners (fig. 4.12). The

FIGURE 4.12. A novelty projection by Arden-Close, averaging rectangular coordinates of a hemisphere of the cylindrical equal-area and its transverse projections, although not equal-area itself. Central meridian 70° E; 10° graticule.

poles are straight lines across the top and bottom; the sides are also straight lines representing the equator at meridians 90° E and W of center. Quadrants of circles form the four "corners." The combined projection is not equal-area and is only a curiosity.

A special oblique equidistant cylindrical projection of the ellipsoid was devised by G. T. McCaw (1870–1942) for the survey of some of the Fiji Islands (McCaw 1917, 161; ref. in Royal Society 1966, 45). Called a Cassini-hyperbolic projection, it involved a double projection from the ellipsoid to a hyperboloid and then to a plane.

M. D. Solov'ev (b. 1887) presented three variations of oblique perspective-cylindrical projections in 1937, the first used for several political and physical maps of the Soviet Union.[80] For the general case, the formulas used by Solov'ev (with symbols revised for consistency) are

$$x = R \lambda' \cos \phi_s', \text{ and}$$

$$y = R (P + \cos \phi_s') \sin \phi'/(P + \cos \phi'),$$

where x, y, and R are as before, (ϕ', λ') are transformed latitude and longitude, respectively, on the base perspective cylindrical projection after conversion from geographic latitude and longitude for the oblique aspect, and $\pm \phi_s'$ are the transformed latitudes of secancy of the oblique cylinder. Constant P is the distance of the point of perspective, divided by the radius of the sphere, from the center of the sphere in the plane of the transformed equator ($\phi' = 0$) and at or above the transformed meridian opposite that being plotted.

On his first variant, Solov'ev placed the north pole of the base cylindrical projection at 75° N, 80° W and the geographic North Pole along the central meridian ($\lambda' = 0°$), with $P = 1$ and $\phi_s' = 45°$. On his second (fig. 4.13), these numbers are

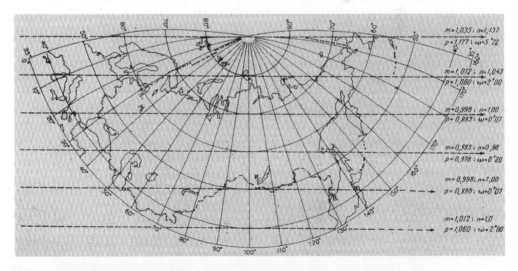

FIGURE 4.13. Solov'ev's oblique perspective cylindrical projection, variant 2, shown for an outline map of the Soviet Union. Lines of constant scale are marked with scale factors m along and n perpendicular to the lines, area scale p ($m \times n$ here), and maximum angular deformation ω; some needed digits of n are missing. Reproduced from Solov'ev (1969, 150).

respectively 30° N, 80° W, 1.3, and on his third variant, 50° N, 80° W, 1.2, and 30° (Solov'ev 1969, 149, 151). As a result, the lines with true scale *along* them are straight lines on the maps (at $\pm\phi_s'$) but small circles on the globe, crossing central meridian 100° E perpendicularly at 60° N and 30° S, 70° and 50° N, and 70° and 10° N, respectively, spanning the breadth of the Soviet Union and not following a geographic parallel or a great circle. The projections are neither equal-area nor conformal. Another variant by T. D. Salmanova in 1948 was used for other published maps of the same nation, with successive constants of 25° N, 80° W, 3.0, and 10°, and lines of true scale crossing the central meridian of 100° E at 55° and 75° N (Solov'ev 1969, 152).

Byron Adams (1984) of Shell Oil Company in Texas proposed in effect an ellipsoidal transverse aspect of the tangent stereographic cylindrical projection devised by Braun (1867). Since this projection is neither equal-area nor conformal, its utility in the ellipsoidal form is obscure, although Adams pointed to its distortion as a compromise between the transverse Mercator and polyconic projections.

The Space Oblique Mercator Projection   The launching of an earth-mapping satellite by the U.S. National Aeronautics and Space Administration (NASA) in 1972 led to another special map projection. The satellite was first called ERTS-1 but was renamed Landsat 1 in 1975. It circled the earth in a nearly circular orbit inclined about 99° to the equator and scanning a swath about 185 km wide from a nominal altitude of 919 km. Two subsequent satellites were launched with the same orbit, followed by Landsat 4 and 5 in the early 1980s with orbits at a nominal altitude of 705 km. The groundtracks were set to repeat every 18 days with 251 orbits for Landsats 1–3, and every 16 days with 233 orbits for the later satellites, after covering the entire earth except for extreme polar regions.

Alden P. Colvocoresses (1974) of the U.S. Geological Survey conceived a map projection that allowed mapping of the scanned swath throughout all orbit cycles, with the groundtrack continuously at correct scale and the swath on a conformal projection with minimal scale variation. If the orbit were about a stationary earth, an oblique Mercator projection would be appropriate, but the earth rotation introduced time. He called the projection the space oblique Mercator (SOM) and appealed for the development of the necessary formulas.

Agencies, other organizations, and individuals under contract did not provide a satisfactory solution, until a human-interest touch was added in 1976 when a practicing chemical engineer without formal cartographic training, but with a pocket calculator and a mathematical interest in map projections as a hobby dating back over thirty years to high school, heard Colvocoresses at an Ohio State University conference (Colvocoresses 1977; Wilford 1981, 351–52). A few months later the engineer (this writer) voluntarily developed a satisfactory solution for the spherical earth and circular orbit as a first step, then for the ellipsoidal earth (in the form finally used), and ultimately for the elliptical orbit. Conformality is within a few millionths throughout the mapping range, but apparently cannot be perfectly achieved. The formulas are impractical without modern computers, but so was Landsat. Visually the projection resembles the oblique Mercator when applied to a world map, but the central line (the groundtrack) is almost a sinusoid (fig. 4.14) (Snyder 1978b; 1981c; 1987b, 214–29).

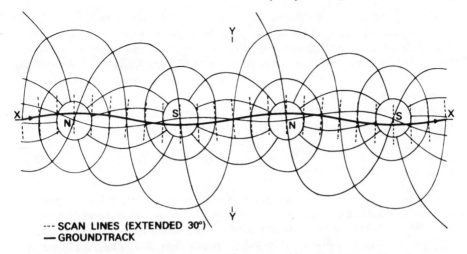

--- SCAN LINES (EXTENDED 30°)
— GROUNDTRACK

FIGURE 4.14. A space oblique Mercator projection of much of the globe with a typical groundtrack as a curved centerline, shown for two orbits of Landsat 1, 2, or 3. 30° graticule. Note the rotation of the graticule from North Pole (N) to North Pole due to earth rotation during the orbit. Reproduced from Snyder (1981c, 14).

Born in Arizona in 1918, Colvocoresses received degrees in mining and civil engineering but became a cartographer as an army officer during the Second World War, retiring as a colonel in 1968. He obtained his doctorate in geodetic science in 1965 from Ohio State University and was a principal investigator for mapping the earth from space for the U.S. Geological Survey from 1968 until his retirement in 1990. He holds two patents on space mapping systems and was president of the American Society for Photogrammetry and Remote Sensing in 1988.

The SOM was incorporated as a standard format for Landsats 4 and 5 and for the launching of Japanese satellite MOS(Marine Observation Satellite)-1 in 1987. The other principal format available was the Universal Transverse Mercator projection, but discontinuities occur at zone changes each 6° of longitude.

New Pseudocylindrical Projections

Although new cylindrical projections were designed during the twentieth century, the constraints set by equidistant straight parallel meridians are strong. On pseudocylindrical projections, meridians are also equally spaced along each parallel of latitude, all of which are straight and parallel to each other. The meridians of pseudocylindrical projections are curved, however, allowing the more polar parallels to be shortened to any desired length, thus decreasing polar distortion in general. Although only two significant pseudocylindrical projections, the sinusoidal and Mollweide, appeared before 1900, scores appeared afterward.[81] Many remained academic presentations, but several became commonly used.

Flat-Polar Pseudocylindricals from Germany, 1900–1920

Poles are shown as lines on some of the early oval projections, such as the one by Agnese and Ortelius in the sixteenth century. These projections, however, are not true pseudocylindricals, since meridians are not equidistant on most parallels. The poles are treated as points on all true pseudocylindrical projections appearing before 1890, when A. M. Nell described a projection of this class with poles shown as lines 0.724 times the length of the equator (fig. 3.15). Inspired by one of Nell's suggestions, Hammer (1900) presented an equal-area projection averaging the x-coordinates of the sinusoidal and cylindrical equal-area projections for a given latitude, thus producing poles half the length of the equator. The y-coordinates were then calculated to maintain equivalence (also see Tobler 1973b):

$$x = R \lambda (1 + \cos \phi)/2;$$
$$y = 2R [\phi - \tan (\phi/2)].$$

Although the Nell and Hammer forms remained relatively unknown, they were quickly followed by two popular pseudocylindricals out of six presented by Max Eckert (1868–1938) (fig. 4.15), then of Kiel, Germany, all with poles and central meridians half the length of the equator (Eckert 1906). Born in Chemnitz, Eckert spent over thirty years, from 1907, as a professor at the Technische Hochschule at Aachen, where he wrote a leading work on cartography published in the 1920s (Geisler 1939;

FIGURE 4.15. Max Eckert (-Greifendorff) (1868–1938), German educator who developed several map projections with line-poles half the length of the equator.

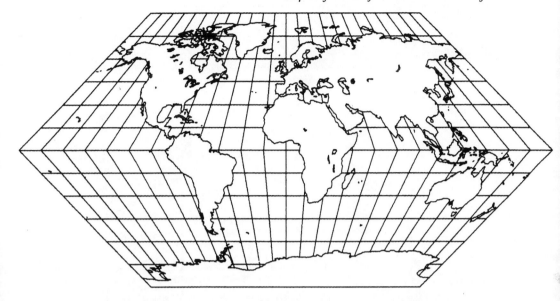

FIGURE 4.16. The Eckert I projection of the world, with equidistant parallels and broken straight meridians. Central meridian 0°; 15° graticule.

Eckert 1921, 1925). The geometric characteristics of his six projections (figs. 4.16 through 4.21) are simply described (using roman numerals consistent with common practice in describing multiple projections by a single inventor):

Eckert II, IV, and VI are equal-area projections, with unequal spacing of parallels.

Eckert I, III, and V have equidistant parallels of latitude and are not equal-area, although the *total* area is normally at correct scale.

On Eckert I and II, the meridians are straight lines, with all but the central meridian broken at the equator.

On Eckert III and IV, the outer (180th) meridians are semicircles, the central meridian is straight, and other meridians are semiellipses.

On Eckert V and VI, the meridians are sinusoids, except for the straight central meridian.

All central meridians are half the length of the equator. As on all true pseudo-cylindricals, meridians are equally spaced along each parallel. Using these principles, the following formulas may be obtained:

I. $x = 2 [2/(3\pi)]^{1/2} R \lambda (1 - |\phi|/\pi);$

   $y = 2 [2/(3\pi)]^{1/2} R\phi.$

II. $x = 2R \lambda (4 - 3 \sin |\phi|)^{1/2}/(6\pi)^{1/2};$

   $y = (2\pi/3)^{1/2} R [2 - (4 - 3 \sin |\phi|)^{1/2}]$ sign $\phi.$

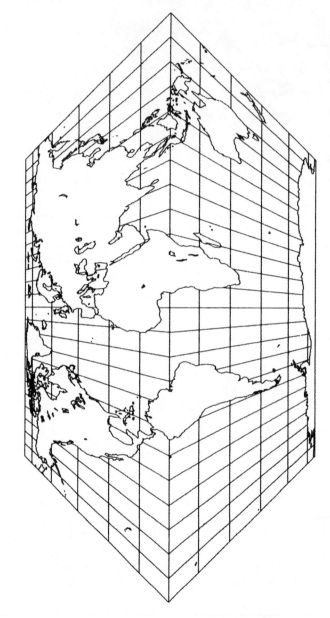

FIGURE 4.17. The Eckert II projection, equal-area with broken straight meridians. Central meridian 0°; 15° graticule.

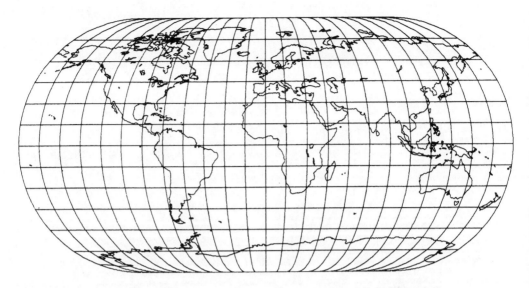

FIGURE 4.18. The Eckert III projection, with equidistant parallels and semiellipses for meridians. Central meridian 0°; 15° graticule. Compare the Agnese-Ortelius oval projection with circular meridians (fig. 1.29).

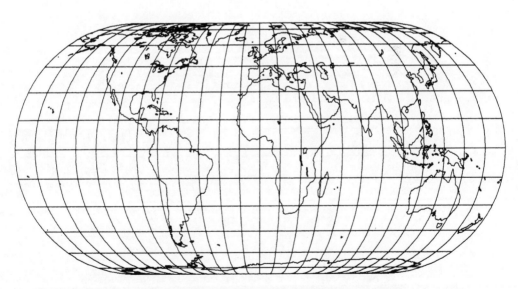

FIGURE 4.19. The Eckert IV projection, equal-area with semiellipses for meridians. Central meridian 0°; 15° graticule. One of Eckert's most popular projections.

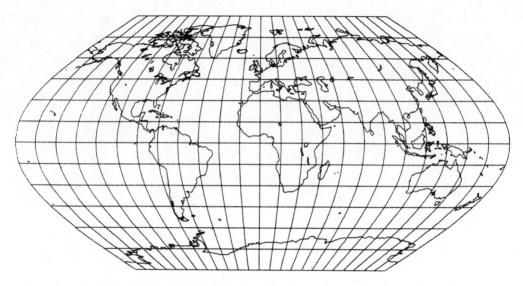

FIGURE 4.20. The Eckert V projection, with equidistant parallels and sinusoidal meridians. Central meridian 0°; 15° graticule.

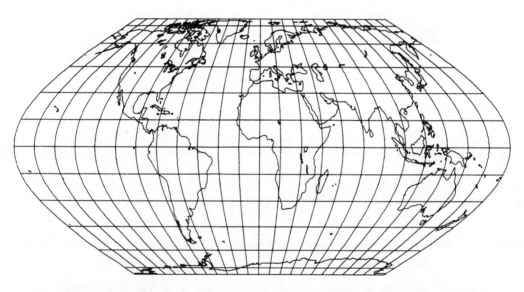

FIGURE 4.21. The popular Eckert VI projection, equal-area with sinusoidal meridians. Central meridian 0°; 15° graticule.

III. $x = 2\{1 + [1 - (2\phi/\pi)^2]^{1/2}\} R \lambda/(4\pi + \pi^2)^{1/2};$

  $y = 4R\phi/(4\pi + \pi^2)^{1/2}.$

IV. $x = 2R \lambda (1 + \cos \theta)/(4\pi + \pi^2)^{1/2},$ and

  $y = 2\pi^{1/2} R \sin \theta/(4 + \pi)^{1/2},$

where $\theta + \sin \theta \cos \theta + 2 \sin \theta = (4 + \pi)(\sin \phi)/2.$

V. $x = R \lambda (1 + \cos \phi)/(2 + \pi)^{1/2};$

  $y = 2R\phi/(2 + \pi)^{1/2}.$

VI. $x = R \lambda (1 + \cos \theta)/(2 + \pi)^{1/2},$ and

  $y = 2R\theta/(2 + \pi)^{1/2},$

where $\theta + \sin \theta = (1 + \pi/2) \sin \phi.$

The formulas for parameter θ in IV and VI require iteration or tabular interpolation.[82] The two equal-area projections IV and VI have been used substantially; the others are nearly dormant, although Eckert III is sometimes wrongly considered identical to the Agnese/Ortelius oval projection (parallels are equally spaced on both, but meridians are ellipses on the former and circular arcs on the latter).[83]

The Eckert IV projection has been used in a few American atlases, in textbooks, in the *National Atlas of Japan,* and for thematic insets on each of the National Geographic Society's world sheet maps for many years.[84] Eckert VI is the basis for climate maps in European-prepared but U.S.-distributed Prentice-Hall world atlases of about 1960; it was a staple, however, for the 1937 Russian *B.S.A.M.,* where in the interrupted and condensed form it almost equally shared the numerous world-distribution maps with the modified Gall stereographic projection.[85] It was called the BSAM pseudocylindrical projection, although credited to Eckert, and was also the basis of the map of the combined Pacific and Indian oceans in the same atlas.[86]

The Eckert V projection is in effect an arithmetic average of both x- and y-coordinates for the plate carrée and sinusoidal projections (the y-coordinates for both are the same). Two of Winkel's (1921) projections are pseudocylindrical. Another presented at the same time (his tripel) is not pseudocylindrical and is described later as a modified azimuthal projection. Winkel I (fig. 4.22), devised in 1914, generalized the Eckert V approach by arithmetically averaging the sinusoidal with the equirectangular having any standard parallels $\pm \phi_1$:

  $x = R \lambda (\cos \phi_1 + \cos \phi)/2;$

  $y = R\phi.$

If $\phi_1 = 0$, Eckert V is obtained. If $\phi_1 = \pm 50°28'$, or arccos $(2/\pi)$, the total area of the map is at correct scale, and this is the example Winkel chose.[87] He had previously (in 1912) used the transverse form of Winkel I, setting ϕ_1 for the base projection at

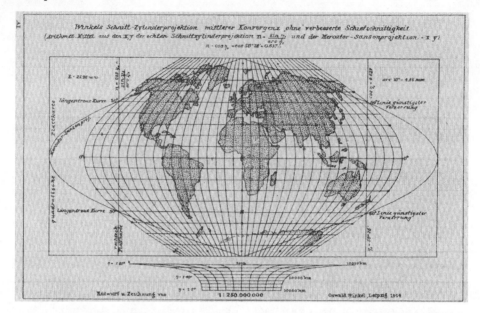

FIGURE 4.22. The Winkel I projection of the world, a generalized form of Eckert V (fig. 4.20), in this case averaging the sinusoidal with the equirectangular projection having standard parallels of 50°28′ N and S. Central meridian 10° E; 10° graticule. Reproduced from Winkel (1921, fig. 4). *U.S. Geological Survey Library.*

±21°51′, or arccos (cos 30°/cos²15°), for a proposed map of Europe and Africa, using a central meridian of 20° E and a latitude of origin of 10° N (Winkel 1921, 249–50).

For his second world map projection, developed in 1918, Winkel proposed a similar arithmetic averaging of the x and y of the equirectangular with a Mollweide modified to have non-equal-area equidistant parallels, and also having the same pole-to-pole dimensions as the equirectangular.[88]

$$x = R \lambda \ [(1 - 4 \ \phi^2/\pi^2)^{1/2} + \cos \ \phi_1]/2;$$
$$y = R\phi.$$

None of Winkel's projections is equal-area.

Two Early American Pseudocylindricals

After interrupting the sinusoidal and then the Mollweide or homolographic projections in 1916 to reduce shape distortion while retaining the equal-area feature, Goode (1925) took an additional step by fusing the two projections in 1923 and interrupting the combination. Thus there was less shearing near the poles and better linear scale near the equator. Calling this new projection the homolosine from the names of the two original projections, Goode obtained it by first superimposing the two uninterrupted projections drawn to the same area scale. He noted that the outer sinusoids extended beyond the outer Mollweide ellipses at the equator, with the reverse near the poles. At one pair of latitudes, the parallels of true scale on the Mollweide, the parallels matched on both projections.

Goode (1925, 123) therefore announced his new equal-area "projection for presenting the earth's surface entire, by fusing the sinusoidal and the homolographic projections, using the sinusoidal up to the latitude of equal scale, 40°44′11.8″ [actually 11.98″], and finishing out the polar cusps on the lobes with the homolographic projection. . . . The projection will be interrupted as in the homolographic of 1916." Linear scale is useful only within the sinusoidal portions. Because the matching parallels are closer to the equator on the sinusoidal than on the Mollweide, the latter portions are moved closer to the equator by 0.05280R. Otherwise the formulas for the portions are the same as those for the respective original projections. The meridians break slightly, near (in practice, at) the 40° parallels, due to the change in projections.

Normally, the homolosine projection is used as Goode prepared it in 1923, with interruptions to preserve landmasses (fig. 4.23a), except for Antarctica, rather than oceans, and with extensions to show Greenland, Iceland, and the eastern tip of Siberia on both the North American and Eurasian lobes. As on his interrupted Mollweide (fig. 4.5), but with two of the six central meridians moved, there are two resulting northern lobes and four southern lobes. The projection has been used in Rand McNally's *Goode's (School) Atlas* series, beginning with the fourth (1932) edition and continuing in every subsequent edition, as well as in other atlases and textbooks.[89] Goode also interrupted the homolosine to show oceans (fig. 4.23b, but with extensions duplicating northern Europe) and included it in several editions of the atlas for ocean-related thematic maps. A further evolution was the condensed interrupted version, deleting much of the Atlantic Ocean to allow larger thematic maps of the continental masses on a given page or sheet (Rand McNally 1986, 6–10, 12–13, 16–49). For several years the projection itself was copyrighted, and few

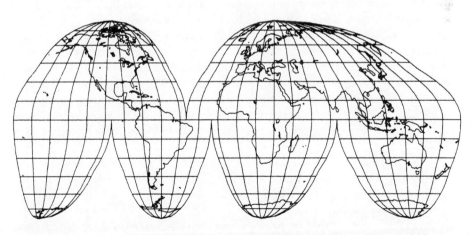

FIGURE 4.23a. The Goode homolosine projection with interruptions used by Goode for landmasses, except for his extensions near Greenland and eastern Siberia. 15° graticule. A fusion of the homolographic (Mollweide) projection toward the poles and the sinusoidal projection toward the equator at 40°44′ N and S latitudes.

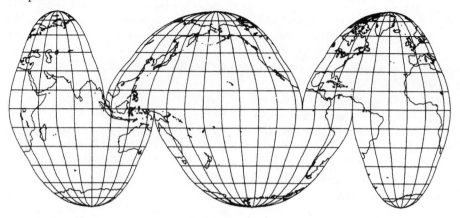

FIGURE 4.23b. The Goode homolosine projection with interruptions used by Goode for oceans, except for his extensions in northern Europe. 15° graticule.

other mapmakers used the projection, but those rights expired, and it is no longer possible to copyright a mathematical design such as a map projection in the United States.

As with the Mercator, Goode's interrupted projections became objects of naïve abuse unintended by the originator. For the Mercator, it was use of the projection for geographic maps. With the interrupted sinusoidal and homolosine, the problem was quite different: the gaps between the interruptions were treated by some mapmakers as merely more ocean, even though all the oceans were already contained within the lobes. The principal such map series, based on the interrupted sinusoidal, was issued in the 1950s by Rand McNally as the Special Ranally Political World map.[90] No meridians were shown, but there was a wide gap between Alaska and Siberia and between Greenland and Iceland on the continuous map.

Richard E. Dahlberg objected in a letter to Rand McNally and in a critical paper not mentioning the company.[91] After first rejecting his complaint, Rand McNally reviewed its use of world map projections and ultimately hired Arthur H. Robinson to design a new projection, which he presented in 1963 and which became known as the Robinson projection. The misuse of interruptions has continued to appear, for example, in a U.S. air-letter postal stamp of the 1980s, in a national news magazine of 1988, and on the cover of a membership flyer of a mapping organization in 1989.[92]

The Denoyer-Geppert Company of Chicago used for many of its world maps a modification of the flat-polar pseudocylindrical in which poles are about 0.31 of the length of the equator, the axes of the map have a 2:1 ratio, and the parallels are equally spaced straight horizontal lines. The meridians are nearly elliptical arcs and are generally equally spaced on parallels, but the spacings in polar regions are narrowed near the central meridian and widened near the outer meridians. Thus this "Denoyer semi-elliptical" projection is not a true pseudocylindrical. Levinus Philip Denoyer developed it about 1920, but the exact design basis is not now known.[93]

A Mushrooming of Pseudocylindricals, 1926–1949

For about two decades beginning in 1926, men from (chronologically) Sweden, the United States, England, the USSR, Latvia, and Germany produced a total of several dozen new pseudocylindrical projections for world maps. Some of the projections are identical, but frequently the originators used a conic section (ellipse, parabola, or hyperbola) to shape meridians. Many of the projections are either equal-area or consist of equidistant parallels, but some have neither feature. Most have remained just proposals.

The first, by Rosén, unlike projections by Eckert and others, uses only a portion of the sinusoid. In the 1926 *Nordisk världsatlas,* there are five world maps on an equal-area flat-polar sinusoidal projection with the poles following the parallels on the regular sinusoidal at 53°08' N and S (± arcsin 0.8) and with the 40th parallels at true scale.[94] This would be equivalent to using Urmayev's formulas (below) with $a = 0.89319$ and $b = 0.8$, and eliminates the iteration required for Eckert's full sinusoid.

The second, devised by Samuel Whittemore Boggs (1889–1954) and with the mathematical development by O. S. Adams, appeared in the British *Geographical Journal.* Boggs served, with variations in the title, as geographer of the U.S. Department of State from 1924 until his death.[95] As Boggs (1929, 241) said in presenting the projection,

> The projection herein described is an equal-area projection which is an arithmetic mean [of y-coordinates only] between the sinusoidal equal-area projection of Sanson and the elliptical equal-area or homolographic projection of Mollweide.
>
> In computing the meridian of 90° from the centre of construction, values of y are initially computed as the arithmetical mean between the values of y, for any latitude, in the Sanson and Mollweide projections. A formula is then derived for x such that the projection will be strictly equal-area, and the area of the curve exactly the same as in the Sanson and Mollweide projections, viz. 4π for the entire surface of a sphere of unit radius. The expressions thus determined are:
>
> $$x = 2\sqrt{2}\ \pi \cos\theta \cos\phi/(2\sqrt{2}\cos\theta + \pi\cos\phi),$$
> $$y = (1/2)(\phi + \sqrt{2}\sin\theta)$$
>
> ϕ being the latitude, and θ the auxiliary angle used in the computation of the Mollweide projection. The intersection on the x axis $= 1.488398$, and on the y axis 1.492505.
>
> In order to equalize them the two formulae are multiplied by reciprocal constants, 1.001379 and 0.9986229.

The parallels of latitude are thus unequally spaced straight parallel lines, while the meridians are complex curves. Boggs preferred the interrupted form and named the projection eumorphic (fig. 4.24). To Goode's unsymmetrical approach to interruption Boggs added a variation of the central meridian within one of the lobes: for the Eurasian lobe, the central meridian is 60° E between the equator and 40° N, and

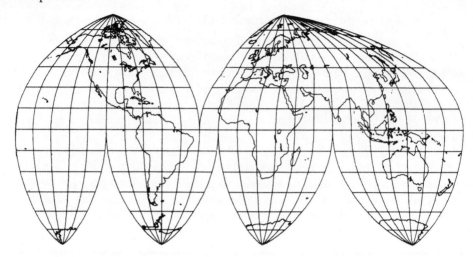

FIGURE 4.24. The Boggs eumorphic projection. Equal-area and usually interrupted as shown, it is a partial averaging of the Mollweide and sinusoidal projections. Boggs also usually added extensions of Greenland and eastern Siberia. 15° graticule.

30° E farther north. The resulting shape is, as expected, intermediate between that of the two original projections, better than the Mollweide near the equator and better than the sinusoidal near the poles. The projection has been used occasionally in textbooks and atlases (Wong 1965, 84, 105–6).

Later the same year and in the same journal, J. E. E. Craster (1873–1942) of England described his experiments with equal-area projections using hyperbolas and ellipses for meridians (Craster 1929), but his final choice was based on parabolas. The projection is calculated more easily than Craster thought, as Deetz and Adams (1934, 165–68, 191) shortly pointed out. The Craster parabolic projection (fig. 4.25) closely resembles the sinusoidal, but it is slightly more rounded and has the following formulas:

$$x = \sqrt{3}\, R\, \lambda\, [2\cos(2\phi/3) - 1]/\pi^{1/2};$$
$$y = (3\pi)^{1/2}\, R\, \sin(\phi/3).$$

It was used in a couple of textbooks of the 1950s and in a symmetrical oblique form in 1971 by the National Geographic Society for one edition of their map of Asia (with the North Pole falling along the central meridian of the normal Craster projection) (Wong 1965, 109; National Geographic 1971).

From 1933 to 1939, Kavrayskiy presented three pseudocylindricals, his fifth through seventh (V–VII) projections. His V is equal-area and uses a sine function for y, but the meridians are not sinusoidal.[96] The scale along the equator is 0.9 of true value, and the scale is correct along the 35th parallels. The formulas are

$$x = (R/ab)\, \lambda\, \cos\phi/\cos(b\phi), \text{ and}$$
$$y = aR\, \sin(b\phi),$$

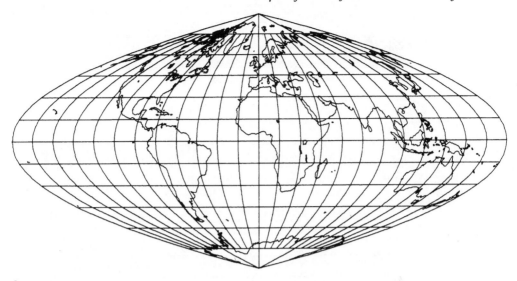

FIGURE 4.25. The Craster parabolic projection. Equal-area, with parabolic meridians. Central meridian 0°; 15° graticule.

where $a = 1/0.9b = 1.504875$ and $b = (\text{arccos } 0.9)/(35\pi/180) = 0.738341$. It was used for a map of the Pacific Ocean in the *Geograficheskiy atlas* (Geographic atlas) of 1940.

Kavrayskiy VI, presented in 1936, is equal-area and has sinusoidal meridians. It had been, however, already presented by Wagner (I) in 1932 and is discussed below.[97] It appeared in the second volume of the *Morskoy atlas* (Sea atlas) in 1953.

Kavrayskiy's third pseudocylindrical, projection VII introduced in 1939, is distinguished by a design repeatedly stated as "so 120-gradusnym krugovym meridianom," that is, with a circular meridian 120° east and west of the straight central meridian (fig. 4.26).[98] It is not equal-area (or conformal), poles are lines half the length of the equator, parallels are equally spaced, and all other meridians are equally spaced arcs of ellipses, not semiellipses. The equator is $\sqrt{3}$ or 1.732 times rather than twice the length of the central meridian, so the scale on the equator differs from that on the central meridian:

$$x = 1.5R \, \lambda \, (\pi^2/3 - \phi^2)^{1/2}/\pi;$$
$$y = R\phi.$$

One other Russian near-pseudocylindrical developed before 1949 is G. A. Ginzburg's arbitrary (non-equal-area and nonconformal) projection of 1944, his VIII, usually called the TsNIIGAiK pseudocylindrical projection in the Soviet Union after the initials of his agency (Tsentral'nyy Nauchno-issledovatel'skiy Institut Geodezii, Aerofotos'emki i Kartografii, or Central Scientific Research Institute of Geodesy, Aerial Photo Survey and Cartography).[99] Like Urmayev's cylindrical projections previously described, this one is defined with a modified series, in this case with coefficients to improve the representation of the Soviet Union on a world map, with the

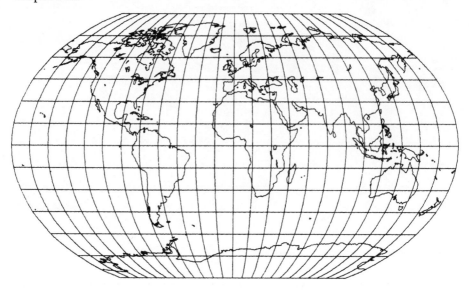

FIGURE 4.26. The Kavrayskiy VII arbitrary pseudocylindrical projection with a circular meridian 120° from the central meridian (0° here). 15° graticule.

central meridian at 70° E, and on which meridians are not quite equally spaced. After setting limits to angular and area distortion, Ginzburg obtained the following equations:

$$x = R \lambda (1 - 0.162388\phi^2)(0.87 - 0.000952426\lambda^4);$$
$$y = R (\phi + \phi^3/12).$$

The projection was used for a six-sheet 1:15,000,000-scale political world map in 1948, as well as for climatic, vegetation, and zoogeographical world maps of the same era (Graur 1956, 239, 267).

Preceding and following these Soviet contributions were a half dozen by Karlheinz (or Karl-Heinrich or Karl Heinrich, as he also signed technical papers) Wagner (fig. 4.27). Born in Leipzig in 1906 to the co-owner of the cartographic firm H. Wagner and E. Debes, he was educated at the Universities of Breslau and Hamburg before resuming work at intervals in his father's firm, becoming head in 1945. In 1951 he started his own firm in West Berlin, where he died in 1985 (Ferschke 1985; Mittelstaedt 1976).

Numbered in the order Wagner (1932, 1949) presented them in his textbook of 1949, four projections appeared in 1932—his I, III, IV and VI. Wagner I (fig. 4.28) is equal-area and has sinusoidal meridians. Each pole and the central meridian are half the length of the equator.[100] The projection is almost identical with Eckert VI in both design and appearance; the only design difference is that Eckert VI uses the full sinusoid quadrant between the equator and poles, while Wagner used only a 60°

FIGURE 4.27. Karlheinz Wagner (1906–1985), a German cartographic businessman who developed at least nine pseudocylindrical and modified azimuthal projections. *Photo courtesy of Ulrich Freitag.*

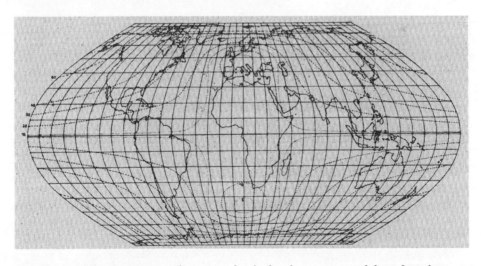

FIGURE 4.28. The Wagner I equal-area pseudocylindrical projection, with lines for poles, partial sinusoids for meridians. Reproduced from K.-H. Wagner (1949, 181). Very similar to Eckert VI with full sinusoids (fig. 4.21). Dashed lines show lines of constant maximum angular distortion. Central meridian 10° E; 10° graticule. *U.S. Geological Survey Library.*

portion and thereby, like Rosén in the 1926 *Nordisk världsatlas,* eliminated the iteration necessary with the Eckert (see n. 94). The formulas are

$$x = [2/3^{3/4}] \ R \ \lambda \ \cos \theta, \text{ and}$$
$$y = 3^{1/4}R\theta,$$

where $\sin \theta = (\sqrt{3}/2) \sin \phi$. As stated above, the later Kavrayskiy VI is identical.

On Wagner III, the parallels are equidistant straight lines, poles are half the length of the equator, and the meridians are equidistant sinusoids, using the same 60° portion as that of his first projection. Wagner adjusted the spacing of meridians to produce true scale along a selected parallel ϕ_1 as well as along the central meridian.[101] The projection is not equal-area:

$$x = [\cos \ \phi_1/\cos \ (2\phi_1/3)] \ R \ \lambda \ \cos \ (2\phi/3);$$
$$y = R\phi.$$

On Wagner IV (fig. 4.29), each meridian is less than half an ellipse, unlike Mollweide's and Eckert's semiellipses. The poles of the map are only $\sqrt{3}/2$ or 0.866 of the distance from the equator to the vertical limits of the ellipse if it were completed. The poles are half the length of the equator, and the parallels are straight lines spaced to preserve areas (K.-H. Wagner 1932, 28; 1949, 190–94):

$$x = 0.86310R \ \lambda \ \cos \theta, \text{ and}$$
$$y = 1.56548R \ \sin \theta,$$

where $2\theta + \sin 2\theta = [(4\pi + 3\sqrt{3})/6] \sin \phi$. The meridian $180°/\sqrt{3}$ or 103°55′ from the central meridian is a circular arc. This projection requires iteration.

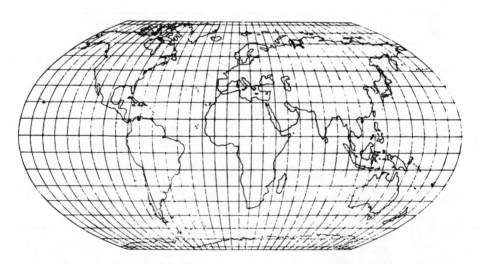

FIGURE 4.29. The Wagner IV equal-area pseudocylindrical projection, with lines for poles and less than semiellipses for meridians. Reproduced from K.-H. Wagner (1949, 192). Independently developed by Putniņš as his P′$_2$ projection. Central meridian 10° E; 10° graticule. *U.S. Geological Survey Library.*

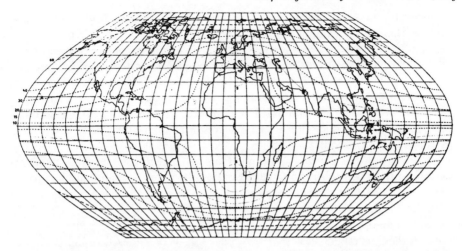

FIGURE 4.30. The Wagner II pseudocylindrical projection, with sinusoidal meridians; developed from Wagner I (fig. 4.28) by shifting parallels toward the equator to vary the area scale. Lines of constant maximum angular distortion are dashed. Central meridian 10° E; 10° graticule. Reproduced from K.-H. Wagner (1949, 187). *U.S. Geological Survey Library.*

Wagner VI, not equal-area, uses equally spaced straight parallels, poles half as long as the equator, and elliptical meridians identical in shape with those of Wagner IV (K.-H. Wagner 1932, 19; 1949, 197):

$$x = R \lambda (1 - 3\phi^2/\pi^2)^{1/2};$$
$$y = R\phi.$$

Wagner's other two pseudocylindrical projections were presented in 1949 and are also not equal-area. His II (fig. 4.30) is prepared in effect by shifting the parallels on Wagner I, other than poles, toward the equator without moving the sinusoidal meridians, and then slightly enlarging the entire map (K.-H. Wagner 1949, 184–89). This was done to achieve an area scale factor gradually increasing from 1.0 at the equator, through 1.2 at 60° latitude, to infinity at the poles. Simplifying constants,

$$x = 0.92483R \lambda \cos \theta, \text{ and}$$
$$y = 1.38725R\theta,$$

where $\sin \theta = 0.88022 \sin (0.88550 \phi)$.

For his V, Wagner applied the same area distortion to his IV that he used to obtain his II (K.-H. Wagner 1949, 194–96). Thus the meridians are the same partial ellipses found on IV. The formulas may be written

$$x = 0.90977R \lambda \cos \theta, \text{ and}$$
$$y = 1.65014R \sin \theta,$$

where $2\theta + \sin 2\theta = 3.00896 \sin (0.88550\phi)$.

In one paper from Riga, Latvia, Reinholds V. Putniņš (1934) offered twelve pseudocylindrical projections with simply described relationships resembling those of Eckert in 1906 and Wagner in 1932. On each the central meridian is half the length of the equator. Putniņš distinguished them with P (for projection or Putniņš?) followed by subscripts 1–6, those without primes having pointed poles, and those with primes having poles half the length of the equator. Odd-numbered subscripts indicate equidistant spacing of parallels (not equal-area). Even numbers indicate equal-area projections with meridians identical to those on the preceding odd-numbered projection, but with parallels appropriately moved. Finally, subscripts 1 and 2 indicate elliptical meridians, 3 and 4 are for parabolic meridians, and 5 and 6 are for hyperbolic meridians.

There is only one curve for a given parabolic meridian, established by the intersections with the two poles and equator, but the other conic sections must be further defined. For the four elliptical projections, Putniņš used the same portion of the ellipse as Wagner did in his IV–VI. The meridians on the four hyperbolic projections are correspondingly prescribed. Any hyperbolas with symmetrical foci along the x-axis fit the general equation $x^2/a^2 - y^2/b^2 = 1$, while similarly oriented ellipses satisfy $x^2/a^2 + y^2/b^2 = 1$. Putniņš used the same a and b for a given meridian on corresponding projections in both of his groups of four ellipticals and hyperbolics. The hyperbolic projections do little more than complete Putniņš's conic section series and are not detailed further because of greater distortion (for formulas, see Snyder 1977, 69).

The first eight Putniņš projections are attractive proposals, although some had already been presented by others (P_4 by Craster in 1929, as Putniņš noted, and P_1' and P_2' by Wagner in 1932, as Putniņš apparently did not realize). The formulas may be written as follows:

P_1 (equidistant parallels, pointed poles, elliptical meridians):

$$x = 1.89490R \lambda [(1 - 3\phi^2/\pi^2)^{1/2} - 0.5];$$
$$y = 0.94745R\phi.$$

P_1' (equidistant parallels, pole-lines, elliptical meridians): Same as the Wagner VI, except that on the Putniņš P_1' the linear scale is multiplied by a factor of 0.94745 to obtain true total area. It is also an arithmetic mean (x and y) of P_1 with the plate carrée of the same width (and therefore height).

P_2 (equal-area, pointed poles, elliptical meridians):

$$x = 1.89490R \lambda (\cos \theta - 0.5), \text{ and}$$
$$y = 1.71848R \sin \theta,$$

where $2\theta + \sin 2\theta - 2 \sin \theta = [(4\pi - 3\sqrt{3})/6] \sin \phi$. It is visually almost identical with the Boggs eumorphic projection, but the meridian at $90°/\sqrt{3}$ or $51°58'$ from the central meridian is a circular arc.

P_2' (equal-area, pole-lines, elliptical meridians): Same as the Wagner IV.[102]

P_3 (equidistant parallels, pointed poles, parabolic meridians):

$$x = (3/\pi)^{1/2} R \lambda (1 - 4\phi^2/\pi^2);$$
$$y = (3/\pi)^{1/2} R\phi.$$

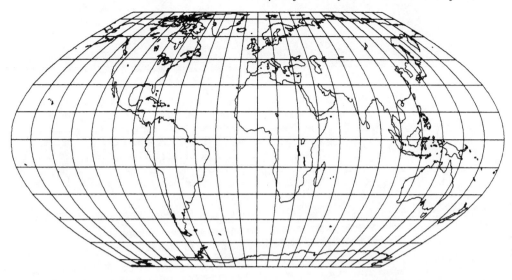

FIGURE 4.31. The Putniņš P′₄ projection, equal-area, with pole-lines and parabolic meridians. One of twelve presented by Putniņš, three of them reinventions. Central meridian 0°; 15° graticule.

P₃′ (equidistant parallels, pole-lines, parabolic meridians):

$$x = (3/\pi)^{1/2} \, R \, \lambda \, (1 - 2\phi^2/\pi^2);$$
$$y = (3/\pi)^{1/2} \, R\phi.$$

P₃′ is an average of P₃ with a plate carrée of the same width and height.

P₄ (equal-area, pointed poles, parabolic meridians): Same as the Craster parabolic projection.

P₄′ (equal-area, pole-lines, parabolic meridians) (fig. 4.31):

$$x = 2 \, (0.6/\pi)^{1/2} \, R \, \lambda \, \cos\theta/\cos(\theta/3), \text{ and}$$
$$y = 2 \, (1.2\pi)^{1/2} \, R \, \sin(\theta/3),$$

where $\sin\theta = (5\sqrt{2}/8) \sin\phi$.

Another German contributor of map projections during the 1930s was Karl Siemon (d. 1937), a professor in Erfurt-Hochheim, who proposed a special-purpose projection (Siemon 1935) and, with a posthumous paper, three general world maps (Siemon 1937), each published in a journal of the Nazi bureau for topographic surveys. All his projections are pseudocylindrical, and two were later independently reinvented in the United States. His first was designed so that straight lines from a chosen center are loxodromes or rhumb lines of true length, path, and direction from the center. While all loxodromes are straight on the Mercator projection, the scale varies widely, but Siemon's Wegtreue Ortskurskarte (fig. 4.32) preserves scale, provided a different map is prepared for each center. For most centers, this world map is somewhat pumpkin-shaped, symmetrical only about the central meridian; if

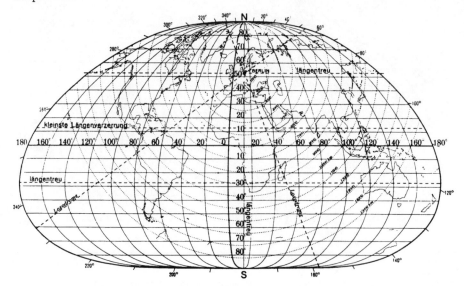

FIGURE 4.32. The loximuthal projection. Reproduced from Siemon (1935, 89). Tobler independently presented it later and provided the name used here. All rhumb lines from the focal point (Berlin here) are straight and at true length. 10° graticule for parallels, 20° for meridians. *U.S. Geological Survey Library.*

a point on the equator is the center, the map is also symmetrical about the equator and resembles the Bordone oval projection or the Mollweide (figs. 1.28, 3.12), if the latter is altered to have equidistant parallels. Siemon's projection became more widely known when Waldo R. Tobler (1930–) of the University of Michigan independently presented it and named it loximuthal (Tobler 1966b).

The formulas, with ϕ_1 the latitude of the central point, are

$$x = R \lambda (\phi - \phi_1)/[\ln \tan (\pi/4 + \phi/2) - \ln \tan (\pi/4 + \phi_1/2)], \text{ and}$$
$$y = R (\phi - \phi_1).$$

If $\phi = \phi_1$, the above equation for x is indeterminate, but $x = R \lambda \cos \phi_1$. Axes cross at the central point.

Siemon's three projections of 1937 are equal-area and more typical of other pseudocylindricals of the period. His II, which he called a Sanson projection with a pole-line, is the same as Wagner I of 1932 (above).[103] His III (fig. 4.33) is original and has fourth-order algebraic curves for the meridians. The parallels are spaced just as they are at the central meridian of the equatorial aspect of the Lambert azimuthal equal-area projection, but the equator becomes $\pi/\sqrt{2}$ or 2.22144 times as long as the central meridian (Siemon 1937, 98–100):

$$x = R \lambda \cos \phi/\cos (\phi/2);$$
$$y = 2R \sin (\phi/2).$$

This projection, however, became known as the quartic authalic because of its fourth-order curves and because O. S. Adams reintroduced it in 1944 in a govern-

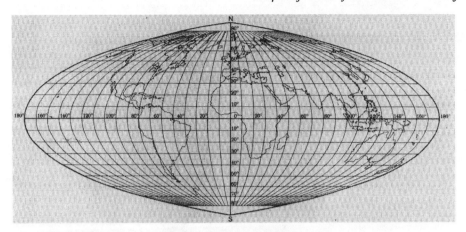

FIGURE 4.33. The quartic authalic projection, reproduced from Siemon (1937, 96). O. S. Adams independently presented it later. Meridians are fourth-order (quartic) curves on this equal-area pseudocylindrical. Central meridian 0°; 10° graticule. *U.S. Geological Survey Library.*

ment agency on the opposite side of the war, commenting that "as far as we know, no such projection has been heretofore proposed." Adams added that it is the Lambert azimuthal equal-area counterpart of Foucaut's stereographic equal-area projection.[104] Like the Putniņš P_2 closely resembling the Boggs eumorphic, except for the axis ratio, it was little used except as an inspiration for the McBryde-Thomas flat-polar quartic of 1949 (discussed later). Siemon's (1937, 99, 101) final projection IV is merely his quartic III with axes adjusted to a 2:1 ratio by dividing his x-values and multiplying y by the same constant $(\pi/2\sqrt{2})^{1/2}$ or 1.05391.

Werner Werenskiold (1945) of Norway proposed three pseudocylindrical projections but all had appeared before, except for a change in scale. He suggested three equal-area projections, but with the area uniformly enlarged to achieve correct linear scale along the equator. Again the poles and central meridians are each lines half the length of the equator. His first projection has parabolic meridians and, as Werenskiold stated, is identical to the Putniņš P_4' except for enlarging the latter uniformly in all directions by a linear factor $(\pi/2.4)^{1/2}$ or 1.14411. His other projections were not credited: Werenskiold II has sinusoidal meridians and is identical to the earlier Wagner I and Kavrayskiy VI except for a scale enlargement of $3/2(3^{1/4})$ or 1.13975. Werenskiold III has elliptical meridians, making it identical to Wagner IV and Putniņš P_2', except for a linear enlargement of 1.15861.

This 1926–49 period is completed not only by some of Wagner's entries discussed above but also by five equal-area pseudocylindrical projections presented by McBryde and Thomas (1949) for world statistical maps in a U.S. Coast and Geodetic Survey publication. F. Webster McBryde (1908–) (fig. 4.34) provided the concepts and Paul D. Thomas the mathematical development. Numbers 1 and 2 meet relatively complicated requirements: no. 1 has pointed poles, with the 80°-latitude lines ⅓ the length of the equator and the x-coordinate at the equator for the meridian 20° from center at 0.85 of the y-value for the 20th parallel (McBryde and Thomas

FIGURE 4.34. F. Webster McBryde (1908–), about the time the cartographic consultant and educator devised the first five of some ten map projections with flat-poles usually 1/3 the length of the equator. *Photo by Al Schwartz, 1952, courtesy of F. Webster McBryde.*

1949, 23). In addition, the projection was to fit a general sine (not sinusoidal) series of equal-area pseudocylindrical projections with the following formulas: [105]

$$x = (q/p) \ R \ \lambda \ \cos \phi/\cos (\phi/q), \text{ and}$$
$$y = Rp \ \sin (\phi/q),$$

in which the constants for no. 1 become $p = 1.48875$ and $q = 1.36509$.

No. 2 is a flat-polar or pole-line modification of no. 1, with the poles ¼ and the parallels for $\theta = 80°$ ½ the length of the equator and the x-value for the 20° meridian at the equator set at 0.85 of the y-value for $\theta = 20°$ (McBryde and Thomas 1949, 28). Consolidating constants,

$$x = 0.22248R \ \lambda \ [1 + 3 \cos \theta/\cos (\theta/1.36509)], \text{ and}$$
$$y = 1.44492R \ \sin (\theta/1.36509),$$

where $0.45503 \sin (\theta/1.36509) + \sin \theta = 1.41546 \sin \phi$.

McBryde-Thomas 3, which the inventors named the flat-polar sinusoidal projection, uses, like the regular sinusoidal and Eckert VI, the full sinusoid and a central meridian half the length of the equator, but the poles were made ⅓ the length of the equator, instead of Eckert's ½ (McBryde and Thomas 1949, 32). Therefore, as with Eckert and not Wagner, iteration is required:

$$x = R \ [6/(4 + \pi)]^{1/2} \ \lambda \ (0.5 + \cos \theta)/1.5, \text{ and}$$
$$y = R \ [6/(4 + \pi)]^{1/2} \theta,$$

where $\theta/2 + \sin \theta = (1 + \pi/4) \sin \phi$.

No. 4, the McBryde-Thomas flat-polar quartic projection (fig. 4.35), is similarly related to the above quartic authalic. The ratio of axes is left at the $\pi/\sqrt{2}$ or 2.22144:1 ratio of the latter projection. With the poles stipulated at ⅓ the length of the equator, the formulas, requiring iteration, become (McBryde and Thomas 1949, 33, 36)

$$x = R \ \lambda \ [1 + 2 \cos \theta/\cos (\theta/2)]/[3\sqrt{2} + 6]^{1/2}, \text{ and}$$
$$y = 2\sqrt{3} \ R \ \sin (\theta/2)/(2 + \sqrt{2})^{1/2},$$

where $\sin (\theta/2) + \sin \theta = (1 + \sqrt{2}/2) \sin \phi$. This is the only projection of the five receiving much attention from others. It captured the interest of Robinson as an interruptible projection in both textbook and paper, while Finch, Trewartha, and associates replaced the Hammer projection with the flat-polar quartic in the fourth edition of their basic geography text.[106]

The last of the McBryde-Thomas series, no. 5 or the flat-polar parabolic projection, is correspondingly related to the Craster parabolic projection, but with an axis ratio of $2\pi/3$ or 2.09440:1 instead of Craster's 2:1. The poles are again ⅓ the length

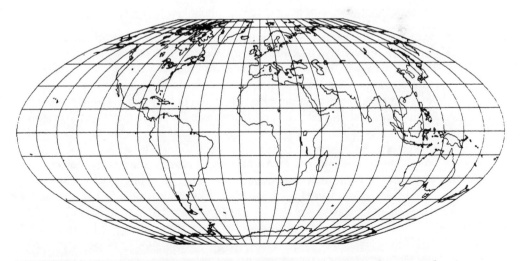

FIGURE 4.35. The McBryde-Thomas flat-polar quartic projection. An equal-area pseudocylindrical with pole-lines 1/3 as long as the equator, derived from the quartic authalic (fig. 4.33). Central meridian 0°; 15° graticule.

of the equator (McBryde and Thomas 1949, 37, 38). Noniterative formulas equivalent to those of Thomas but slightly simpler are as follows:

$$x = \sqrt{6} \, R \, \lambda \, [2 \cos (2\theta/3) - 1]/\sqrt{7}, \text{ and}$$
$$y = 9R \, [\sin (\theta/3)]/\sqrt{7},$$

where $\theta = \arcsin [(7/3)(\sin \phi)/\sqrt{6}]$.

In 1977 McBryde would "fuse" three of these with older projections to create still others, but this is discussed later. McBryde was born in Lynchburg, Virginia, and during and after obtaining his doctorate in geography at the University of California at Berkeley in 1940 he taught at several universities, especially at Ohio State University in 1937–42. A geographical consultant to several federal agencies, he also traveled extensively to Central and South America and was president of the American Society for Geographical Research. Since 1969 he has operated a consulting firm for "human ecology" in Maryland and has written on his research in a variety of geographical fields (Marquis Who's Who 1981–82, 510).

As McBryde and Thomas point out, crediting others who in turn lead to Germain in 1866, there are relatively simple formulas permitting the development of an almost endless set of new equal-area pseudocylindrical projections as well as allowing proof of the equivalency of such existing projections.[107] If y is any function of ϕ multiplied by R, or $y = Rf(\phi)$, then

$$x = R^2 \, \lambda \, \cos \, \phi/(dy/d\phi),$$

where x and y provide an equal-area pseudocylindrical. There can be specific subsets, such as the sine and tangent series discussed by Baar and by McBryde and Thomas (see n. 175).

Pseudocylindricals with Unconventional Parameters, 1950 and Later

Some of those making new contributions to map projections after 1950 took less conventional approaches. The many proposals for new pseudocylindricals before 1950 had generally fit into neat categories: often the projections were equal-area or had equidistant parallels, with elliptic arcs or other common curves for meridians. After that period, inventors more often broke out of these bounds and offered proposals that caused some observers, more than ever before, to question the proposals' suitability. It was also true that by then, most standard approaches had been worked and reworked.

In 1950 Urmayev extended his 1947 studies of cylindrical projections to pseudocylindricals.[108] First, for a class of equal-area sinusoidal projections using various portions of the sinusoid and having line-poles of various lengths, he generalized the approach taken for Kavrayskiy VI (preceded by Wagner I), so that

$$x = aR \, \lambda \, \cos \, \theta, \text{ and}$$
$$y = R\theta/ab,$$

where $\sin \theta = b \sin \phi$, $a = 2/3^{3/4}$, and b can vary from 0 (the cylindrical equal-area) through $\sqrt{3}/2$ (Kavrayskiy VI) to 1 (the sinusoidal). The poles are lines $(1 - b^2)^{1/2}$ as long as the equator, and the area scale varies with b. No iteration is involved.

Urmayev also modified this concept with a short series to produce projections neither equal-area nor sinusoidal by adopting a varying area scale of $(1 + K\theta^2)$ and a desired pair of standard parallels. The spacing of the parallels increases away from the equator: [109]

$x = aR \lambda \cos \theta$, and

$y = (R\theta/ab)(1 + K\theta^2/3)$,

where θ is found from ϕ as just above. Specific forms of this type were used for climatic world maps in the 1947 *Atlas ofitsera* and for the physical-political map of the Pacific and Indian ocean regions in the *Atlas mira* of 1954 and 1967.[110] Constants for the latter are $a = 2/3^{3/4}$, $b = 0.8$, and $K = 0.414524$, giving an area scale at latitude 70° just 1.3 times that of the equator, and a vertical shape distortion of 1.299 at the equator.

Two arbitrary or non-equal-area pseudocylindricals appeared about this time in England. Apparently no details were published. One, in *The American Oxford Atlas* of 1951, was developed by Bomford for this atlas (Lewis and Campbell 1951, 3, 6, pls. I–V). Calling the projection a modified Gall, the text only states that for world climate maps at a scale of 1:110M, "Gall's [stereographic] projection has . . . been . . . modified by some lateral compression to reduce enlargement in high latitudes" (Lewis and Campbell 1951, 6). The spacing of the straight parallels is the same as on Gall's; the curved meridians are equally spaced along parallels, but at about $(1 - 0.04\phi^4)$ of the spacing at the equator.

The Times Atlas counterpart of this modified Gall is called The Times projection (fig. 4.36) and was devised for Bartholomew by John Moir (1906–1988) of Edinburgh, their chief draftsman, in 1965 to replace various interrupted projections devised by John Bartholomew (discussed later).[111] With its straight parallels also spaced as those of the Gall stereographic, the sinusoidal meridians are curved more than those of Bomford, but were "designed to reduce the distortions in area and shape which are inherent in cylindrical projections, whilst, at the same time, achieving an approximately rectangular shape overall."[112]

Moir's 180th meridian is plotted so that the length of each parallel is decreased from that of the equator by 1.0866 times $\sin^2(\theta/2)$ for the unit sphere, where θ is 90° times the ratio of the map distance of the parallel from the equator to the map distance of the pole-line from the equator. True scale is maintained at latitudes 45° N and S on the International ellipsoid, and the sphere used has as its radius the average of the major and minor semiaxes of the ellipsoid. The formulas are equivalent to

$x = R \lambda \{0.74482 - 0.34588 \sin^2[45° \tan (\phi/2)]\}$, and

$y = 1.70711 R \tan (\phi/2)$.

The Times projection has been used for several of the world maps beginning with the 1967 second ("Comprehensive") edition of the *Atlas*.

A third modified Gall was designed by Lawrence Fahey of the United States in 1975.[113] The world is enclosed in an ellipse with an axis ratio of about 1.42 to 1. Other meridians are equidistant ellipses. The parallels are spaced for a Gall stereographic based on a cylinder secant at about 35° instead of 45°.

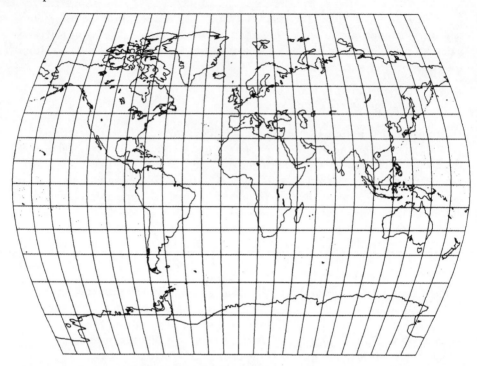

FIGURE 4.36. The Times projection, developed for Bartholomew and used in *The Times Atlas*. A modified Gall projection, the parallels are spaced like those of the Gall stereographic (fig. 3.10). Central meridian 0°; 15° graticule.

During the 1960s, approaches still more empirical than those of Urmayev and others were taken by two separate cartographers in the United States and Hungary. The first was Arthur H. Robinson (fig. 4.37). Born in Montreal in 1915 of American parents, he received his Ph.D. in geography from Ohio State University in 1947. During the Second World War he was in charge of mapping at the U.S. Office of Strategic Services, predecessor of the Central Intelligence Agency. In 1946 he began to teach geography and cartography at the University of Wisconsin, a career from which he retired as professor emeritus in 1980, after establishing one of the leading U.S. academic programs and writing several editions of the standard American textbook on cartography, *Elements of Cartography*. He was president of the Association of American Geographers (1963–64) and of the International Cartographic Association (1972–76) (Ormeling 1988, 110–13; Robinson 1953a).

Following the problems with the Ranally map of Rand McNally, Robinson was asked to select a world map projection that would fulfill a list of nine requirements, including an uninterrupted format, least possible shearing, least possible apparent area-scale distortion for major continents, a simple graticule, and a system resulting in a map suitable for readers of all ages. He could not find a suitable existing projection and therefore designed a new one in 1963 which was briefly called the orthophanic (right-appearing) projection, but soon became known as the Robinson projection (fig. 4.38). The details were later published by Robinson (1974).

FIGURE 4.37. Arthur H. Robinson (1915–), a leading educator in cartography shown about the time he developed the Robinson projection used by Rand McNally, the National Geographic Society, and others. *Photo by Chase Ltd., Washington, 1970, courtesy of Arthur H. Robinson.*

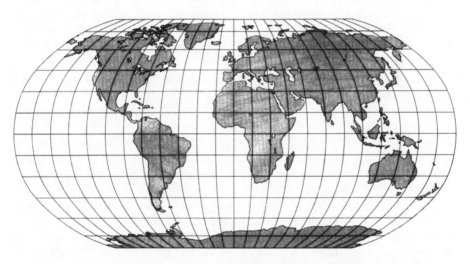

FIGURE 4.38. The Robinson projection, a compromise pseudocylindrical developed for Rand McNally. Reproduced from Robinson (1974, fig. 3). Central meridian 0°; 15° graticule. *Plotted by the University of Wisconsin Cartographic Laboratory. Reproduced by permission of Universitätsverlag Ulm GmbH, publisher of the* International Yearbook of Cartography.

Human computers and plotters assisted in its trial-and-error development, but the projection broke from the tradition of computing coordinates by mathematical formulas or using simple geometric construction. It was derived to make the world map "look" right, minimizing the exaggerated polar areas of the Mercator, Miller, or Gall stereographic cylindricals, and the polar compression of the flat-polar, equal-area projections. The projection has flat poles 0.5322 as long as the equator, which is 1.9716 times the length of the central meridian. As usual, the curved meridians are equally spaced along the straight parallels, but the parallels are equally spaced between latitudes 38° N and S, beyond which spacing gradually is reduced; the area scale gradually changes away from 40° latitude, 22% low at the equator and 30% high at 60°, but becoming infinite at the poles.[114] The coordinates for points along the meridians 90° from center are interpolated from a table for each 5° of latitude, and other meridians are plotted proportionally.[115]

Rand McNally began to use the projection in various atlases[116] and as a wall map, but purchasers preferred the Mercator to such an extent that the Robinson sheet map was withdrawn from popular distribution and left only in the more expensive educational series. Some European publishers also used the projection. In 1987, however, John B. Garver (1988), chief cartographer of the National Geographic Society, arranged a seminar consisting of its senior cartographic staff and outside consultants to select a world map projection to replace the Van der Grinten used since 1922. Of some twenty possibilities considered, the Robinson was selected, and the new sheet map was issued at a press conference as part of the hundredth anniversary of the society in the fall of 1988 and as an insert with the twelve million copies of the *National Geographic*. As a result of the press conference, "more than 550 newspapers and magazines with a combined circulation of over 51 million" carried a story about the change, perhaps a record for a single map projection event (Robinson 1989, 20), and several other cartographic firms began to use the same projection. Rand McNally immediately issued an inexpensive version, and the National Geographic replaced the Van der Grinten with the Robinson for physical and political world maps in their 1990 world atlas.[117]

Hungarian János Baranyi (1968) took another nonformula approach in presenting seven examples of non-equal-area pseudocylindrical-like projections. In these the parallels are spaced at slightly increasing (on three examples), decreasing (on one), or increasing and decreasing (on three) distances from each other as the latitude varies from equator to pole. The poles are points on five examples, and lines on the others are ½ (on one) or ⅔ (on one) of the length of the equator. The meridians are equally spaced on two but become gradually closer away from the central meridian on the rest. The 180th meridians are circular arc segments tangent to each other or to straight lines; the central arc is centered on the equator, while other arcs are centered on the central meridian or its extension. The other meridians are spaced on each parallel in the same proportions as the equatorial spacing to fit the outer meridian, and all projections are symmetrical about the equator and about the central meridian.

For example, in Baranyi's projection IV (fig. 4.39), using his scale, the distance between 10° parallels is 12 mm from 0° to 30°, 13 mm from 30° to 60°, and 12 mm from 60° to 90°, totaling 222 mm pole to pole. The equator is marked with 12 mm

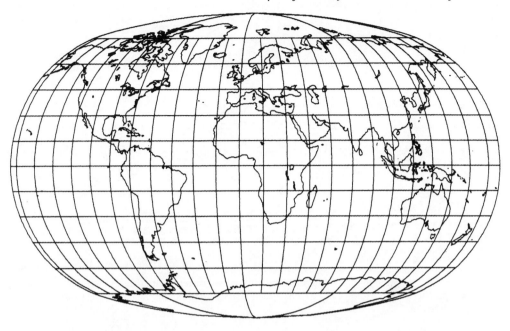

FIGURE 4.39. The Baranyi IV projection, an arbitrary pseudocylindrical with varied tabular spacings; used in Hungarian atlases. Central meridian 10° E as used in atlases; 15° graticule.

for each 10° longitude from 0° to 20° from the central meridian, 11 mm from 20° to 60°, 10 mm from 60° to 140°, and 9 mm from 140° to 180°, totaling 368 mm. Each 180th meridian consists of a circular arc of 100 mm radius centered on the equator, tangent to another arc centered on an extension of the central meridian and passing through the pointed pole.

Although these projections may be encoded in their original forms, Baranyi's meridian curves and their variable spacings add complications. Therefore, closely fitting empirical series formulas were subsequently developed in Hungary for the two of his projections that have been used in Hungarian atlases: Baranyi II was used in historical atlases "after the 70's," and his IV replaced the Soviet TsNIIGAiK modified polyconic for thematic maps in the world atlas beginning 1985.[118] Baranyi and Földi (1976) also published comparisons between his projections and that of Robinson.

With Baranyi's paper was one by György Érdi-Krausz (1968), also of Hungary, who provided his own pseudocylindrical (fig. 4.40) with a more classic construction. Reminiscent of Goode's homolosine, it is a fusion of the Mollweide with a flat-polar sinusoidal. With, in effect, a modification of Urmayev's unmentioned 1950 approach, Érdi-Krausz used a portion of the sinusoidal curve to obtain a flat-polar sinusoidal with central meridian 0.4 and each pole 0.6 of the length of the equator. He utilized this projection only between 60° N and S latitude. Poleward he used the Mollweide enlarged in scale so that its 60° parallels matched those of the truncated sinusoidal. The combined projection, looking much like a Mollweide but with a slight break in

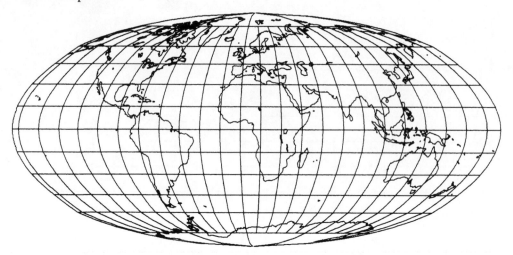

FIGURE 4.40. The Érdi-Krausz projection. Used in Hungarian atlases, this is equal-area between and beyond 60° N and S latitude, but at two different area scales. A fusion at those parallels of the Mollweide and a flat-polar sinusoidal projection. Central meridian 10° E; 15° graticule.

meridians at 60° latitude, is equal-area only within each of the two projections. The sinusoidal portion has only 0.70768 of the area scale of the Mollweide. For the formulas between 0 and ±60° latitude,

$$x = 0.96042R \; \lambda \; \cos \theta, \text{ and}$$

$$y = 1.30152R \; \theta,$$

where $\sin \theta = 0.8 \sin \phi$. From $\phi = \pm 60°$ to the respective pole,

$$x = 1.07023R \; \lambda \; \cos \theta, \text{ and}$$

$$y = R \; (1.68111 \sin \theta - 0.28549 \text{ sign } \phi),$$

where $2\theta + \sin 2\theta = \pi \sin \phi$.

The Mollweide portion requires iteration. The larger polar area scale, although abruptly changed at 60°, reflects the common thread through some of the projections of Gall, Wagner, Miller, Urmayev, Robinson, and others to have area exaggeration as the poles are approached, although to a far less extent than that on the Mercator projection. The Érdi-Krausz projection was used in Hungarian historical atlases during the 1970s, for a world map in Hungary's 1967 *National Atlas,* and for the principal political and physical world maps in the *Nagy világatlasz* (Great world atlas), beginning in 1985.[119]

A few years after Érdi-Krausz's presentation, fusion was again applied to normal pseudocylindrical projections. McBryde (1978), who had, with Thomas, described some flat-polar alternatives in 1949, fused three of their earlier projections and also

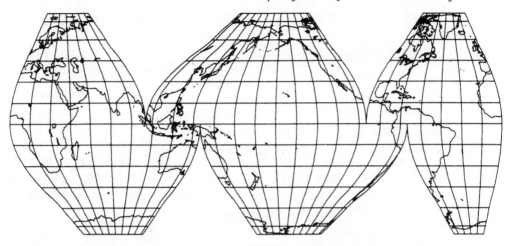

FIGURE 4.41. The McBryde S3 projection, an equal-area fusion of the sinusoidal and the McBryde-Thomas flat-polar sinusoidal at 55°51′ N and S latitude. Here McBryde's interruptions are used to show oceans. 15° graticule. *Patent by McBryde (1982a). Used by permission of F. Webster McBryde.*

Eckert VI with the pointed-polar sources to create four new world maps. All remain equal-area and contain the pointed-polar portion between the matching parallels and the flat-polar portions poleward. Thus all have flat poles. Using the names he applied, each starting with McBryde, Q3 is the quartic authalic plus the McBryde-Thomas 4 (flat-polar quartic), matching at 52°09′ N and S latitudes. P3 is the Craster parabolic plus the no. 5 (flat-polar parabolic), matching at 49°20′; S2 is the sinusoidal plus Eckert VI, matching at 49°16′; and S3 (fig. 4.41) is the sinusoidal plus the 1949 no. 3 (flat-polar sinusoidal), matching at 55°51′. The poles on the fused projections are 0.306 to 0.316 times the length of the equator, except on S2 where, with the longer Eckert pole, the ratio is 0.441. The formulas are the same as those for the respective base projections, except that y for the flat-polar portion must have subtracted from it the sphere radius R times a constant taking the sign of ϕ, with numerical values (Q3) 0.042686, (P3) 0.082818, (S2) 0.084398, and (S3) 0.069065. McBryde added codes for the uninterrupted (A1) and more common interrupted (B1) forms.

McBryde (1982a, 1982b) patented this series, especially describing S3, and has used the latter for a published map interrupted to show oceans, highlighting two hundred–mile offshore limits as they appear worldwide.

Applying the concept of partial ellipses used by Wagner, Putniņš, and Kavrayskiy in the 1930s, as well as fusion, but fusing at the equator, Masataka Hatano (1972) of Japan obtained an asymmetric elliptical pseudocylindrical in which the North Pole is ⅔ the length of the equator, while the South Pole is ¾ its length.[120] In addition, he made the scale factor along the equator 0.85, and the vertical scale was adjusted to preserve the correct area scale. The consolidated formulas become as follows:

$$x = 0.85R \; \lambda \; \cos \theta, \text{ and}$$
$$y = a \; R \; \sin \theta,$$

where $2\theta + \sin 2\theta = b \sin \phi$. For the northern hemisphere, $a = 1.75859$, $b = 2.67595$; for the southern, $a = 1.93052$, $b = 2.43763$. The meridians are thus elliptical arcs changing eccentricity when crossing the equator, and circular at longitude $118°32'$ from center (northern) and $130°08'$ (southern hemisphere).

Another pseudocylindrical projection presented twice was first proposed by Franz Mayr (1964). By spacing meridians in proportion to the square root of the cosine of the latitude and preserving equal-area, Mayr produced a pointed-polar projection requiring numerical integration for values of y:

$$x = R \; \lambda \; \cos^{1/2}\phi;$$
$$y = R \int_0^\phi \cos^{1/2}\phi \; d\phi.$$

Tobler (1973b) reintroduced this projection, but as an equal-area geometric mean of the x-values of the cylindrical equal-area and sinusoidal projections. It was only one of several projections Tobler suggested, inspired in part by Foucaut's weighted arithmetic averages of the same projections (p. 113). Tobler pointed out that, using geometric means of y-coordinates instead for an equal-area projection,

$$x = 2R \; \lambda \; \cos \phi \; (\phi \sin \phi)^{1/2}/(\sin \phi + \phi \cos \phi), \text{ and}$$
$$y = R \; (\phi \sin \phi)^{1/2}.$$

Unequally weighting the coordinates of the two projections leads to other combinations. Tobler stressed in his paper an equal-area pseudocylindrical he called the hyperelliptical (fig. 4.42), with meridians on the base projection following the curve $x^k + y^k = \gamma^k$. The basic projection is then averaged with the cylindrical equal-area to produce a series of equal-area projections, including some common ones, depending on constants chosen. Numerical integration is required for the general case. Working with the equations

$$x = \lambda \; [\alpha + (1 - \alpha)(\gamma^k - y^k)^{1/k}/\gamma], \text{ and}$$
$$\alpha y - [(\alpha - 1)/\gamma] \int_0^y (\gamma^k - y^k)^{1/k} \; dy = \sin \phi,$$

Tobler showed several examples, but selected $\alpha = 0$, $\gamma = 1.183136$, and $k = 2.5$ for the preferred projection, multiplying x and dividing y by $(2\gamma/\pi)^{1/2}$ or 0.8679 to obtain $2:1$ axes, and then multiplying each by R. These meridian curves lie between the Mollweide ellipses and the corresponding rectangles. The poles are points.

While nearly all of the many specifically presented pseudocylindricals of the twentieth century are mentioned above, the list is not exhaustive. Atlases have also noted modified cylindrical or pseudocylindrical projections not otherwise identified, and cartographic studies have mentioned others in passing to show the versatility of

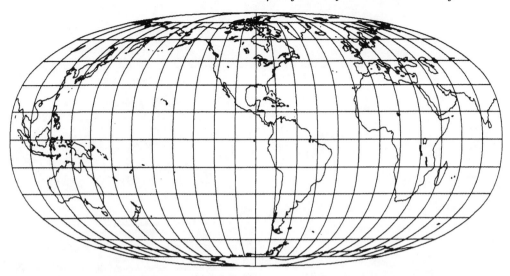

FIGURE 4.42. The Tobler hyperelliptical projection. An equal-area pseudocylindrical with complex meridian curves falling between Mollweide ellipses and corresponding rectangles. Central meridian 90° W; 15° graticule.

a certain approach.[121] Specially used interrupted forms have been noted above, but all pseudocylindricals can be interrupted in various ways. It is appropriate now to describe the use of some of these projections in oblique and transverse forms, a tedious construction until the advent of modern computers.

Oblique and Transverse Pseudocylindrical Projections

Although there are now computer programs that can routinely plot any oblique or transverse aspect of almost any known projection regardless of type, this was not routine until about 1980. For pseudocylindrical projections, apparently the first transverse aspect appeared in 1908 and the first oblique view in 1928, both applied to the Mollweide and both in England. The former was prepared by Charles Close, and the result (fig. 4.43) appeared as the frontispiece of a 1908 *Catalogue of Maps* issued by the general staff of the British Topographic Section.[122] Using a central meridian of 70° E, Close showed the usual eastern hemisphere bounded by the circle representing the 90th meridians east and west of center on the normal Mollweide, but half the western hemisphere abuts the northern and half abuts the southern semicircles rather than abutting eastern and western semicircles, and polar regions are much less distorted than those of the normal aspect.

On the oblique aspect shown by J. Fairgrieve (1928), the geographic North Pole is placed at latitude 45° north and longitude 90° west of center on the base Mollweide, with a curved Greenwich meridian passing through the center (see also Close 1929, 248–49). In 1932 Wagner experimented with offset transverse (often called

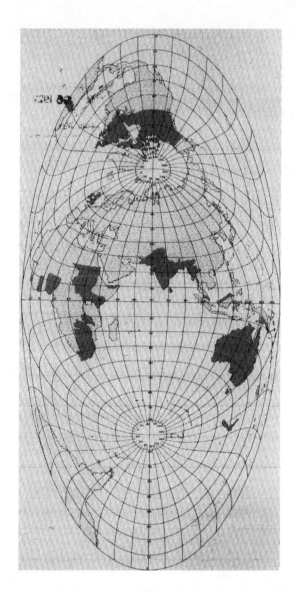

FIGURE 4.43. The first transverse pseudocylindrical projection published. Prepared by Close in 1908 and reproduced here from Hinks (1912; 1921 ed., frontispiece). An equal-area transverse Mollweide with the eastern hemisphere enclosed in a circle. Central meridians 70° E, 110° W; 10° graticule.

transverse oblique) aspects of his own flat-polar sinusoidal projections and an oblique Eckert VI to show the Atlantic Ocean and the Americas.[123]

Still another precomputer example is a transverse oblique Mollweide named the Atlantis (reconstructed in fig. 4.44) by John Bartholomew (1948, 1958), who published it as the first plate of the *Regional Atlas* of 1948 and ten years later as the frontispiece in the first of the five volumes of *The Times Atlas*. The name was based on the central position of the Atlantic Ocean, with longitude 30° W along the equator of the base Mollweide and latitude 45° N passing through the center of the elliptical world map. A symmetrical oblique Mollweide world map appears in Japan's *National Atlas* (Geographical, Japan 1977, 206).

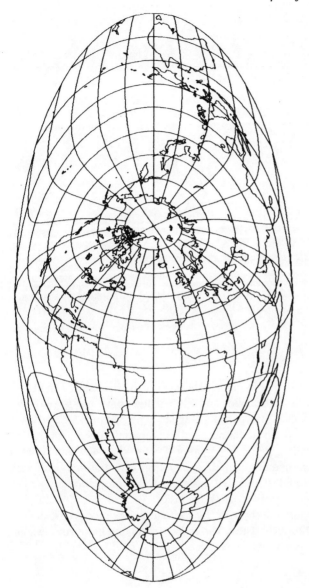

FIGURE 4.44. The Atlantis projection, prepared and used by Bartholomew in atlases. An equal-area transverse Mollweide with 30° W and 150° E as the central meridians and the North Pole displaced 45° from the center to allow focus on the Atlantic Ocean. 15° graticule.

Thus the aesthetically attractive bounding ellipse of the Mollweide was retained in many of the early oblique and transverse pseudocylindrical studies. Even the oblique Hammer used in several other cases (see below) is bounded by the same outer ellipse.

Much more complicated is Allen K. Philbrick's (1914–) interrupted oblique Sinu-Mollweide projection (fig. 4.45). Altering Goode's approach with the homolosine, Philbrick (1953), then of the University of Chicago, retained the Mollweide everywhere north of 40°44′ S and fused the sinusoidal at that parallel, for use to the

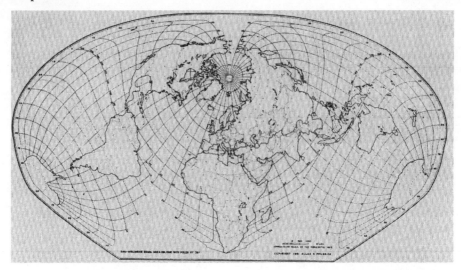

FIGURE 4.45. The Sinu-Mollweide projection, an equal-area interrupted oblique fusing of the Mollweide and sinusoidal projections. Reproduced from Philbrick (1953, 206). Central meridian 20° E; 10° graticule. *By permission of Allen K. Philbrick. U.S. Geological Survey Library.*

South Pole. He then rotated the globe to place 55° N and 20° E at the center of the original projection and the North Pole along the central vertical axis. Finally he interrupted the lower half into three lobes and the portion above the level of the actual North Pole into two lobes. This was a challenge to construct in 1953, and even now it would be a challenge to program for automatic computation, but "only Africa and Antarctica have areas with angular deformations" more than 40°, and he recommended the attractive arrangement for showing world distributions (Philbrick 1953, 208–9). Use of a symmetrical oblique Craster parabolic projection for a 1971 *National Geographic* map of Asia was mentioned earlier.

Many pseudocylindrical and other projections with arbitrary properties, such as the Ginzburg, Robinson, Miller cylindrical, Van der Grinten, and globular, have essentially no value in the oblique or transverse aspect.

New Azimuthal Projections

The new azimuthal projections of the twentieth century are few when contrasted with new pseudocylindricals, but as with cylindrical projections, this is largely due to the paucity of design variables available to adjust. For example, there is only one azimuthal equal-area projection (unless the center point is badly distorted), and even that cannot be expanded and contracted like the cylindrical equal-area projection and remain azimuthal. The variations have therefore been arbitrary or compromise azimuthal projections.

The first twentieth-century form to be developed was named for Arthur Breusing, who developed its prototype in the 1880s. This is the projection named by its developer Young (1920, 7–8) as the Breusing (harmonic) minimum-error projection, because it uses the harmonic mean of the same base projections as those used by Breusing for his geometric mean, namely, the stereographic and the Lambert azimuthal equal-area projections. In appearance, it strongly resembles Airy's minimum-error azimuthal projection of 1861–62 (fig. 3.23).

The principal development of new azimuthal projections occurred in the Soviet Union. In 1949 Solov'ev applied double projection to azimuthal projections (double projection is described above for the development of ellipsoidal versions of projections).[124] His work did not produce specific new projections, but the basic technique involves projecting the globe onto a sphere tangent at the projection center and with twice the radius of the base globe, and then projecting from this sphere onto a plane tangent at the same point. If both projection steps are carried out stereographically with the base globe and outer sphere, respectively, the Breusing harmonic, which Solov'ev independently presented, results; if the first step is stereographic and the second is orthographic, the Lambert azimuthal equal-area projection results.

In the same year Ginzburg (1949a) modified the Lambert azimuthal equal-area projection for less shape distortion on school maps of hemispheres. He added a series term to the formula for the projection radius to reduce angular distortion at the expense of creating a slight area distortion.[125] The projections remain azimuthal. In terms of the angular distance z in radians from the center, the radius ρ is found thus:

$$\rho = R \left[2 \sin (z/2) + az^q \right].$$

For Ginzburg I, $a = 0.00066$ and $q = 9$; for Ginzburg II, $a = 0.00025$ and $q = 10$.

Urmayev's (1950) first two projections are also azimuthal with arbitrary properties. The formulas for the radius are given in integral form:

$$\rho^2 = 2R \int_0^z \left[1 \pm (z°/K)^2 \right] \sin z \, dz,$$

where K is 100 and the sign is plus for Urmayev I and 130 and minus, respectively, for Urmayev II, z being in degrees.[126] The expression may be analytically integrated. Maling (1960, 208) states that I "has angular and areal distortions which are very similar to those of the minimum error projection of Sir Henry James," but is nonperspective; Urmayev II, however, "appears to have no particular purpose or value."

Emphasizing the central portion of a map by enlarging it attracted cartographers in Australia, Hungary, Israel, Poland, Sweden, and the United States.[127] Only two of the projection types developed are azimuthal: Torsten Hägerstrand (1957, 73, 74) showed logarithmic enlargements of Asby, Sweden, on maps limited to Sweden and showing migration patterns. Crediting Edgar Kant with the method, Hägerstrand (1957, 74) details the projection only as one "in which the distance from the centre shrinks proportionally to the logarithm of the real distance. . . . The rule obviously cannot be applied to the shortest distances." There are approaches, however, that allow the projection to function at any distance (fig. 4.46).[128]

This writer (Snyder 1987a) has presented a series of "magnifying-glass" azimuthal projections which permit the direct enlargement of a desired central portion

FIGURE 4.46. A logarithmic azimuthal projection, based on the principles presented by Hägerstrand (1957), with an enlarged region centered on Stockholm. Range 45°; 10° graticule.

of an azimuthal map whether equal-area, equidistant, stereographic, or gnomonic, and the continuation of the rest of a given region (or even the world) with a tapering of radial spacing or, optionally in the case of an equal-area or equidistant map, a single step down in area or distance scale, respectively (fig. 4.47). The central portion preserves its original characteristics, just as a magnifying glass would allow, but there is no loss of adjacent features as there is when holding a glass over a map.

New Modified Azimuthal Projections

While only the radius can vary on a true azimuthal projection, the projections loosely termed here modified azimuthals have no such restriction. As a result, numerous new projections may be described under this general heading, with special characteristics offsetting the loss of true azimuths.

FIGURE 4.47. A "magnifying-glass" azimuthal equal-area projection, centered at 85° W, 45° N, with the inner circle (radius 7°) enclosing a Lambert azimuthal equal-area projection and the band between outer (radius 20°) and inner circles enclosing an azimuthal equal-area projection with 0.25 the area scale of the inner circle. 5° graticule. Reproduced from Snyder (1987a, 64).

Retroazimuthal Projections

In contrast to an azimuthal projection, a retroazimuthal projection provides the true azimuth *of* the center *from* all other points. Prior to 1909, the only such projection developed was by Littrow in 1833, and it was not known to have this feature until 1890. The Littrow and some other retroazimuthals do not resemble an azimuthal in appearance; others do in part.

The first and best-known retroazimuthal designed as such is often called the Mecca projection because James I. Craig (1868–1952) of the Survey Department in Cairo, Egypt, designed it in 1909 with Mecca as the center (Craig 1910). Limited

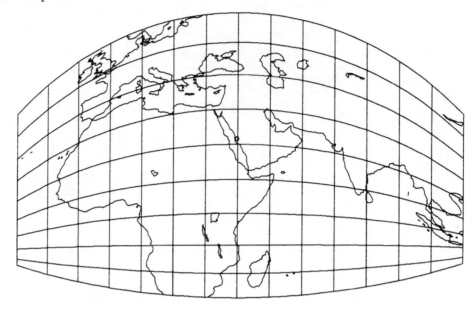

FIGURE 4.48. The Craig Mecca projection, retroazimuthal centered on Mecca, giving the direction relative to north from any other point to Mecca as the angle that a straight line connecting the two points makes with the vertical. Extending the map much beyond this range leads to great crowding and overlapping. 10° graticule.

to the predominantly Moslem world, it enables a worshiper to determine the direction of Mecca, so that the appropriate direction can be faced during prayers. The meridians (fig. 4.48) are equidistant vertical straight lines, and the parallels are curves. The true direction to the center is the angle between the meridian of the user and a straight line connecting the user's location with the central point. Edgar N. Gilbert (ca. 1973) of Bell Telephone Laboratories in New Jersey satirically used the projection for a map centered on Wall Street in New York City, as the Mecca of the financial world (see also Gardner 1975). The formulas are as follows:

$$x = R \lambda, \text{ and}$$
$$y = RB (\sin \phi \cos \lambda - \cos \phi \tan \phi_1),$$

where $B = \lambda/\sin \lambda$. If $\lambda = 0$, the equation for B is indeterminate, but $B = 1$. Axes pass through the center (ϕ_1, $\lambda = 0$). The projection is almost useless much beyond the range shown in figure 4.48, and beyond a hemisphere centered on the focus there is overlapping.

Almost simultaneous with Craig's work was Hammer's (1910) presentation of a very different retroazimuthal projection with curved meridians and parallels and correct distance along the retroazimuthal line. In this case the retroazimuth is the angle between a vertical line through the user's point (not along a meridian) and the straight line to the center. Hammer's illustration was limited to 4,000 km from center, but Hinks (1929) and E. A. Reeves (1929) presented a retroazimuthal projection of the entire world (replotted in fig. 4.49) using the same projection independent of Hammer's paper, although mentioning Craig. Hinks and Reeves chose as their center Rugby, England, the source of a powerful radio signal, thus helping the user to point an antenna to receive the signal. Beyond 90° in longitude (not distance) from the center, this projection is plotted backward and, unless the center is on the equator, partially overlaps the nearer hemisphere. The formulas for the Hammer retroazimuthal resemble those for the azimuthal equidistant projection and may be written

$$x = RK \cos \phi_1 \sin \lambda, \text{ and}$$

$$y = -RK (\sin \phi_1 \cos \phi - \cos \phi_1 \sin \phi \cos \lambda),$$

where $K = z/\sin z$, and

$$\cos z = \sin \phi_1 \sin \phi + \cos \phi_1 \cos \phi \cos \lambda.$$

If $\cos z = 1$, then $K = 1$. If $\cos z = -1$, for the antipode of the center, the point is plotted as a bounding circle of radius πR.

In 1913 Carl Schoy of Germany introduced a projection somewhat like Craig's but not retroazimuthal. While it shows correct azimuths from the center, the map distance from the center is affected by azimuth as well as globe distance.[129] On the Schoy projection all meridians are equidistant parallel straight lines. For formulas,

$$x = R \lambda, \text{ and}$$

$$y = RB (\cos \phi_1 \tan \phi - \sin \phi_1 \cos \lambda),$$

where $B = \lambda/\sin \lambda$. But if $\lambda = 0$, $B = 1$.

Shortly afterward, in 1919, Maurer, who invented the two-point azimuthal projection in 1914 (see below), invented a two-point retroazimuthal projection but required that both focal points be on the equator and less than 90° apart.[130] Limited to half a hemisphere bounded by the poles and two meridians 90° apart, the projection has straight parallel meridians and shows parallels of latitude as semiellipses that would pass through the two focal points if completed:

$$x = \sin 2\lambda/\sin 2\lambda_1, \text{ and}$$

$$y = 2 \sin \phi (\cos^2\lambda - \sin^2\lambda)/\sin 2\lambda_1,$$

where retroazimuths are correct to two points on the equator with longitudes λ_1 to either side of the central meridian.

The retroazimuthal concept was applied to a polar azimuthal projection by Arden-Close (1938). With meridians drawn as straight lines radiating at their true angles from the pole, the retroazimuthal center is marked on one of the meridians at the true-scale distance from the pole. The parallels are then placed on the other meridians so that their retroazimuths relative to the actual north direction of the meridian are correct. Beyond 90° from the retroazimuthal center the projection folds back, as it does on Craig's. J. E. Jackson (1968) pointed to new retroazimuthals with other properties, such as one like Craig's but with meridians spaced from the

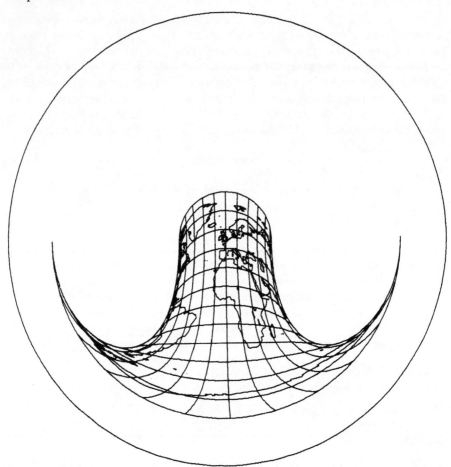

FIGURE 4.49. A reconstruction of the (*a*) inner and (*b*) outer hemispheres of the Hammer retroazimuthal projection as applied independently by Hinks (1929) with Reeves (1929) to a world map centered on Rugby, England, from which all distances are true to scale. Directions from north to point antennas to receive Rugby radio transmissions can be determined as angles from vertical for the straight connecting lines. Maps (*a*) and (*b*) should be superimposed; note that (*b*) has a reversed map. 15° graticule.

central meridian in proportion to the sine of the longitude difference rather than equidistantly. The parallels become ellipses. He also described another that is azimuthal and a third that is stereographic.

More recently, Yasuchi Hagiwara (1984) of Japan devised a retroazimuthal projection in somewhat the form of an oblique equidistant cylindrical projection with the retroazimuthal center along the north pole-line of the base map. The retroazimuth and distance can be obtained from the rectangular coordinates, as Botley had done for azimuth and distance with the actual oblique equidistant cylindrical some forty years earlier.

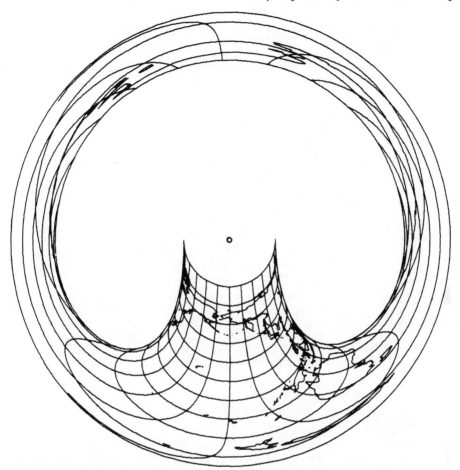

FIGURE 4.49. Continued.

Retroazimuthal maps are little more than curiosities. The ordinary stereographic projection, centered on the point to which azimuths are desired, can serve the same purpose, except that the azimuth to the center must be measured with respect to the inclined meridians, rather than from the vertical, and the distance is not correct as it is on Hammer's retroazimuthal.

Winkel's Tripel Projection

When Winkel described his two combination pseudocylindrical projections in 1921, he also included a third world map which he called tripel, normally meaning triple (fig. 4.50). Like his others, this projection is an arithmetic mean between two older projections. In this case, the mean applies to both x- and y-coordinates of an equirectangular projection and the Aitoff projection introduced in 1889. For his example, Winkel made the standard parallel of the equirectangular the 50°28′ (arccos

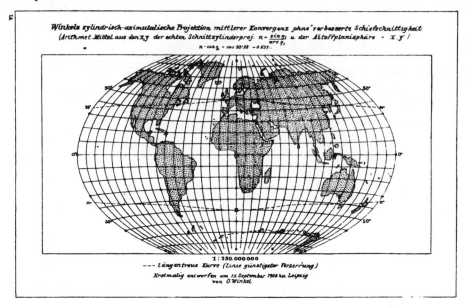

FIGURE 4.50. The Winkel tripel projection used in *The Times Atlas*. Reproduced from Winkel (1921, fig. VI). An averaging of the Aitoff (fig. 3.26) and equirectangular (fig. 1.4) projections, the latter with standard parallels 50°28′ N and S (40° in *The Times Atlas*). Central meridian 10° E; 10° graticule. *U.S. Geological Survey Library.*

$(2/\pi)$) used in his other examples, but the tripel projection itself does not have a standard parallel.

This is the Aitoff projection based on modifying the azimuthal equidistant projection, not the Hammer or Hammer-Aitoff formed by modifying the Lambert azimuthal equal-area. Winkel makes this clear, but *The Times Atlas,* a principal user since its first edition of 1958, wrongly states that the Lambert and the Hammer are involved, even though the atlas maps are correctly constructed, using 40° as the standard parallel for the equirectangular.[131]

The tripel projection is neither equal-area nor conformal. The parallels are curved except for the straight poles and equator. For an equirectangular standard parallel ϕ_1, the poles are $\cos \phi_1/(1 + \cos \phi_1)$ as long as the equator, or 0.389 for Winkel's example and 0.434 for *The Times Atlas* form. The equator and central meridian are equidistantly marked for meridians and parallels, respectively, but at two different scales. (These spacings would not be uniform if the Hammer were used.) The tripel projection also appeared in wall maps of the world published by Wenschow Map Company of Germany and for several climate maps in a high school textbook of 1959.[132] For the formulas:

$$x = (R/2)[\lambda \cos \phi_1 + 2B \cos \phi \sin (\lambda/2)], \text{ and}$$
$$y = (R/2)(\phi + B \sin \phi),$$

where $B = a/\sin a$, and $\cos a = \cos \phi \cos (\lambda/2)$, but if $a = 0$, $B = 1$.

Projections with Two or More Focal Points

Both Maurer of Germany and Close of England realized at about the same time that a projection could have either correct azimuths or correct distances from two points, rather than being azimuthal *and* equidistant from just one point. Maurer was first with each type, but the years of invention were 1914–22, when technical communications between the two countries were understandably limited.

The two-point azimuthal projection (fig. 4.51) was developed first, and Maurer's (1914) initial paper was followed by several other German commentaries in 1919 discussing it. Unaware of all these, Close (1922) independently reinvented the projection.[133] Maurer pointed out that the two-point azimuthal is equivalent to a tilted gnomonic projection: After drawing an oblique gnomonic with a projection center halfway between the two focal points, the gnomonic may then be compressed in a direction parallel to the line connecting the two points so that the distances in that direction are multiplied by the cosine of half the great-circle distance between the

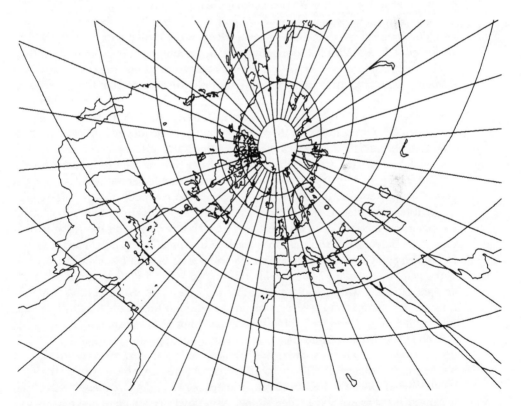

FIGURE 4.51. A two-point azimuthal projection of less than a hemisphere giving true azimuths from either Washington or Moscow to any other point as the angle between a straight connecting line and the north meridian through the initial city. All great-circle paths are shown as straight lines between any points shown. Equivalent to an oblique gnomonic projection compressed in one direction. 10° graticule.

focal points, from which all azimuths are to be correct. The compressed form is a two-point azimuthal projection, all great-circle paths remain straight lines, and the projection is limited to less than one hemisphere.

The two-point equidistant projection, in which all distances are true from two chosen points, is not so easily related to other projections. Maurer's (1919a) description was soon followed by that of Close (1921). As an example of its utility, Maurer said the projection could be used "to read on a map, true to scale, by how many sea-miles the distances from one place in the Atlantic to a place in the Pacific differ when traveling via Panama, Cape Horn or the Magellan Strait."[134] The National Geographic Society has used the projection, said to be "pioneered by the Society," for most of their maps of Asia since 1959, although the focal points at 35° N, 40° E and 35° N, 140° E are chosen near the extremities of Asia to reduce overall distortion, not to serve as origins for distance measurement (National Geographic 1959; 1990, pls. 71, 72). The straight lines of true scale provide great-circle distances but are almost never along great-circle paths.

Close (1934) extended the two-point equidistant projection to the entire world, placing his two focal points 90° apart along the equator (reconstructed in fig. 4.52). The two points may be placed anywhere, but, as Tobler (1966b) showed, the map is bounded by an ellipse of increasing eccentricity as the focal points are more widely separated. Close (1935) subsequently gave an example of a two-point azimuthal-equidistant projection, which he had mentioned earlier (Close 1922) and on which distances are correct from one point and azimuths are correct from the other point.

The principle of the two-point equidistant projection has led to two other projections with long-term commercial usage by the organizations in which they were developed. The first of these is the Chamberlin trimetric, devised in 1946 by Wellman Chamberlin (1908–1976), a member of the National Geographic Society staff beginning in 1935. He was chief cartographer from 1964 until his retirement in 1971 (*New York Times* 1976). The geometric design is simple. Three widely separated points are chosen near the edges of the region to be mapped, and the three great-circle distances connecting the points are drawn to scale as straight lines to connect the points on the map. The true distance to each other point on the globe is then plotted on the map as a short arc from each of the three base points. The three arcs for a given point form a small near-triangle, of which the center is made the location of the point on the map.[135] Graphically, this was done just for intersections of meridians and parallels, with shorelines and other features interpolated, but this principle is used mathematically for every point in computer plotting, using lengthy formulas such as those developed by George Bynum (1978) of Mobil Corporation.

The Chamberlin trimetric projection is thus an approximate three-point equidistant projection (an exact one is impossible) and is a low-error compromise used because of its balance of distortion, which has only recently been studied in detail (Bretterbauer 1989; Christensen 1992). Except between the three base points, true distance cannot reasonably be determined. First used for a map of northern North America in 1947, it has become the standard projection for the National Geographic *Atlas* and wall maps of all continents except Asia (two-point equidistant) and Antarctica (azimuthal equidistant), and for sectional maps of South America.[136] The Na-

FIGURE 4.52. A reconstruction of a two-point equidistant projection of the world with the same 15° graticule presented by Close (1934) and giving correct distances from two points on the equator at 0° and 90° E longitude to any other point as the length of a straight connecting line. These lines are not great-circle paths, except along the equator here.

tional Geographic map of North America served as the base for two sectional maps of the continent appearing in all the editions of *The Times Atlas*.[137]

A direct but more complicated outgrowth of the two-point equidistant projection for the sphere is a form developed for the earth as an ellipsoid by Jay K. Donald (1957) of American Telephone and Telegraph. The Donald elliptic projection was used by the Bell Telephone system in grid form to establish the distance component of long-distance rates in southern Canada and the forty-eight contiguous United States. The Clarke 1866 ellipsoid in use for North America was first transformed onto an authalic or equal-area sphere (instead of a conformal sphere discussed above), but the latitude adjustment was altered empirically to maintain true distance from the equator along the meridians. A map of southern Canada and the forty-eight states was then prepared on the two-point equidistant projection of this special sphere with focal points in the eastern and western United States. After other adjustments, a central point in each city was given a pair of coordinates, and the calculated diagonals provided distances for grouping into rate zones.

The principle of multiple focal points for distance measurement was carried further by Tobler (1977) and Aribert Peters (1984), son of Arno, using least squares and the speed of modern computers to devise an "empirical" minimum-error projection of the United States and the world, respectively. Tobler minimized the error in measured distances between sixty-five points located at every 5° of latitude and longitude in the region; Peters used thirty thousand points worldwide. Finally there is the polyfocal projection of Shlomi and Kadmon, but this is discussed under cartograms.

Modifications of the Hammer Projection

Hammer's (1892) stretching of the equatorial Lambert azimuthal equal-area projection may not have been the basis of the Winkel tripel projection, but it was the springboard for several other innovations. For two elliptical maps each covering 205°43′ (8/7 of 180°) of longitude, or slightly more than an equatorially centered hemisphere, Rosén in 1926 modified the equatorial Lambert so that the meridians are given 8/7 instead of Hammer's two times their original values before elongating horizontal coordinates by the same ratio.[138] Eckert, who by then used the hyphenated surname Eckert-Greifendorff (1935), proposed 4 instead of Hammer's 2 as the multiplying factor. This produced an equal-area world map (for example, fig. 4.53) with almost straight parallels, looking more like the quartic authalic projection, which is actually the limiting case when the factor is made infinity.

Besides his six new pseudocylindrical projections, Karlheinz Wagner developed two new projections from the Hammer and one from the Aitoff. In the order he

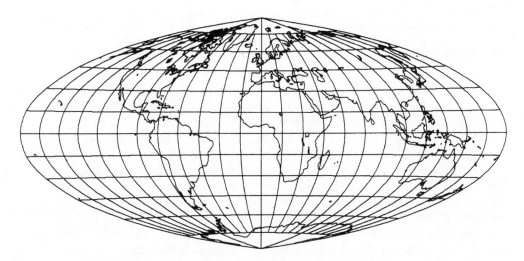

FIGURE 4.53. The Eckert-Greifendorff modified Hammer projection. Equal-area with nearly straight parallels, after a 2:1 expansion of part of the Hammer projection (fig. 3.27). Central meridian 0°; 15° graticule.

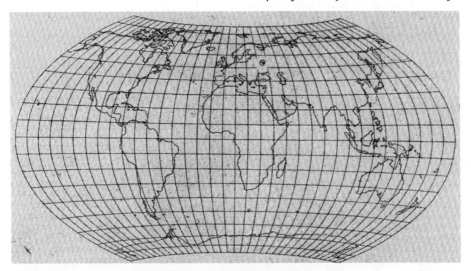

FIGURE 4.54. The Wagner VII or Hammer-Wagner projection, an equal-area modification of the portion of the Hammer projection (fig. 3.27) between 65° N and S latitudes. Reproduced from K.-H. Wagner (1949, 211). Central meridian 10° E; 10° graticule. *U.S. Geological Survey Library.*

listed all nine in 1949, his VII (fig. 4.54), first presented in 1941, is equal-area and is titled the Hammer projection with pole-line (equal-area) by Wagner and Hammer-Wagner by Maling.[139] It has curved poles corresponding to the 65° parallels on the Hammer (or Lambert), and the poles are therefore convex toward the equator. Each meridian is plotted along the Lambert meridian that corresponds to ⅓ of the longitude difference (rather than Hammer's ½), with horizontal stretching such that the resulting equator is made twice as long as the central meridian; neither is equally marked off.

Wagner (1941) preferred 65° and ⅓ as parameters over three alternatives which he also illustrated, together with transverse and oblique examples of each, in his paper.[140] It may be constructed with the following formulas:

$$x = 2.66723 \ R \cos \theta \sin (\lambda/3)/\cos (\alpha/2), \text{ and}$$
$$y = 1.24104 \ R \sin \theta/\cos (\alpha/2),$$

where $\sin \theta = \sin 65° \sin \phi$, and $\cos \alpha = \cos \theta \cos (\lambda/3)$.

The U.S. Environmental Science Services Administration (a one-time name of essentially the Coast and Geodetic Survey) used this projection for climatic world maps, and several commercial mapmakers appeared to be using this type for wall maps.[141]

For VIII, introduced in 1949, Wagner (1949, 208–12) applied a specified area distortion to VII, very much as he had done with two of his pseudocylindricals (II and V), to obtain the "Hammer projection with a pole-line and prescribed area distortion." This modification similarly shifts the parallels toward the equator without moving the poles or meridians.

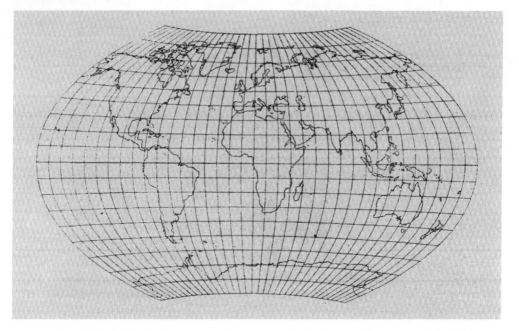

FIGURE 4.55. The Wagner IX or Aitoff-Wagner projection, modifying the portion of the Aitoff projection (fig. 3.26) between 70° N and S latitudes. Not equal-area. Reproduced from K.-H. Wagner (1949, 216). Central meridian 10° E; 10° graticule. *U.S. Geological Survey Library.*

For his IX (fig. 4.55) Wagner (1949, 214–17) adapted Aitoff's modification of the azimuthal equidistant projection to provide curved poles convex toward the equator. Wagner in effect used as his poles the 70° parallels of the equatorial azimuthal equidistant and placed his other parallels at ⅞ of the corresponding original latitudes. For meridians, Wagner multiplied those of the azimuthal equidistant by ¹⁸⁄₅. To retain true scale along the equator and central meridian, Wagner then multiplied x- and y-coordinates by ¹⁸⁄₅ and ⁹⁄₇, respectively. He called it the Aitoff projection with pole-line, and Maling (1973, 245) refers to it as the Aitoff-Wagner projection.

The oblique Hammer projection itself has appealed to several cartographers with its elliptical outline and equal-area feature. Athelstan Spilhaus (1911–) employed a transverse oblique aspect to show oceans of the world (Spilhaus 1942). He then developed a three-petaled version centered at the South Pole, with one petal centered on each of the three major oceans, the Arctic attached to the Atlantic, and the petals cut to follow continental shorelines.[142] The resulting map was called equal-area, but it was only approximately so in the far southern portions where the three petals did not precisely merge without alteration; a 1989 version left appropriate gaps to make it truly equal-area (Spilhaus and Snyder 1991). John Bartholomew's version is a symmetrical oblique Hammer, centered at latitude 45° N with the

Greenwich meridian along the vertical centerline, and called the Nordic projection; it first appeared in a 1950 atlas.[143] Bomford used both symmetrical and nonsymmetrical oblique aspects in the *American Oxford Atlas* of 1951 (Lewis and Campbell 1951, 16, A1).

William Briesemeister (1953, 1959) of the American Geographical Society, with a simple modification, turned a symmetrical oblique Hammer projection into a form subsequently called the Briesemeister projection (fig. 4.56) and touted by the inventor as "a world equal-area projection for the future." The oblique Hammer base has as its center 45° N, 10° E, with the true North Pole vertically above the center. To change the oval shapes of the parallels of latitude in the north polar regions nearly to circles, Briesemeister stretched the vertical y-coordinates and compressed the horizontal x-coordinates so that the ellipse bounding the world map has a 1.75:1 ratio of axes instead of Hammer's 2:1. The continental masses (except Antarctica) are grouped well about the projection center. First applied to a map of 1948, the projection was used for an *Atlas of Diseases* published by the American Geographical Society beginning in 1950.[144] It also appeared in several textbooks of the late

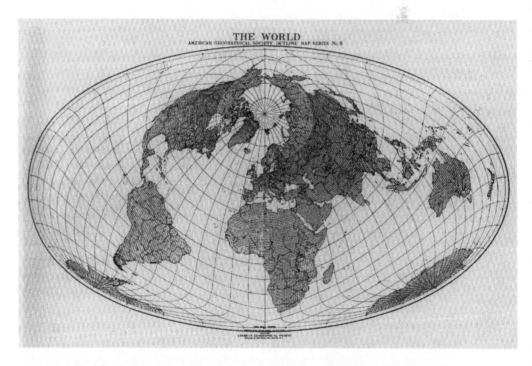

FIGURE 4.56. The Briesemeister projection, reproduced from Briesemeister (1953, 261). An equal-area oblique Hammer projection reshaped to provide axes with a 1.75:1 ratio instead of the Hammer's 2:1. The landmasses are grouped without interruption except for Antarctica. North Pole at 45° N on the central meridian of the base Hammer projection. Central meridian 10° E; 10° graticule. *U.S. Geological Survey Library, by permission of the American Geographical Society.*

1950s and replaced the Mercator as a world map projection in Hammond atlases by 1970.[145] The equations may be expressed as

$$x = R \ (3.5D)^{1/2}\cos \ \phi'\sin \ (\lambda'/2), \text{ and}$$
$$y = R \ (2D)^{1/2}\sin \ \phi'/1.75^{1/2},$$

where $D = 2/[1 + \cos \ \phi'\cos \ (\lambda'/2)]$,
$\sin \ \phi' = [\sin \ \phi - \cos \ \phi \cos \ (\lambda - 10°)]/2^{1/2}$,
and $\cos \ \lambda' = [\sin \ \phi + \cos \ \phi \cos \ (\lambda - 10°)]/(2^{1/2}\cos \ \phi')$.

The axes cross at the projection center. The world map is bounded by an ellipse having $R\sqrt{7}$ as the semimajor axis and $4R/\sqrt{7}$ as the semiminor axis.

The concept of expanding and contracting the Lambert azimuthal equal-area and Hammer projections to obtain equal-area projections with rearranged distortion patterns and elliptical boundaries of various eccentricities seemed to have considerable attraction. It also appealed to Solov'ev by 1946[146] and later to Maling (1962), Tobler (1962b), and O. M. Miller (1965), with Tobler, for example, calling attention to the fact that a world map based on the Lambert may be made elliptical rather than circular and the elliptical Mollweide may be made circular.

Other Modifications of the Lambert Azimuthal Equal-Area and Azimuthal Equidistant Projections

Although the derivatives of the Hammer owed their beginnings to Lambert's 1772 azimuthal equal-area projection, they were actually derivatives of a derivative; there were several other projections relating directly to the Lambert. Some used a hemisphere of it as the centerpiece of a world projection, with all or part of the other hemisphere as interrupted points or lobes, and others altered the projection itself.

Under the first category, Maurer produced an equal-area star projection in 1935, inspired by the star projections of the nineteenth century, but using six equal lobes with curved sides appended to a north polar Lambert hemisphere.[147] On the lobes, the parallels of latitude continue as concentric circular arcs with radii spacing in mirror image to the spacing of the corresponding northern parallels, and meridians are spaced along each southern parallel at the same distances as those used along the corresponding northern parallel. Maurer also produced a star with equidistantly spaced parallels and six equal points, but consisting entirely of broken straight meridians and parallels, with a hexagon for the northern hemisphere; this is not based on the Lambert, but is more like Jäger's 1865 star projection.[148]

C. B. Fawcett (1883–1952) of England combined the Lambert as an inner hemisphere with an equal-area nonazimuthal projection for lobes of the other hemisphere (Fawcett 1949). In the polar form, Fawcett's construction of the projection was identical to that of Maurer's equal-area star, except that Fawcett used a total of three unequal lobes to portray southern continents to 60° S; he used a south polar Lambert for the more southern regions. Fawcett added an example centered on London, with an oblique Lambert for the inner hemisphere. In this case only two lobes were needed for South America and Australia, the spacing on each lobe an oblique transformation of the spacing on the lobes used in the polar form. Oceans are incomplete in both versions. Fawcett's equal-area oblique map resembles in shape Frye's equidistant world map of 1895.

Proposing a polar projection only, William William-Olsson (1903–1990) of Sweden produced another lobed modification (William-Olsson 1968). Using the polar Lambert azimuthal equal-area from the North Pole only to latitude 20° N, he appended four identical lobes extending to the South Pole, with central meridians starting from longitude 65° W. For parallels, the lobes have equidistantly spaced circular arcs centered on the North Pole, with spacing equal to that between parallels at 20° N on the Lambert, or 0.81915 (cosine (70°/2)) of true spacing. To retain equality of area and a fit to the polar projection, the meridians on the lobes are spaced from the four central meridians at secant (70°/2) or 1.22077 times true spacing. Thus the lobes are constructed using a projection resembling the Bonne, but centered on the North Pole, compressed north to south and expanded east to west.

A more complicated alteration of the Lambert azimuthal equal-area projection was developed by this writer (Snyder 1988), who had been seeking an equal-area counterpart of some of the modified stereographic conformal projections developed previously. Calling the series oblated equal-area (after the Miller oblated stereographic, below), this writer obtained either oval, rectangular (fig. 4.57), or rhombic lines of constant distortion, depending on the chosen values of two constants, but with rounded corners on the rectangles and rhombi. Whereas the oblique Lambert azimuthal equal-area projection is ideal for maps of circular regions, the oblated equal-area series has less overall distortion for oblong or rectangular regions. A further much more complicated modification was principally developed in 1984 prior to the above by John A. Dyer, professor of mathematics at Southern University in Louisiana, while working with this writer, who developed applications (Dyer and Snyder 1989). This projection type involves multiple equal-area compressions, ro-

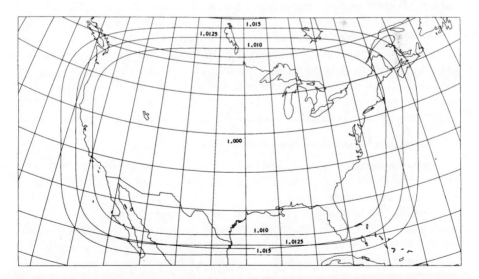

FIGURE 4.57. An oblated equal-area projection of the United States with three near-rectangular lines of constant distortion. Reproduced from Snyder (1988, 352). A modification of the Lambert azimuthal equal-area projection, centered at 39° N, 96° W, to reduce overall distortion in the rectangular region. 5° graticule.

tations, and shearing of the base projection in conjunction with least squares to achieve a potentially minimum-error equal-area projection of a given region with lines of constant distortion tending to emulate the shape of the region. This projection was applied specifically to an outline map of the irregularly shaped Alaska, beginning with the oblique Lambert azimuthal equal-area as a base.

Still more remotely resembling the Lambert, but using it as a starting point, is a series of projections suitable for displaying the surfaces of irregularly shaped satellites of some of the planets, such as Phobos and Deimos of Mars. Philip J. Stooke of Canada developed the projections somewhat empirically, but they are superior to the commonly used orthographic projection (Stooke 1989; Stooke and Keller 1990).

Beginning with the azimuthal equidistant projection, John Bartholomew's (1942, 2–3) so-called tetrahedral projection was used as two small insets for a main world map on the Kite projection (discussed later), to show landmasses in a northern form and oceans in a southern version. The central portion, from the pole to about the tropic line (23 ½° N or S) of the same hemisphere, is drawn to the polar azimuthal equidistant projection. Beyond the Tropic and extending to the opposite pole are three equal lobes based on interrupted Werner projections, adjusted to match the meridian spacing at the Tropic.

One world map for displaying routes of the air age was called the "Matter-most map," presented by S. C. Gilfillan (1946) and James Wray of the University of Chicago. The central portion consists of an oblique azimuthal equidistant projection centered on Paris, but extending only 75° in radius. Australia and the remainder of South America were appended with two legs called equidistant equal-area, probably an adapted Bonne; much of the Pacific Ocean and all of Antarctica were omitted.

Arthur Davies (1949) produced another interrupted modification of the oblique azimuthal equidistant projection, this one centered on the Balearic Islands. He too stressed landmasses (again most of the Pacific Ocean is missing, and Antarctica is separated but shown). Continental masses more than 90° from this center were appended with two legs having, in the north polar base first prepared, equidistant circular arcs for southern latitudes and meridians drawn as straight lines between their locations on the equator of the polar base and on the southernmost parallel of the extension. After the oblique graticule was plotted graphically from the polar base, true distance from the Balearics was preserved throughout the contiguous portion.

New Pseudoazimuthal Projections

The rare pseudoazimuthal classification of map projections is characterized in the polar aspect by concentric circles for parallels of latitude, but curved instead of straight meridians. Only the Wiechel (1879) equal-area pseudoazimuthal projection is known to have appeared prior to 1950. In 1951, however, the *American Oxford Atlas* displayed a "modified zenithal [or azimuthal] equidistant projection" for maps of each of the three largest oceans (Lewis and Campbell 1951, 3, 7, 12–15). Designed by Bomford, the projection results from curving the meridians on a polar azimuthal equidistant projection "so as to give approximately equal scale error all round a bounding oval" (Lewis and Campbell 1951, 7). Distance from the center remains correct, but the scale along the circular parallels of latitude, normally constant at a given latitude, is now constant along an oval touching the center of each of the four sides of a rectangular map. The adjusted polar aspect fitting a given rectangle is then given a standard transformation to produce the equatorial aspect used for each of the oceans. Mathematical details were not given, but equations could be similar to those just below for the Ginzburg pseudoazimuthal, except that $\rho = Rz$, while $q = 2$ and C and z_n are selected to suit.

A pseudoazimuthal projection series resembling Bomford's in principle but more general and mathematically explicit was developed by Ginzburg (1952a).[149] Called an oblique TsNIIGAiK projection with oval isocols or "kosaya proyektsiya TsNIIGAiK s oval'nymi izokolami," where *isocol* is a commonly used Russian term for lines of constant distortion, the projection (fig. 4.58) was used in one form for a combined map of the Atlantic and Arctic oceans in the various editions of *Atlas mira* beginning in 1954.[150] In its general form, polar coordinates (ρ, θ) may be expressed as

$$\rho = KR \sin (z/K), \text{ and}$$
$$\theta = Az - C(z/z_n) \sin (q\ Az),$$

where z and Az are great-circle distance and azimuth from south, respectively, of points relative to the center of the projection. For the above ocean map, $C = 0.1$,

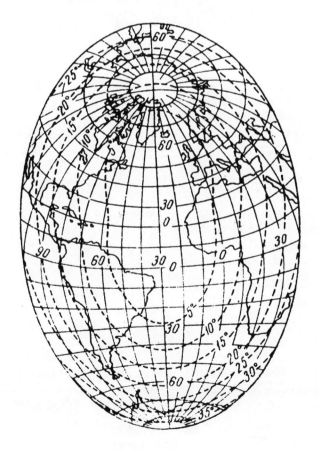

FIGURE 4.58. Ginzburg's oblique TsNIIGAiK projection with oval isocols or lines of constant distortion, applied to a map of the Atlantic and Arctic oceans. Reproduced from Grauer (1956, 66). A pseudoazimuthal projection centered at 20° N, 30° W; 10° graticule. *U.S. Geological Survey Library.*

$q = 2$, $K = 3$, central lat./long. = 20° N/30° W, and range $z_n = 120°$.[151] The properties are arbitrary, neither equal-area nor conformal, but the center has no distortion.

Modifications of the Stereographic Projection

The two short formulas known as the Cauchy-Riemann equations, which provide necessary and sufficient conditions to establish conformality (p. 65), led to a means of modifying the stereographic (or any other conformal) projection so that conformality is maintained but the resulting projection provides less distortion for a non-circular region. Since these are partial differential equations,

$$\partial x/\partial x' = \partial y/\partial y', \text{ and}$$
$$\partial x/\partial y' = -\partial y/\partial x',$$

where (x', y') are the old and (x, y) are the new rectangular coordinates, an integrated solution is useful for practical applications. One such solution was presented by Ludovic Driencourt and Laborde (1932, 4:202). It is a compactly stated series involving complex algebra (i.e., both real and imaginary numbers) that fully satisfies the Cauchy-Riemann equations and permits the formation of an endless number of new conformal map projections when the constants are changed. It may be written

$$x + iy = \sum_{j=1}^{n} (A_j + iB_j)(x' + iy')^j,$$

where n is any chosen positive integer indicating the highest order or power of the equation, i is $\sqrt{-1}$, Σ is a summation, and (A_j, B_j) is a set of coefficients ranging from $j = 1$ to $j = n$. This is not an infinite series, but is exactly conformal even if n is 2 or 3. (If n is 1, the original projection is retained but may be rotated.)

Apparently the first practical application of this series was by Laborde himself in 1926 in developing the projection for Madagascar, mathematically a modified transverse Mercator using $n = 3$ in the above equation and letting (x', y') be the coordinates of the transverse Mercator base (p. 162). The first such modification of the stereographic was made by O. M. Miller (1953) to develop a lower-distortion conformal map combining Europe and Africa (fig. 4.59). Also using a third-order ($n = 3$) form of the above equation, Miller in effect chose $A_1 = 0.9245$, $A_3 = 0.01943$, and zero for A_2 and all B coefficients. The complex equation then simplifies to the pair presented by Miller (with symbol changes):

$$x = Kx'\{1 - (Q/12)[3(y')^2 - (x')^2]\}, \text{ and}$$
$$y = Ky'\{1 + (Q/12)[3(x')^2 - (y')^2]\},$$

where $K = 0.9245$ and $Q = 0.2522$. Miller's base projection for which (x', y') were calculated was the oblique stereographic projection with its center at latitude 18° N and longitude 20° E. As a result, the lines of constant scale are ovals, following the general combined shapes of the two continents, and reducing overall scale variation below that of the base stereographic projection. When the map was prepared on contract to the Army Map Service (AMS), the central meridian was shifted from 20° to 18° E.

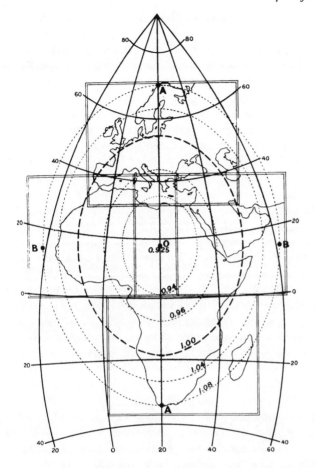

FIGURE 4.59. The Miller oblated stereographic projection. Reproduced from Miller (1953, 407). He presented it to reduce overall distortion for a map of Africa and Europe. A conformal modification of the oblique stereographic projection centered at 18° N, 20° E; 20° graticule. Scale is constant along dashed ovals, with scale factor as shown. *U.S. Geological Survey Library, by permission of the American Geographical Society.*

The AMS then asked Miller to extend the mapping to include the rest of the continental masses in the eastern hemisphere, but without altering the map of Europe and Africa. This imposed a considerable restraint on him, but Miller (1955) soon described in detail the very complicated solution to the problem. He decided to use three separate conformal projections, each with oval lines of constant scale to suit respectively Europe and Africa as before, Central Asia, and Australasia, the latter two regions having oblique ovals. Because these projections are distinct, they do not fit together conformally, so Miller connected them with four nonconformal fill-in zones, making seven zones in all. Much of the former Soviet Union, Iran, and Indonesia are in the nonconformal sections. The maps were prepared in sheets, all labeled Miller oblated stereographic projection, at a scale of 1:5,000,000, and were used as bases for many other maps, so that by the 1980s computer programming was needed and supplied for this potpourri of distortion characteristics (Sprinsky and Snyder 1986). The same projection, but labeled Miller's prolated stereographic, is used in *The Times Atlas* for the map of Africa, and with its more common name in the Rand McNally *New International Atlas* for Europe and sectional maps of Africa.[152]

Lee (1974) of New Zealand followed Miller's third-order approach with oblique ovals for a map of the Pacific Ocean. A decade later, this writer (Snyder 1984b), utilizing the speed of computer plotting, determined complex coefficients for conformal map projections that are modified stereographics with nearly regular polygons, ovals, or rectangles of constant scale near the desired limits of the map. All the polygons and rectangles have rounded corners, but the enclosed map content is shown with minimum overall error for a conformal map in accordance with the Chebyshev-Grave principle (pp. 140–41).

A more complicated use of the Driencourt-Laborde series, practical only in an electronic-computer age, was described by W. I. Reilly (1973) of New Zealand and this writer (Snyder 1984a, 1986).[153] Both used least squares to determine optimum conformal projections based on the ellipsoid for irregularly shaped regions. Reilly in effect began with the regular Mercator projection and obtained sixth-order coefficients for the new official New Zealand Map Grid. This writer began with the oblique stereographic and calculated tenth-order coefficients for a map of the fifty United States in their correct relative locations having a scale range of less than 4% (rather than a range of 15% using a standard Lambert conformal conic projection). He then determined sixth-order coefficients for a conformal map of Alaska with less than 0.6% scale range (versus 2.4% using the conic). The U.S. Geological Survey continues to use a so-called modified transverse Mercator projection for the sheet map of Alaska; neither conformal nor equal-area, it more nearly resembles an equidistant conic projection (Snyder 1987b, 64–65).

In 1992, Hammond Inc. (1993; press release; pers. com., C. Dean Hammond, 1992) announced "a revolutionary new projection: The Hammond Optimal Conformal," developed by Mitchell J. Feigenbaum (1944–), professor of physics at Rockefeller University, New York, and renowned for work on the predictability of chaos and fractal geometry. Said to produce "the most accurate maps that can *ever* be made," the projection is used for continental maps in Hammond's new world atlas. With oval or other closed loops of constant scale, one of which follows a smoothed shape of the particular continent, the Hammond optimal conformal projection is a more thorough extension of the work of Miller and this writer discussed above on the modified stereographic. The arbitrary degree of smoothing of continental shapes for the lines of constant scale was needed to minimize the effects of singularities and to make computation feasible. The projection is an encouraging computer-age advance in commercial projection choice.

New Conic Projections

Although most twentieth-century developments of conic projections established criteria to determine optimum standard parallels for common projections (pp. 175–77), a small number of basically different conic projections have been proposed. Except for the modified polyconics, none of them has apparently been used for finished maps.

First was a minimum-error conic described by Young (1920, 22–23).[154] Young had comprehensively studied several low-error map projections, especially azimuthal and conic. The formulas for his minimum-error conic are too difficult to justify regular use, especially since there is almost negligible difference from the equidistant conic. Inspired by Airy's (1861) approach, Young developed formulas requiring numerical integration for the computation of the radius of each parallel, after computation of the cone constant for a given pair of limiting parallels. Young acknowledged the impracticality and suggested Murdoch III as the best low-error conic.[155]

When this writer developed the cylindrical satellite-tracking projections in 1977, he also included a series of conic projections with the same purpose, that of showing Landsat mapping-satellite groundtracks as straight lines. Instead of two standard parallels at true scale, there are two parallels at which conformality is maintained, but only one can also have true scale.

Likewise, Byron Adams (1984), in presenting his transverse cylindrical stereographic projection of the ellipsoid, also included a stereographic conic projection applied to the ellipsoid. Unlike Braun's (1867) conic projection of the sphere from the South Pole onto a cone tangent at latitude 30° N for a world map, Adams proposed a regional map (befitting the use of the ellipsoid) projected from the point on the ellipsoid opposite the point of tangency of the cone suitable to the region. Like his transverse cylindrical, this conic is neither equal-area nor conformal.

Modified Polyconic Projections

All three of the projections that formally include the name *polyconic* and that were developed before 1900 originated in the U.S. Coast Survey. In this century the name *modified polyconic* has been applied to projections developed in several other countries. The first and best known was devised by Charles Lallemand of France. In 1909 it was adopted by the International Map Committee in London as the basis for the 1:1,000,000-scale International Map of the World (IMW) series.[156] Used for sheets 6° of longitude by 4° of latitude between latitudes 60° N and S, and for 12°- and 24°-wide sheets more poleward, the projection differs from the ordinary polyconic projection in two principal features: all meridians are straight, and there are two meridians (2° east and west of the central meridian on the sheets between 60° N and S) that are made true to scale. All parallels of latitude are constructed as nonconcentric circular arcs with the same radii they would have on the ordinary polyconic. The meridians are marked off true to scale on the top and bottom parallels of each sheet, and the true-scale markings along the two "standard" meridians determine the spacing of all parallels.

The 1909 conference said that in view of the scale,

> several suitable projections differ but little from each other, and that the contraction and expansion of the paper on which the map is printed affect all lengths on the map, and prevent it from being in fact exactly either orthomorphic or equivalent, it is not necessary to lay great stress upon the selection of a projection which has the best properties as to conformity or equivalence. (United Nations 1954, 35)

Therefore an easily constructed, edge-matching (but not mosaickable) projection was selected. This compromise projection was used for all maps prepared until 1962, when the Lambert conformal conic with two standard parallels replaced the modified polyconic as the recommended but not required projection (United Nations 1963, 9–10). Meanwhile, the projection had been the topic for several papers, including an intense international debate on its merits in 1927 between Hinks (English) and Antoni Łomnicki of Poland.[157] The formulas were available in outline form for many years and more recently have been given in detail for computer programming.[158]

McCaw's (1921) modified rectangular polyconic is approximately conformal, with scale errors comparable to those of the transverse Mercator. The concept of equal-area polyconics was mathematically analyzed by O. S. Adams (1919a, 114–18), but he found nothing worthy of further consideration. Using a general formula by Tissot, Maurer presented an equal-area polyconic in 1935 with formulas requiring iteration; like other polyconics it has a central meridian true to scale and distortion-free, while each parallel of latitude is a circular arc of true curvature.[159]

Albert H. Bumstead, the first chief cartographer of the National Geographic Society, developed another equal-area polyconic that was used for maps of South America in 1937 and 1942 (in the next revision of 1950, the projection was the Chamberlin trimetric). Details are unclear, but it was described as

> a new projection upon which the outlines of the [South American] continent and the countries appear more nearly as they would on a globe than on any other projection so far devised. . . . The continent . . . was divided into zones running east and west, each having a width of five degrees of latitude. Each zone is given its own two standard meridians, placed to reduce the distortion of that zone to a minimum. (Grosvenor 1937, 810)

Parallels of latitude have true curvature, and spacings of meridians *and* parallels were adjusted to provide correct area scale.

There are numerous Russian modified polyconics. Some eight are symmetrical about the equator, including seven with nonconcentric circular arcs for all parallels. Two of these were presented in 1937, one by the Khar'kov Engineering-Construction Institute (KhISI) and the other by V. D. Taich.[160] Four others have been used by TsNIIGAiK for maps of the world on projections first described by Ginzburg in 1949 and 1951, and another for school maps of the Soviet Union on a projection presented by Salmanova in 1951.[161] Ginzburg and Salmanova used Urmayev's method of calculation (p. 179) for cylindrical and other projections with prescribed scale errors at certain points. For definition, equations for the polyconic projections need only give the location of each circular parallel along the central meridian (at true scale on some of the projections) and along the meridian 180° from center. The other meridians are equally spaced along each parallel.

The four polyconic projections of Ginzburg, using Maling's numbering, are successive versions used for world maps in the *Geograficheskiy atlas* for secondary-school teachers (1939–49) (Ginzburg IV), for school maps in 1950 (V), in the *Bol'shoy sovetskiy entsiklopediya* (*BSE*) (2d ed. 1950) (VI) (fig. 4.60), and in the

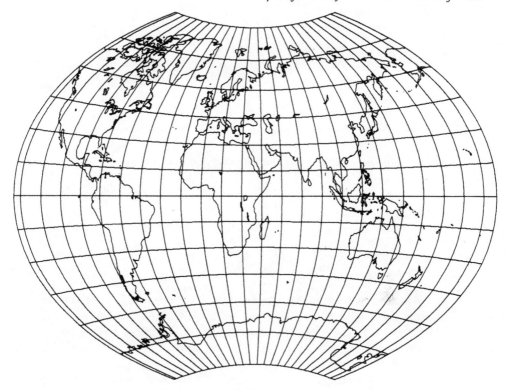

FIGURE 4.60. The Ginzburg VI or BSE modified polyconic projection of the world. Central meridian 40° E as used in the Soviet Union; 15° graticule.

Atlas mira of 1954 (VII, the only one of these projections not also symmetrical about the central meridian); a fifth presented in 1966 has been used for various thematic world maps.[162] Formulas provided for the *BSE* projection (VI), but with radians instead of the degrees of the Russian formulas, are as follows:[163] For latitude ϕ at the central meridian,

$$x = 0, \text{ and}$$
$$y = R(\phi + 0.045\phi^3).$$

For the 180th meridian east of center,

$$x = R(5\pi/6 - 0.62636\phi^2 - 0.0344\phi^4), \text{ and}$$
$$y = R(1.3493\phi - 0.05524\phi^3).$$

For the other Ginzburg polyconic projections, additional series terms appear to be needed to match coordinates provided. Philip Voxland (pers. com., 1991), using least squares and omitting asymmetric projection VII, has derived the following coefficients (rounded to five significant figures) to fit published tables of coordinates:

	IV	V	(1966)
a	2.8284	2.5838	2.6516
b	− 1.6988	− 0.83584	− 0.76534
c	0.75432	0.17037	0.19123
d	− 0.18071	− 0.038097	− 0.047094
e	1.76003	1.54331	1.36289
f	− 0.38914	− 0.41143	− 0.13965
g	0.042555	0.082742	0.031762

These coefficients are for use in the following equations: Along the central meridian,

$$x = 0, \text{ and}$$
$$y = R\phi.$$

For the 180th meridian from the center,

$$x = R(a + b\phi^2 + c\phi^4 + d\phi^6), \text{ and}$$
$$y = R(e\phi + f\phi^3 + g\phi^5).$$

Five other polyconics are asymmetrical about a curved equator and were apparently also presented by Ginzburg (1952b, 68–72, 81–82) between 1934 and 1944. The first two have a pointed North Pole, as has a similar-looking 1949 projection with noncircular curved parallels (thus not strictly polyconic) by A. I. Mikhaylov (Ginzburg 1952b, 68, 72–73).

A modified rectangular polyconic projection was designed about 1950 by G. E. Bousfield of the former Energy, Mines and Resources Canada for the portions of Canada north of latitude 80° N.[164] Initially given that name, it has also been called a modified polyconic or the Bousfield projection. It was designed to match the Lambert conformal conic with standard parallels at 49° and 77° N used for the general mapping of Canada (on the Clarke 1866 ellipsoid), in order to reduce excessive scale variations that would occur if the latter were used in the extreme north. The lengthy analytical formulas were recently derived by G. V. Haines (1981, 1987) of the Geological Survey of Canada.

Oblique and Transverse Conic Projections

Oblique conic projections have the same theoretical advantages as oblique aspects of other projections, namely, moving lines of low distortion to follow the regions of special interest. Like other oblique aspects, they are complicated, especially without automatic computers. Few have been formally promoted, but a double one became important through its use by the American Geographical Society.

The first to appear was a transverse polyconic developed by Charles H. Deetz (1864–1946) of the U.S. Coast and Geodetic Survey and apparently first described in 1919.[165] The poles on the base polyconic projection lie on the equator of the globe, and the poles of the globe fall on the equator of the polyconic base. For his example, Deetz chose longitude 160° W as his central meridian coinciding with the equator of the polyconic base, and the central meridian of the polyconic base was made the great circle perpendicular to this meridian at latitude 45° N, thus providing a map recommended for the northern Pacific Ocean. The National Geographic Society used the projection for a combined map of Eurasia and the West Pacific in 1942, with a map center of 95° E and 40° N instead of the one above; they used it more recently, with a different center, for their maps of the Soviet Union.[166] As a compromise projection, neither area nor local shape is preserved, and linear scale is preserved only along lines that are not easily discerned.

Proposed oblique conics have generally been either conformal or equal-area. Craster (1938) proposed an oblique conformal conic projection with its two standard parallels rotated to follow the general curve of the two principal islands of New Zealand, while Solov'ev (1937; 2d ed., 1946, 202, 210) proposed a single standard transformed parallel with an oblique conformal conic projection for the Soviet Union. These were not adopted, but the same type of projection (with two standard transformed parallels) is used for many of the maps of Japan in that country's *National Atlas*.[167]

A more complicated oblique conic appeared double as the bipolar oblique conic conformal projection (figs. 4.61, 4.62), developed by O. M. Miller (1941) and Briesemeister of the American Geographical Society. Prepared strictly for a combined map of North and South America, it accommodated the tendency of North America to curve toward the east as one proceeds from north to south, while South America tends to curve in the opposite direction. To construct the map, a great circle arc 104° long was first selected, beginning at latitude 20° S and longitude 110° W, cutting through Central America from southwest to northeast, and terminating at 45° N and the resulting longitude a bit east of 20° W.

The initial point is the pole and the center of transformed parallels of latitude for an oblique conformal conic projection, with two standard transformed parallels at distances from this pole of 31° and 73°, for all the land in the Americas southeast of the 104° great-circle arc. The other end of the 104° arc serves as the pole and center of transformed parallels for an identically shaped oblique conic projection for all land northwest of the same arc. The inner and outer standard transformed parallels of the northwest portion are therefore tangent to the outer and inner standard transformed parallels, respectively, of the southeast portion, touching at the dividing line. The rest of the two conic projections do not match exactly, so a smoothing was applied, resulting in a narrow nonconformal strip along the 104° arc.

Because of the high quality of the map of the Americas prepared in sheets at a scale of 1:5,000,000 by the American Geographical Society, others used these as bases, like the later maps on the Miller oblated stereographic projection discussed above, for the preparation of other maps; analytical formulas were developed in the late 1970s (Snyder 1987b, 117–23). The U.S. Geological Survey used the base maps and the projection for several maps of North America, while maps of South America were prepared by the Pan-American Institute of Geography and History and by Rand McNally, the latter for sectional maps of the continent in their *New International Atlas;* the National Geographic used the base maps for a 1979 map of bird migration in the Americas.[168] When this projection is used for the map of only one continent, it would be more precise to use a "mono"-polar oblique conic, thus avoiding the fitting problem along the 104° arc, but the quality and availability of the American Geographical Society maps were overriding factors.

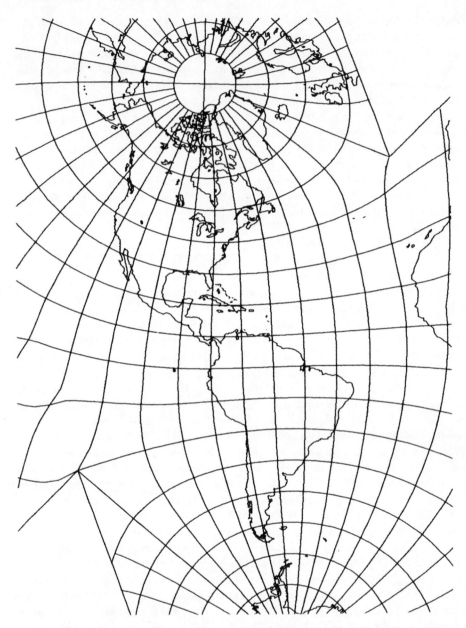

FIGURE 4.61. The bipolar oblique conic conformal projection devised by Miller and Briese-meister for a map of the Americas, in outline form and including the two poles of the conic projections in the upper right and lower left. 10° graticule.

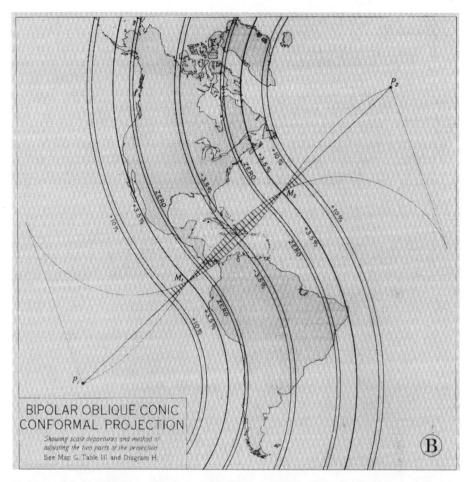

FIGURE 4.62. Scale patterns for the bipolar oblique conic conformal projection reproduced from O. M. Miller (1941, facing 104). The narrow band of nonconformality is shown with diagonal lines across the center. *U.S. Geological Survey Library, by permission of the American Geographical Society.*

Some oblique equal-area conic (oblique Albers) projections have been described in addition to Hammer's (1889) earlier use. One was by M. Hunter (Tobler 1963b) for a map of the China coast, and another was by John E. Westfall (1970) to show the ancient *oikumene* of Europe, the Near East, India, and the Orient with reduced scale variation.

Pseudoconic Projections

The only significant pseudoconic projections have been equal-area—the so-called Bonne projection and its earlier polar-limiting form, the Werner, with their concentric circular arcs for parallels and curved meridians. Several of the twentieth-century pseudoconic developments have included the Bonne or Werner in a specially interrupted form, often in conjunction with other projections. William-Olsson (1968) has already been mentioned.

The first innovator was Wilhelm Schjerning (1862–1917), a geographer who taught in and near Berlin, finally serving as director of Kaiser Wilhelm-Realgymnasium, Berlin (*Petermanns* 1917). In 1904 he proposed six projections that he developed as modifications of the azimuthal equidistant projection of the entire world.[169] Most are pseudoconical, and all are world maps with true linear scale from the projection center. For each projection in its polar aspect, Schjerning kept the parallels of latitude as equidistant concentric circular arcs; the meridians, however, instead of being the equidistant radial straight lines of the polar azimuthal equidistant projection, are spaced along the parallels equidistantly, but at intervals depending on the projection.

For Schjerning's first projection (actually first presented in 1882), the spacing of meridians is half that of the azimuthal equidistant, so the projection is an equidistant conic with standard parallels at 90° N and 18°36′ S, a world in a semicircle and a specific case of Mendeleev's later 1907 approach. In projection II, the boundary of the map on the azimuthal equidistant base in the first step of the transformation is an ellipse with the North Pole at the center, the minor axis extending vertically to the equator, and the major axis horizontally to the South Pole. The northern hemisphere remains a polar azimuthal equidistant projection, while the meridians in the southern hemisphere are spaced so that 180° of longitude are included in the arcs of the parallels within each half of the ellipse. Finally the graticule is rotated on the globe to move the map center from the pole to the equator on the Greenwich meridian; the right half of the ellipse is enlarged with an unspecified delineation to improve the shape of Asia.

In projection III (fig. 4.63), the map boundary is a pair of circles, each with a

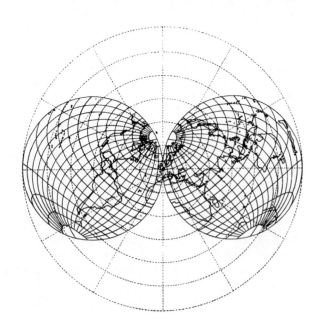

FIGURE 4.63. The Schjerning III projection, showing the world bounded by two circles. This is an adaptation of the azimuthal equidistant projection centered at London. Scale but not direction is correct from London to all other points on the map. Reproduced from Zöppritz and Bludau (1912, 195). 10° graticule. *Library of Congress.*

diameter initially extending horizontally from the North Pole of the polar azimuthal equidistant projection to the South Pole at the left and right edges, respectively. After meridians are spaced on each parallel so that 180° fits inside each circle, the center of the projection is shifted to London with an oblique transformation, and the Greenwich and 180th meridians form the outline of each circle.

In projection IV, the meridians are initially spaced at their true distances along each parallel, starting from a central meridian straight down from the North Pole of the polar azimuthal equidistant. This is the same as the Werner projection, but Schjerning then moved London to the pole of the projection with a central Greenwich meridian, and thus ended with an oblique heart-shaped equal-area Werner. Projection V is the same as IV, except that the meridians are spaced at *half* their true distances and the graticule remains centered at the North Pole. The map is thus beet-shaped and at a constant area scale, but is nowhere free of local shape distortion.

Schjerning's projection VI (fig. 4.64) is like the normal Werner of IV, but centered at the South Pole and interrupted to consist of three continuous, symmetrical, "flower-petal" lobes extending to the North Pole and interrupted at longitudes 10° E, 150° E, and 70° W to emphasize the three major oceans. The map is extended in places to complete ocean outlines. Some of Schjerning's projections are clever and attractive, but the complexity of construction in spite of simple principles helped negate their practicality.[170]

After interrupting pseudocylindrical projections as described above, Goode (1929a, 1929b) adapted the Werner projection in 1928 to a polar petallike equal-area projection more complicated than Schjerning's VI. The parallels are all equally

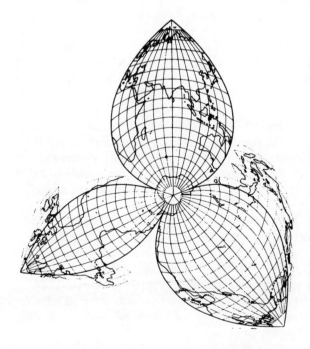

FIGURE 4.64. The Schjerning VI equal-area projection of the world, an interrupted three-lobed Werner projection centered at the South Pole to emphasize oceans. 10° graticule. Reproduced from Zöppritz and Bludau (1912, 201). *Library of Congress.*

spaced circular arcs centered on the North Pole. North America and Eurasia have separate central meridians, and other meridians are marked off true to scale on the parallels, with interruptions at longitudes 30° W and 170° W. At about latitude 10° N, North America's central meridian of 90° W ends and South America's 60° W begins. At 30° S, the central meridian shifts to 70° W. Similar shifts occur for Africa and Australia. Parallels then are marked off for meridians with respect to the new central meridians, and meridian curves are completed. The continents are thus emphasized. This projection was used not in the Rand McNally *Goode's Atlas,* but instead for land distribution and outline maps in various geography texts by P. E. James.[171]

John Bartholomew introduced four combination projections of the complete or nearly complete globe near midcentury, including the tetrahedral described above. This John was the fourth generation of John Bartholomew mapmakers. The first John (1805–1861) trained his son John (1831–1893) at his Edinburgh firm. John, Jr., then worked for Petermann in Gotha, Germany, before returning to assume management of his father's firm in 1856 and to build its reputation to an unexcelled level in Great Britain. His son John George (1860–1920) succeeded him in 1888, soon named the firm the Edinburgh Geographical Institute, and then in 1919 reorganized it as John Bartholomew and Son, Ltd. Meanwhile he developed personally and corporately an extraordinary international profile with joint publishing ventures and worldwide contacts. John George broadened the use of existing projections, but it was *his* son John (1890–1962) who developed new projections as well.[172] Fifth generation John Christopher (1923–) succeeded him, oversaw the development of The Times projection, and used a half dozen different world projections, including a new recentered Nordic, in a 1984 atlas.

Three of the Bartholomew projections are combinations of the equidistant conic with two standard parallels and the Bonne projection in interrupted form. On the "Kite" and "Regional" projections emphasizing landmasses, the standard parallels for the conic are 22°30′ N, near the Tropic of Cancer, and 67°30′ N, near the Arctic Circle (Royal Society 1966, 43). On the "Lotus" projection, emphasizing oceans, the standard parallels for the conic are 22°30′ S and 67°30′ S. As a result, the complete graticule for the conic encloses about 249° of the full circle, with the gap at longitude 165° W on the Kite, 160° W on the Regional, and 15° E on the Lotus.

On the Kite projection, first shown in the *Citizen's Atlas* of 1942, the equidistant conic is employed only between the standard parallels (J. Bartholomew 1942, 2–3). North and south of this range, the Bonne projection, its parallels concentric with those of the conic, is used with three central meridians and interruptions at 15° W, 45° E, and 165° W to the North Pole, and interruptions of 60° E, 30° W, and 165° W to the South Pole.

The Regional projection was introduced in the *Regional Atlas* of 1948 with four lobes and in *The Times Atlas* of 1958 with three lobes limited to continental coverage (J. Bartholomew 1948, 2–3; 1958, pl. 1). On it the equidistant conic extends north to 67°30′ N in the 1948 version with the Bonne to the pole, but it extends to 80° N in the 1958 form, where the region north of 80° is omitted. South of 22°30′ N, the Bonne is used with concentric parallels and the three or four interruptions.

The Lotus projection, introduced with the revised Regional projection in 1958, has the equidistant conic extending from 22°30′ S to 80° S (J. Bartholomew 1958, pl. 2). North of the conic, three Bonne-type lobes extend to the North Pole with interruptions at 15° E, 105° E, and 75° W and variations at the Gulf of Mexico and Mediterranean Sea.

D. G. Watts (1970) of England probably holds the record for the most new projections presented in one paper. All fifty-five of these are combinations of earlier projections, with Watts fully crediting the sources. Forty are modifications of pseudocylindricals to reduce distortion at the outer meridians. They are interrupted in the manner suggested for pseudocylindricals by Goode, but they cannot be readily interrupted into more than two sections in the northern hemisphere without dividing the landmasses themselves. To offset some of the remaining distortion in northwestern North America and in northeastern Asia, Watts proposed bending the normally straight parallels in those areas upward from the equator, beginning say 80° east and west of the central meridian, into concentric circular arcs. Some parallel such as 30° N is made the parallel of true curvature, and the meridians are marked off along the curved parallels at the same uniform spacings that were used along the straight parallels.

If the original projection is equal-area, the modification is also equal-area. If the original projection is the sinusoidal, the wings are Bonne projections, as in Watts's no. 16. The southern hemisphere can be a mirror image of the northern hemisphere, except for different meridians of interruption, thus requiring interruptions along the equator, as in no. 16. The southern hemisphere can also be an extension of the northern hemisphere without interruptions along the equator, but with changes in central meridians and meridians of interruptions in the southern hemisphere, as in his no. 36.

Watts began with thirteen existing pseudocylindricals. He also developed several new pseudocylindricals, some by shortening the poles of some of the Eckerts from half to a quarter of the length of the equator. The wings for the sinusoidal and Eckert I and II are existing projections, such as the Bonne. The others provide fifteen new pseudoconic projections. With the choice of base projection, curvature of outer parallels, and location of interruptions in meridians and the equator, there is considerable flexibility in producing equal-area and compromise projections with varying distortion patterns.

In 1937–38 Solov'ev proposed a modified Bonne projection for maps of the Soviet Union.[173] As the result of introducing two additional constants, this projection is no longer equal-area. It has apparently not been used in atlases; the Kavrayskiy IV form of the equidistant conic, Solov'ev's oblique perspective cylindrical, and the Salmanova modified polyconic have been used instead. The Solov'ev formulas may be written

$$x = R\rho \sin \theta, \text{ and}$$
$$y = R(\cot \phi_1 - \rho \cos \theta),$$

where $\rho = \cot \phi_1 + C_1(\phi_1 - \phi)$, $\theta = C_2 \lambda (\cos \phi)/\rho$, and ϕ_1 is the central latitude. The Bonne projection is obtained if C_1 and $C_2 = 1$. For Solov'ev's modification, $\phi_1 = 52°18′$, $C_1 = 1.02$, and $C_2 = 0.95$.

An adaptation of pseudocylindrical and pseudoconic principles is an "approximately Equal-area" (for land only) world map projection called the Oxford, devised by Bomford about 1950 and used for numerous thematic world maps in Oxford atlases.[174] There are two straight central meridians, 70° W and 40° E pole to pole, with pole-to-pole interruption and condensing in the Atlantic Ocean. The parallels between these meridians are straight, but in the "outer parts of the map the parallels have been curved away from the Equator in order to reduce distortion and fit the most distant land masses into the page" (Lewis and Campbell 1951, 6). Other meridians are curved, and the spacing is adjusted to reduce area distortion to a maximum of about 20%.

Other Projections

Most projections can be placed in the various standard categories (cylindrical, conic, pseudocylindrical, etc., or modifications) listed above, but several significant projections, as well as novelties, remain to be described separately.

Van der Grinten's Projections of the World in a Circle

One of the most popular new projections of the twentieth century was announced by Alphons J. van der Grinten (1852–1921) of Chicago in a German journal in 1904 and in an American journal the following year (van der Grinten 1904a, 1905a, 1905b). Inventing the projection in 1898, van der Grinten received patents for it in the United States and three other countries.[175] Commonly called the Van der Grinten projection (fig. 4.65) (in Russian it is regularly called the projection of Grinten), this is the first of two conventional or arbitrary projections he presented at the same time. In 1912 Alois Bludau described the two but presented two modifications of the first, which he called van der Grinten's circular (*kreisförmige*) world maps 2 (fig. 4.66) and 3 (fig. 4.67).[176] These were later changed to roman numerals, and the second original 1904 projection was given the numeral IV (fig. 4.68). No. III has been reinvented at least twice with the name *modified Van der Grinten;* it has been used moderately.[177] Geometric constructions were originally provided; formulas for all four have been recently derived for computer plotting.[178]

The first Van der Grinten projection was appealing because (1) the meridians and parallels are entirely circular arcs or straight lines, (2) the world is enclosed in a circle, and (3) the general appearance of the Mercator projection is preserved with somewhat reduced area distortion. Van der Grinten emphasized that the projection combines the Mercator appearance with the roundness of the Mollweide projection. Van der Grinten projections I, II, and IV at first glance closely resemble some of the Lagrange projection variants, but the Van der Grinten projections are not conformal, and the meridians and parallels on I and IV do not generally intersect at right angles. The ordinary Lagrange projection of the world in a circle (fig. 2.8), described first by Lambert in 1772, was discussed in some detail by van der Grinten as an introduction to his own projection. Van der Grinten III has straight parallels of latitude and is almost a pseudocylindrical projection, except that the circular meridians do not intersect most parallels at equal intervals.

On all four variants, the central meridian and equator are straight, and all other meridians are circular arcs equidistantly spaced along the equator and intersecting

FIGURE 4.65. The Van der Grinten (I) projection, showing the world in a circle, with all meridians and parallels as circular arcs, reproduced from van der Grinten (1904a, pl. 10). By far his most popular projection, used for wall maps throughout the twentieth century, partly because of its resemblance to the Mercator (fig. 1.37). It more closely resembles the "Lagrange" projection (fig. 2.8), but van der Grinten's is not conformal. Central meridian 10° E; 10° graticule. *U.S. Geological Survey Library.*

at the poles. On I, II, and III, the entire world is enclosed in a circle, the meridians are identical on all three, and the poles are points at the intersections of the central meridian with the outer circle.

On I, the parallels other than the equator are circular arcs symmetrical about the central meridian and with increasing spacing away from the equator along all meridians. The 75th parallels are only slightly more than halfway between the equator and poles on the central and outer meridians, so regions more polar than about 80° N or S are normally not included in the world map.

The parallels on both II and III intersect the central meridian at the same points that they do on I, but on II they are circular arcs curved with smaller radii to cross

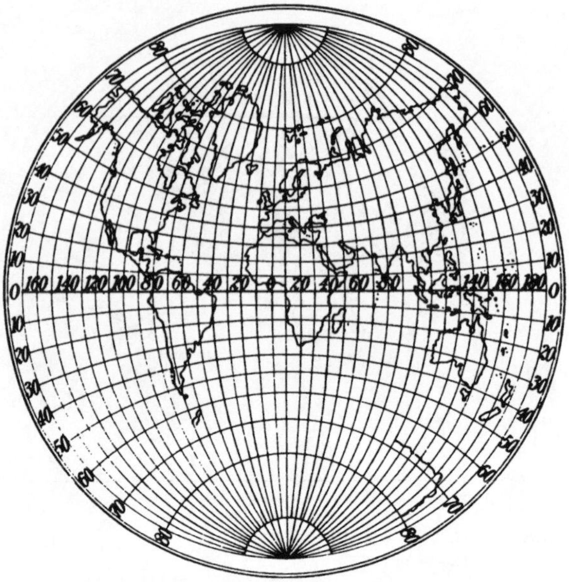

FIGURE 4.66. The Van der Grinten II projection, a modification by Bludau and similar to Van der Grinten I (fig. 4.65), but the parallels cross the meridians at right angles. Reproduced from Zöppritz and Bludau (1912, 183). Central meridian 10° E; 10° graticule. *Library of Congress.*

FIGURE 4.67. The Van der Grinten III projection, a modification by Bludau with meridians like those of Van der Grinten I (fig. 4.65) but straight lines for parallels intersecting the central meridian (10° E here) as they do on I. Reproduced from Zöppritz and Bludau (1912, 183). 10° graticule. *Library of Congress.*

each meridian at right angles, thus making the nonconformal II resemble the conformal Lagrange projection more than does I, the most notable difference being that the Lagrange meridians are not equally spaced along the equator. On III, the parallels of latitude are straight parallel lines.

On Van der Grinten IV, which the inventor called *apfelschnittförmige* or "apple-

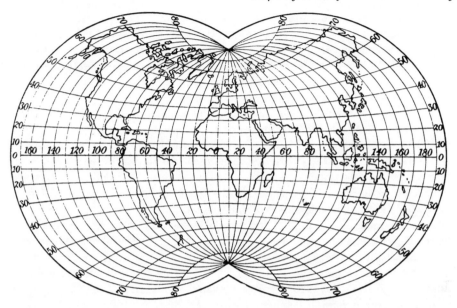

FIGURE 4.68. The Van der Grinten IV, his apple-shaped projection, with an all-circular graticule. Central meridian 10° E; 10° graticule. Reproduced from Zöppritz and Bludau (1912,
185). *Library of Congress.*

shaped," the world is enclosed in two equal intersecting circles (van der Grinten
1905b). The poles are at the intersections of these two circles, and the equator is
perpendicular to the central meridian through the center. The two circles are so
located that the central meridian is half the length of the equator; both lines are
then marked equally for parallels and meridians, respectively, and thus are true to
scale. The parallels are circular arcs. Bludau also showed how IV can be modified to
provide right-angle intersections of parallels on the same meridians (Zöppritz and
Bludau 1899; 1912 3d ed., 185).

When it was first proposed, *the* Van der Grinten (I) projection promptly pro-
voked generally favorable comment (Goode gave it "very high value") from cartographers in England, France, and the United States.[179] One of the earliest and most
persistent and prominent users was the National Geographic Society. The projection's adoption followed a request by the society's magazine editor Gilbert H. Grosvenor to staff cartographer Bumstead in 1918 for a new world map projection, saying
the "Mercator Projection is atrocious. In my office I have a MAP of the WORLD on
another projection by a Chicago man whose name [van der Grinten] I forget at this
moment, but while this is an improvement over Mercator, it is not satisfactory."[180]
Nevertheless, the Van der Grinten was adopted in 1922 for the society's official new
world map, which was revised every few years, the last revision occurring in 1985.[181]
Only polar azimuthal equidistant and equatorial Lambert azimuthal equal-area
maps, issued as adjacent hemispheres to constitute interrupted world maps, devi-
ated from this adoption, and only during the 1930s and 1940s. In 1988, the society
chose the Robinson projection to replace the Van der Grinten.

The U.S. Department of Agriculture adopted the Van der Grinten by 1949 as the base map for economic data. Consequently, between 1940 and 1960 the projection became the second most heavily used for world maps in textbooks (after the Goode homolosine). But 82% of its use, according to Frank Wong's tabulation, was due to one author, geographer Richard M. Highsmith.[182] The projection was used in atlases of 1940, 1942, and 1958 and in the first five editions of the National Geographic *Atlas of the World;* it was available in wall maps from at least five commercial mapmakers in 1988, second only to Mercator.[183] In the Soviet Union, the projection was used for numerous political and mineral world maps at scales of from 1: 32,000,000 to 1:100,000,000 between 1940 and 1950, and on other maps for over forty years.[184] In many ways, the high level of usage of the Van der Grinten projection was an extension of the acceptability to the public, even the geographic community, of the gross area distortion of the Mercator.

The "modified" Van der Grinten, his III, is the only other one of the four to have received commercial attention, mainly because of the availability of base maps from McKnight and McKnight Map Co., which apparently independently developed this projection from Van der Grinten I.[185] This projection was found next most often in the textbooks of 1940–60 after the Van der Grinten I projection, and it was principally used for land-distribution maps.

Maurer proposed in 1935 a modification of Van der Grinten II, retaining the same meridians but relocating the parallels so that the geographical poles became circular arcs falling along the 72d parallels N and S on Van der Grinten II.[186] Other parallels were correspondingly, but not proportionately, placed along former parallels closer to the equator. The right-angle crossing of meridians and parallels was thus retained, but the large polar expanse on II was substantially reduced.

A very recent projection considerably resembling Van der Grinten I, but having somewhat less distortion and no circular arcs, is by Léo Larrivée (1988). Designed for the Canadian International Development Agency, the graticule consists of complex curves, and the poles are curved lines. For its equations,

$$x = R \lambda (1 + \sqrt{\cos \phi})/2, \text{ and}$$
$$y = R \phi/[\cos (\phi/2) \cos (\lambda/6)].$$

Cartograms

Although the preservation of small shapes or of area scales can be placed in neat (if often complicated) mathematical packages for conformal or equal-area projections, respectively, there is an important category of transformation that is frequently used for practical maps of a different type. These projections, called cartograms, ordinarily display particular statistical data so that a measurable characteristic is in true proportion to some demographic, economic, or other set of data. For example, the area of a region may be shown proportional to population or to gross national product, or the distance from a center can be shown proportional to shipping costs. As with equal-area projections, there are numerous cartogram designs for any given set of data.

Cartograms as special map projections are primarily twentieth-century developments. The term *cartogram* was coined about 1860 with a variously shaded map

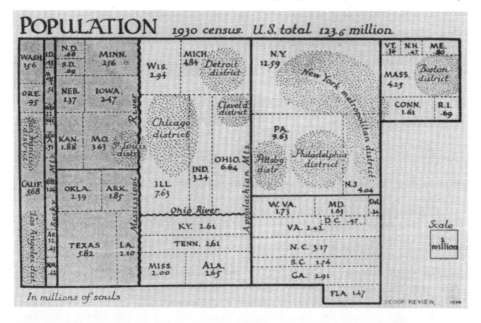

FIGURE 4.69. One of the earliest cartograms published in the United States with areas of regions in proportion to a non-land-area variable, in this case the 1930 populations of states. Reproduced from Raisz (1934, 293). *U.S. Geological Survey Library, by permission of the American Geographical Society.*

of illiteracy in France by Charles Dupin, and furthered in 1874 by Georg Mayr of Bavaria, who used the term in a paper about the graphical display of statistics. All the cartograms described involved maps with shadings, isolines, added symbols, and the like, rather than reconfigured map projections.[187]

However, Maurer (1919a) referred to the two-point equidistant projection as a cartogram. Erwin Raisz (1934) initiated the development in the United States with a paper showing the U.S. states as contiguous rectangles having areas proportional to population, wealth, manufacturing production, or farm products. The states were located approximately in the proper position relative to each other (fig. 4.69), but shape was ignored, and Raisz (1934, 292) emphasized that "the statistical cartogram is not a map."

Tobler made cartograms a scientific form capable of iteration by computer into optimum maplike statistical tools. He challenged Raisz's comment, saying in 1963 that "the diagrams can be regarded as maps based on some unknown projection," and he presented a mathematical analysis of area cartograms by adapting equations for geographically equal-area map projections (Tobler 1963a, 64, 67–68, 75–78). Tobler applied these and related principles to specific programs for developing, after numerous iterations, cartograms with continuous reshaped boundaries (fig. 4.70) (Tobler 1973a, 1974a).

In the late 1970s, Eli Shlomi (1977) and Naftali Kadmon (Kadmon and Shlomi 1978) of Israel devised the polyfocal projection for similar purposes. Here, the rect-

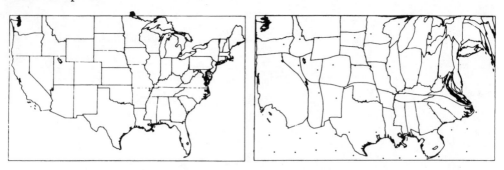

FIGURE 4.70. A map showing the contiguous United States (left) on an equirectangular projection, and with areas in proportion to their populations (right) with more realistic boundaries than those of fig. 4.69, after several iterations. Reproduced from originals for Tobler (1973a, 217). *By permission of Waldo R. Tobler and the New York Academy of Sciences.*

angular coordinates of a given standard map projection are modified to present a new map with various enlarged ("fish-eye") centers in any selected locations and with any selected enlargement, separate for each center.

Although the two-dimensional quality of area has predominated in cartograms, distance cartograms have also been employed, often including correct azimuths, to show, for example, true or perceived directions from a central point and distances proportional to actual road mileage or parcel post rates (Monmonier 1990, 17). Carrying perception still further, comical area cartograms have depicted the United States as viewed by a stereotyped Texan or New Yorker,[188] and serious distance cartograms are used to make subway routes more unstandable to users, with both distance and direction considerably distorted and with few nearby features.[189]

Octant Maps

Leonardo da Vinci and others in the sixteenth century had constructed world maps using curved equilateral triangular octants of the globe, each bounded by two meridians 90° apart from pole to equator and by the equator. Use of octants was revived for this purpose in the twentieth century, especially by California architect Bernard J. S. Cahill. In 1909 he presented and in 1913 patented a world map with the octants arranged to resemble a butterfly (Cahill 1909, 1913a, 1913b). Thus began a thirty-year effort through numerous papers to promote this "butterfly map" arrangement as "a world map to end world maps," quoting the title of one of Cahill's (1934) papers; he especially stressed the potential for weather maps.[190] Bounding his lobes longitudinally with 22°30′ W, and with each 90° interval from there minimizing interruptions of land, he used his own arbitrary projection for the principal version, on which the bounding meridians and equator were combinations of straight, circular, and otherwise curved lines, not quite filling out the equilateral triangles of an octahedron. He also published variants filling these triangles, and onto which the globe was projected conformally or gnomonically (fig. 4.71); for the conformal variant he credited O. S. Adams with the mathematical design, calling the variant "after 25 years' study, perfected for the use of meteorologists. The author firmly believes

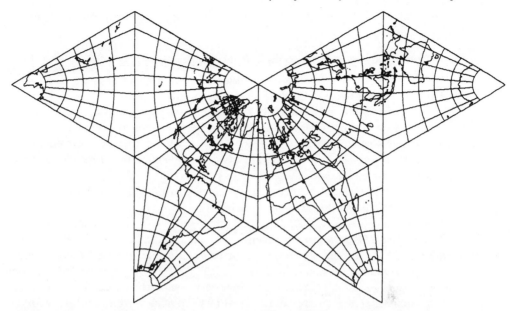

FIGURE 4.71. An outline reconstruction of the Cahill butterfly map of the world in its simplest variant, using the gnomonic projection on each of the eight triangular faces of an octahedron. 15° graticule.

that no better world map is possible for this purpose."[191] In spite of Cahill's confidence and long-term efforts, the projection was almost unused.

A second revival of the octant map was less belabored, but was accompanied by similar claims. In 1943, L. E. Pitner published a small atlas with octants closely resembling Lambert azimuthal equal-area projections centered at the octant center, or latitude 35°16′ N or S (arccot $\sqrt{2}$) (Ward 1943). There were two patterns: on one the octants were clustered about the nearest pole (like those of Leonardo, fig. 1.30), and on the other they straddled a straight line along the equator. Applying the name Octovue projection, Pitner claimed that "size, shape, position, distance, and area are shown more nearly correct than ever before on a flat map of the world. . . . practically free from distortion" (Ward 1943, 2).

Maurer's Globular Projections

In 1922, at a time when globular projections were declining in popularity, Maurer proposed some more.[192] Two may be used for the entire world, with the straight central meridian and equator true to scale and all other meridians and parallels arcs of circles, spaced equally on the central meridian and equator. The maps are bordered by portions of two circles, identical to the bounds of Van der Grinten IV. In Maurer's "full-globular" projection, the parallels intersect the outer meridian at equal intervals between the equator and high and low points of the circles, the poles becoming curved lines. In his "apparent-globular," the parallels are somewhat larger in radius, striking outer meridians at ¾ of the distance from the equator to the intersection with the latitude on the "full-globular." Maurer's third globular was called "all-globular."

Conformal World Maps as Mathematical Challenges

The most extensive treatment of the subject of map projections as a mathematical challenge without necessarily a practical outcome was to devise conformal projections for hemispheres or world maps bounded by simple geometric figures such as regular polygons, rhombi, or ellipses. This work was partially undertaken by Peirce, Guyou, and August in the late nineteenth century when they conformally placed hemispheres in squares or the world in a two-cusped epicycloid, using elliptic functions and complex algebra.

During the twentieth century, the principal cartographic practitioners of this esoteric specialty within the field of projections were O. S. Adams of the United States and L. P. Lee of New Zealand, both of whom held similar positions in government agencies. Oscar Sherman Adams was geodetic computer and mathematician with the U.S. Coast and Geodetic Survey from 1910 until his retirement in 1944. Born in 1874 in Mt. Vernon, Ohio, where he died in 1962, he received degrees from Kenyon College in Ohio, including an honorary doctorate in 1922. He served as a school superintendent and principal in several Ohio school systems just before moving to the Washington, D.C., area. In addition to his numerous publications on projections and his work developing the North American Datum of 1927 and the State Plane Coordinate System, he was president of local scientific associations.[193]

First, Adams (1925, 72–111) described the derivation of conformal projections of the sphere within a regular hexagon, a rhombus, a six-pointed star, a rectangle, and an ellipse, as well as projections of the hemisphere in a rhombus, triangle, and square, the latter a different aspect of the projections of Peirce and Guyou. The mathematics included Jacobian and Dixon elliptic functions, Weierstrassian and Abelian functions, the Schwarz integral for a conformal circle mapped in a regular polygon, and Lagrange's conformal sphere within a circle. Adams (1929, 1936) later presented a conformal world map in a square with the poles in the diagonals and then in the middle of opposite sides.

Lee (1965, 1976) carried Adams's work further, showing the conformal world map in an equilateral triangle (fig. 4.72) and ellipse (with due credit to less accurate earlier attempts), a hemisphere in a rectangle, and the world conformally on the faces of all five of the regular Platonic polyhedra (tetrahedron, cube, etc.). Laurence Patrick Lee (1913–1985) was born in England but early moved to New Zealand. His career from 1936 to retirement in 1974 was with the New Zealand Department of Lands and Survey, where he was chief computer during the last ten years and was involved in various geodetic surveys of the country. A prolific and profound writer on the topic of map projections, Lee was also active in several cartographic and survey organizations (*New Zealand* 1985).

Although these projections are truly conformal, they generally have a singular point at each vertex of the bounding polygon and often elsewhere as well, with wide ranges of scale. They principally remain admirable mathematical achievements and artistic curiosities.

Less complicated and more practical was a transverse oblique August conformal

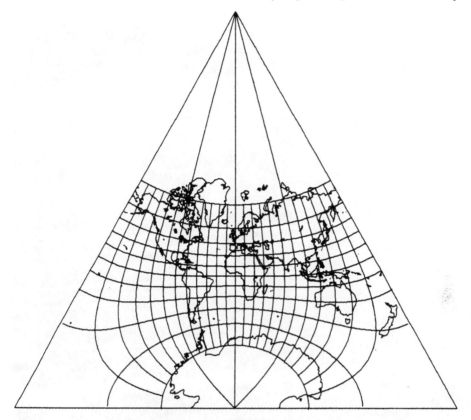

FIGURE 4.72. A reconstruction of Lee's conformal map of the world in an equilateral triangle. A novelty and mathematical exercise. Central meridian 0°; 15° graticule.

ocean map prepared by Spilhaus (1942) to accompany his transverse oblique Hammer equal-area ocean map. Both maps were centered at 70° S, 15° E, with longitudes 15° E and 165° W coinciding with the equator of the base projection, and both thereby showed the oceans with reduced interruption.

Raisz's Orthoapsidal Double Projections

Erwin Josephus Raisz (1893–1968; fig. 4.73) (pronounced "raw' eess"), mentioned above for introducing cartograms to the United States, is especially noted as the author of the first textbook in English on cartography. His *General Cartography* appeared first in 1938. It was revised in 1948 and superseded by his *Principles of Cartography* in 1962. While each edition includes a significant discussion of map projections, Raisz (1943) originated a series of projections that he called orthoapsidal, "derived from 'orthographic' and 'apsidal.'" As he later summarized the principle (Raisz 1962, 181), "we draw a parallel-meridian network on any suitable solid other than a sphere and then make an orthographic projection of that."

Raisz presented examples with the world map on half a tilted ellipsoid, on an

FIGURE 4.73. Erwin J. Raisz (1893–1968), author of the first textbook on cartography in English, pioneer in cartograms, and inventor of orthoapsidal projections, such as the armadillo. *Photo courtesy of Robert B. McMaster.*

interrupted hyperboloid, and on half a tilted torus ring (with no central hole). The latter, with a tilt of 20°, he named the armadillo projection (fig. 4.74) because of the resemblance of its graticule to the animal in a coiled position. The parallels on the torus (but not on the final projection) are equally spaced, with the poles as semicircles running along the top and bottom. The meridians are equally spaced around half the circumference of the torus, normally with longitude 10° E as the central meridian. After the torus is tilted and projected orthographically, the meridians and parallels appear as elliptical arcs. The southern polar region is hidden from view, but New Zealand is occasionally separately appended as a pigtail. The inventor said the armadillo projection has a relatively high land-to-sea ratio for a world map (Raisz 1938; 1948 2d ed., 84). It was used to a limited extent in a 1944 atlas by Raisz, in some textbooks beginning in 1951, and as an inspiration for Jacques Bertin, who devised modifications in 1950–51.[194]

Born in Löcse, Hungary (now Levoča, in Slovakia), Raisz worked as an engineer in Budapest until moving to the United States in 1923. He received a doctorate in geology from Columbia University in 1929 and a year later joined the new Institute of Geographical Exploration at Harvard. There he taught cartography until 1950,

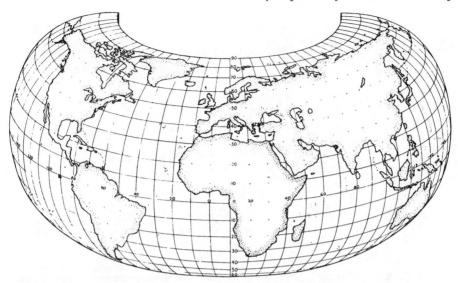

FIGURE 4.74. The armadillo projection as presented and reproduced from Raisz (1943, 133). Given the name because the shape resembles the curled animal, it is an orthographic projection of a world map placed on part of a torus ring. Central meridian 10° E; 10° graticule. *U.S. Geological Survey Library, by permission of the American Geographical Society.*

when the Institute closed, and he continued cartographic experiments and lecturing elsewhere (Robinson 1970).

More Polyhedral Globes

A polyhedron approximating a globe was the subject of two special promotions during the twentieth century, both involving patents of the late 1940s and one remaining commercially available for the rest of the century.[195] Irving Fisher (1867–1947), a noted Yale economist, used a regular icosahedron (with twenty equilateral triangles) for a world map, flat or folded, and based on the gnomonic projection (fig. 4.75). First presented in 1943, it was patented in 1948.[196] The poles were generally placed at opposite vertices, with one edge extending from the North Pole along longitude 5°06' E (Fisher 1943, 610). Construction is relatively simple, but use of the gnomonic projection leads to a 58% range of scale.

With much more intensive promotion, prolific inventor and innovator R. Buckminster Fuller (1943) unveiled his Dymaxion map as a cutout in the widely read *Life* magazine. The polyhedron he used is a cuboctahedron, with eight equilateral triangles and six squares, onto which the globe is projected with a special geometric construction that results in constant scale along all edges. It is described in detail in his patent of 1946, in which he too claims distortion "less than with any form of projection heretofore known"; on the other hand, his *Life* map states that linear scale varies by 50% (Fuller 1943, 1946). He later replaced the cuboctahedron with the icosahedron, adapting the low-distortion projection he had used previously.[197] Devo-

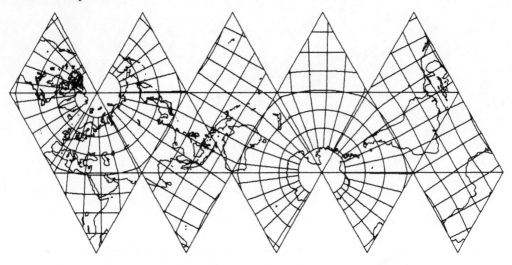

FIGURE 4.75. A reconstruction of Fisher's gnomonic projection of the world onto an icosahedron, with the poles at vertices and meridians as placed by Fisher. 15° graticule.

tees were still touting his claims of minimal distortion in 1989, in spite of the many interruptions of the flattened polyhedron.[198]

Fascination with polyhedral faces was also expressed by Cahill beginning in 1909 and Lee beginning in 1965, as noted earlier, as well as by Hinks (1921) on the cube, J. A. Smith (1939) and Stefania Gurba (1970) on the dodecahedron, several others on the icosahedron, Botley (1949, 1954) on the tetrahedron and octahedron, R. Clark (1977) on the octahedron, A. J. Potter (1925) on the tetrahedron, and Charles Burky (1934–35) on the truncated icosahedron (with thirty-two pentagons and hexagons).[199] Most projections onto the polyhedra were gnomonic, but Cahill, Fuller, and Lee used conformal or arbitrary projections as noted above, and A. D. Bradley and Fisher devised two versions of an equal-area icosahedron.[200]

Gilbert (ca. 1973) used the round globe itself as the basis for projection, but with two complete worlds conformally mapped onto one sphere. The principles developed by Lagrange in 1779 were used, and, as on Lagrange's flat map of the world in a circle, there is serious area distortion in polar regions. Gilbert's work inspired Alan A. DeLucia to present an orthographic view of the double-world globe, tilted so that its center is 5° N and 5° E.[201] Essentially the complete world is shown, reminiscent of the familiar Mercator, but with a globular appearance and reduced polar distortion.

Still Other New Projections

Some teachers of college-level cartography and geography found a need to develop still other world map forms combining several existing projections. The projection names could be lengthy (Paul S. Anderson's 1974 oblique, poly-cylindrical, orthographic azimuthal equidistant projection, or Borden D. Dent's 1987 poly-centered oblique orthographic) or short (R. R. Macdonald's 1968 optimum continental, a combination of a rectangular polyconic, equidistant conic, Bonne, and equirectangular). The connections between multiple projections varied from mathematical contiguity through vague fill-ins to clearly discontinuous map sections.

Bertin (1967, 294–95) developed some innovations between 1950 and 1955. J. G. P. Baker's (1986) dinomic projection consists of a Mercator projection between latitudes 45° N and S, but a more gradual spacing of parallels beyond there to pointed poles, the meridians converging in the latter regions with interruptions to preserve continental forms (except for split Antarctica).

In addition to various cylindrical and polyconic projections described earlier, in which polynomial equations are used to determine spacings, such equations have also been used to develop pseudocylindrical and general world map projections with reduced error.[202]

For numerous other projections, details are obscure or unavailable to this writer. The index of a recent *Bibliography of Map Projections* lists some such projections under the names *geomatic* (by J. R. Schwendeman), Křovák, Karel Kuchař, A. Milne, F. J. Müller, A. P. Petrenko, and *pseudostereographic* (Snyder and Steward 1988, 103–10).

Attempts to Classify Projections

Maurer, in his 1935 *Ebene Kugelbilder*, had as his primary goal the classification of nearly all prior projections into families reminiscent of those developed in the eighteenth century by the Swedish botanist Carolus Linnaeus (1707–1778) for plants and animals (Maurer 1935, 7–16; 1968 trans., 1–22). In the process Maurer found a number of gaps. While this could present a problem to a botanist or biologist looking for species, Maurer often filled such a gap by developing a new projection, which he did in over two dozen cases. He pointed out, however, that he wanted to "ward off the likely reproach that I have suggested a huge number of useless projections" (1935, 8; 1968 trans., 2–3). Except for certain designated items, generally mentioned above, these were not serious proposals, but were intended to complete, elucidate, or explain.

Maurer also felt moved during these early years of Nazi control of Germany to use "German terms . . . for these designations wherever possible, and germanizations for many of the foreign terms have been suggested."[203] He itemized his preferences, for example, for *Kegel, Säule, flächentreu,* and *winkeltreu* instead of *Konus, Zylinder, äquivalent,* and *konform,* respectively. A longtime official in the German navy, Maurer committed suicide in 1945, a few months after the Nazi collapse.[204]

Maurer was only the most detailed of the classifiers of map projections. As the number of projections multiplied, so papers devoted to classification proliferated, aside from sections of other works, such as d'Avezac-Macaya (1863, 473–85). The *Bibliography of Map Projections* lists twenty-two such papers plus Maurer's monograph, all dating from 1897, but only seven predate 1950.[205] Nine are in English and seven in Russian, with four other languages for the remaining seven; the most often referenced papers in English are those by Lee (1944) and Tobler (1962a).

GENERAL WORKS AND JOURNALS

In the nineteenth century, detailed treatises on map projections multiplied relative only to those of earlier centuries; the total for the century was some ten significant works, eight of them in German or French. After 1900, about three dozen book-length studies and a dozen more popular books can be listed, all concentrating on the one subject of map projections and appearing in a combined total of a dozen languages.

In English, eight advanced studies, some with numerous revisions, have been frequently cited. In chronological order of first edition, these are Hinks (1912), *Map Projections* (2 eds., 1912, 1921); Deetz and Adams (1934), *Elements of Map Projection* (5 eds., 1921–44); Steers (1970), *An Introduction to the Study of Map Projections* (15 eds., 1927–70); Melluish (1931), *An Introduction to the Mathematics of Map Projections;* Kellaway (1946), *Map Projections* (2 eds., 1946, 1949); Richardus and Adler (1972), *Map Projections for Geodesists, Cartographers and Geographers;* Maling (1973), *Coordinate Systems and Map Projections* (2 eds., 1973, 1992); and Snyder (1982, 1987b), *Map Projections Used by the U.S. Geological Survey* (2 eds.) plus its revision *Map Projections: A Working Manual.* In addition, works by Hendrikz (1946–47), Pearson (1977), Lauf (1983), Canters and Decleir (1989) and Bugayevskiy and Snyder (1995) have been very recent or have had less impact. More specialized phases of the subject have appeared in lengthy treatises by O. S. Adams (1919a, 1925), Young (1920), Thomas (1952), Lee (1976), and Snyder (1985a).

More popularly oriented or elementary studies in English include those written by Morrison (1901), Garnett (1914), Mainwaring (1942), Fisher and Miller (1944), Chamberlin (1947), Merriman (1947), McDonnell (1979), and Snyder and Voxland (1989). These listings do not include important chapters of cartography textbooks and of other volumes, for example, the detailed exposition by Close and Clarke in the 1911 to 1926 editions of the *Encyclopaedia Britannica.* This entry was abridged in the 1929–60 editions and printings, and was replaced during 1962–72 with a less technical discussion by O. M. Miller. Thereafter the entry was reduced to less than a page written by others.[206]

There were at least three significant French treatises: Duchesne's (1907) *Les projections cartographiques,* Driencourt and Laborde's (1932) *Traité des projections des cartes géographiques,* and Reignier's (1957) *Les systèmes de projection.* German treatises consisted especially of Maurer's (1935) *Ebene Kugelbilder,* Wagner's (1949) *Kartographische Netzentwürfe,* and Hoschek's (1969) *Mathematische Grundlagen der Kartographie.* In addition, Fiala's Czech treatise was translated into German as *Mathematische Kartographische* in 1957, and the third edition of the Zöppritz-Bludau work dating from 1884 appeared in 1912.[207]

From the Soviet Union came numerous treatises, beginning with (translating the titles) Kavrayskiy's (1933, 1934) *Investigation in Mathematical Cartography* and *Mathematical Cartography;* Solov'ev's (1937, 1952, 1969) *Cartographic Projections, Practical Manual for Mathematical Cartography,* and *Mathematical Cartography;* Graur's (1956) *Mathematical Cartography;* Urmayev's (1941, 1962) *Mathematical Cartography* and *Fundamentals of Mathematical Cartography;* Ginzburg and Salmanova's (1957, 1964) *Atlas for Selection of Cartographic Projections* and *Manual of Mathematical Cartography;* Meshcheryakov's (1968) *Theoretical Fundamentals of Mathematical Cartography;* and Bugayevskiy and Vakhrameyeva's (1992) work.

Polish cartographers contributed three works: Łomnicki (1927a) with *Mathematical Cartography,* Biernacki (1949) with *Theory of Representation of Surfaces for Surveyors and Cartographers,* and Churski and Galon (1968) with *Cartographic Projections.* Major works have also appeared in Bulgarian by Khristov (1949, 1961), Chinese by Fang (1934) and Yang (1990), Czech by Fiala (1952), Hungarian by Hazay (1954, 1964), Serbo-Croatian by Borčić (1955), Slovakian by Kuska (1960), and Spanish by Sanchez and Bustamente (1928).

Governmental agencies in the United States and the former Soviet Union have been leading sources of monographs on map projections, including several of the books just mentioned. In the United States the subject was the almost exclusive domain of the Coast and Geodetic Survey and its predecessors from the 1820s, when Hassler originated the polyconic projection, through Thomas Craig's and Schott's work of the 1880s and especially O. S. Adams's prolific publications from 1918 to 1945, until about 1970. The U.S. Geological Survey followed USC&GS recommendations with little or no change, although beginning in 1981 the USGS published several monographs that have assisted the public as well as government agencies in relating computers to old and new projections. The *Works (Trudy)* of the central, Moscow, and Novosibirsk engineering and research institutes of geodesy, aerial photo survey, and cartography in the Soviet Union were an important source of papers (over thirty) on the subject from the 1930s to the 1970s.

Numerous journals served as major conduits for twentieth-century papers on map projections. In Germany, the Soviet Union, England, and the United States, outstanding journals were published: In Germany, the commercially published *Petermanns Mitteilungen* (with *Geographische* inserted in 1938) contained over fifty papers, all in or before 1951, when the Deutsche Gesellschaft für Kartographie began to publish *Kartographische Nachrichten.*[208] The latter journal has carried over thirty more papers. The other German journals publishing large numbers of such papers have been *Annalen der Hydrographie und Maritimen Meteorologie* with over thirty, all before 1944, its demise, and *Zeitschrift für Vermessungswesen* with over fifty, including some in every decade, but half were by Bulgarian geodesist V. K. Khristov (his name germanized to W. K. Hristow).

In the Soviet Union, *Geodeziya i Aerofotos'emka* with about seventy and *Geodeziya i Kartografiya* with three dozen published the bulk of these papers. In England, the *Empire Survey Review* (since 1963 the *Survey Review*) has led with seventy papers, but the *Geographical Journal* formerly was just as prominent in the field with over fifty, since all but two appeared before 1964, when the *Cartographic Journal* of the British Cartographic Society was first published, drawing another twenty-some. A similar shift in journal usage occurred in the United States, where the *Geographical Review* published twenty-four papers and the *Journal of Geography* and *Surveying and Mapping* about eighteen each, but *The American Cartographer* (since 1990 *Cartography and Geographic Information Systems*) has largely supplanted them as a map-projection paper outlet with over thirty since its first issue of 1974, although most of the latter are by Tobler and this writer.

Several other countries have had journals with active records, notably China

(*Acta Geodetica et Cartographica Sinica*), Hungary (*Geodézia és Kartográfia*), Poland (*Geodezja i Kartografia*), and Switzerland (*Schweizerische Zeitschrift für Vermessung, Kulturtechnik und Photogrammetrie*, the latest of its three names). Each of these journals has contributed over twenty papers on map projections.

There is of course much redundancy, especially in the books, but none is all-inclusive. The variety of repositories for such information contributes to the continued enthusiasm for research into both the past and future of the subject.

CONCLUSIONS AND OUTLOOK

The twentieth century produced a great mixture of map projections ranging from trivial to important, depending on one's viewpoint. With an increase in the number of cartographers and related professionals (as well as amateur cartophiles) and of journals available for presenting ideas, the numbers of map projections and books and papers about them continued to rise exponentially.

Sociological factors such as these can be used to explain the quantitative expansion of any continuing scientific field, but what of the quality? Employing relative usage as an index to quality is generally inappropriate here as in other consumer fields, because of political and promotional factors. The very different Gall orthographic and Robinson projections are no better or worse now than they were in 1855 and 1974, respectively, when they were first announced to the scientific public. However, in 1973 via Arno Peters (independently as his own) and in 1988 via the National Geographic Society (with full credit), these projections jumped from relative obscurity to substantial usage by not only their sponsors but others. Also, the nationalities of Gall, O. M. Miller, Ginzburg, and Baranyi undoubtedly influenced the countries in which their world map projections are heavily used.

Claims that one projection shows less distortion than any other have been quoted in this work, and most are probably true using certain criteria stated or not stated by and varying with the inventor. These artificially restricting criteria present others with the same problems in choosing the best projection for a given application. The best equal-area projection for a very long narrow strip following a parallel of latitude is almost certainly the Albers equal-area conic, but almost any actual region is not this or any other idealized shape, and the property required may not be clear. The choice of projection for a region—province, continent, or ocean—should be governed by the shape and extent of the region, the purpose of the map, the likelihood that the map would be one of a series covering several regions or several characteristics of the same region, and the possible need for the maps of the set to fit together.

The selection of a world map projection is more subjective, and the choice generally makes much more of a visual difference. Mathematical analyses or minimization of error is much less helpful in providing a visually satisfactory product, in part because of the diversity of land and water, polar and equatorial portions (Robinson 1986, 1988). For an overall view of the world, it would seem that for the model the designer of a flat map should draw chiefly upon the area, shape, and distance relationships on the globe. Instead, the familiar Mercator projection was used as the model for the commonly seen Gall stereographic, modified Gall, Van der Grinten, Miller cylindrical, and The Times projections—retaining the general appearance, but of course with modifications reducing the area distortion. Even the numerous other cylindrical and pseudocylindrical projections have the property in common with the Mercator that all points at the same distance from the equator on the globe are shown at the same perpendicular distance on the map, a desirable property closely related to climate which is not true of azimuthal, modified azimuthal, and oblique and transverse projections in general. On the other hand, the latter types emphasize the curvature of meridians and parallels and the roundness of the globe, and azimuthals can better relate other locations to the center. The mutual exclusiveness of many of these properties provides impetus to the ongoing developments in the field.

Continuing the summaries shown near the end of chapter 3 by utilizing table 4.2 of this chapter and the *Bibliography of Map Projections* (Snyder and Steward 1988, vi–vii), the numbers of projections are as follows: pre–1670, 16 new projections; 1670–1799, 16; 1800–1899, 53; 1900–1992, 180 (see comment below); and a total of 265.

For books and papers:

Language	English	German	Russian	French	Italian	Polish	Hun-garian	Spanish
pre–1700	2	1	—	2	—	—	—	—
1700–1799	8	6	—	6	2	—	—	—
1800–1899	60	125	2	71	25	—	1	2
1900–1988	1024	459	253	101	45	56	45	34

	Chinese	Bul-garian	Czech	Dutch	Japa-nese	Swedish	Latin	Others	Total
pre–1700	—	—	—	—	—	—	9	2	16
1700–1799	—	—	—	—	—	3	10	—	35
1800–1899	—	—	—	2	—	2	—	2	292
1900–1988	30	27	26	20	21	11	—	56	2208

Because of oversights, errors, and subjective decisions concerning literature entries and the counting of projections, these numbers must be treated as approximations. For example, several series of projections presented by an author in one paper, such as those by Watts and this writer, are treated as single projections. Other series, such as those by Eckert and McBryde/Thomas, are treated as several individual projections. Thus the total of 180 projections (listed above for 1900–1992) could be shown as less than 170 or more than 240.

Of additional interest is the fact that German dominance of the subject, based on language of publications, declined after the First World War and ended with the Second World War. Thereafter, the English language took a leading role, with the period from 1950 to 1988 producing a tally of 648, 223, and 187 papers and books in English, Russian, and German, respectively, of which over half the English-language items originated in the United States. These three languages in turn represented three-fourths of the total of 1,385 publications found for the same period.

The modern computer has been an incentive in developing special low-error projections, but to date these have tended to remain academic rather than become the basis of cartographic application. Those capturing most attention are still those that look distinctive, and pre–computer-age world maps are the more popular twentieth-century projections. This dichotomy of scientific analysis versus acceptability is true of many fields, however, and it helps give human interest to a subject that can otherwise be especially esoteric.

The current reluctance to use projections that are more mathematically complicated, in spite of the need to program the formulas only once for a given type of computer, will hopefully dissipate for medium- and large-scale maps. The increasing availability of both forward plotting (lat/long to x, y) and inverse data-gathering (x, y to lat/long) formulas in a published algorithmic form for numerous projections at least since 1982, as well as in numerous computer software programs, is reducing this reluctance, but working with projections still strikes fear in the hearts of many trained cartographers and geographers because of the mathematical aspects. Just as the available instructions for computer-operated home appliances seem beyond the ability of many adults to follow, even though they need no knowledge of the computer program itself, so map projection programs require more than the routine instruction "enter the standard parallels."

It is more important that cartographers, for whom use of map projections is only one of many skills involved in map work, be brought closer to the projections already available than that still more projections be developed. One avenue for this education is in the field of artificial intelligence (AI). This writer is aware of three recent attempts, all aborted, to develop AI or "expert" systems for map projection selection for a given mapping task.[209] Each was intended to ask the user a number of questions about the application and then provide ranked choices. These systems were abandoned not because of technical impasse but because the principals became involved in other projects.

In practice, the field of map projections will undoubtedly continue in various directions. More projections will be developed by both mathematically and educationally oriented individuals, but more of the latter will try to make the subject digestible to others. The more esoteric work of the past and present will be occasionally picked up by a mapmaker, but generally the accepted choices of the past will dominate future usage as they have dominated the present. The increased emphasis on the computer will not merely make it easier to prepare the printed map; it will also remove the need for many printed maps as Mark Monmonier (1985, 179) has stressed ("A prime casualty of cartography's Electronic Transition will be the attitude that the map is a printed product"). Computers still take a few minutes to plot a map on a different projection, unless the map is already in computer storage, but this time is being reduced so that a series of maps will soon be able to appear rapidly on the screen with varying properties for a set of requirements, and the deficiencies of one projection can be shortly offset with another view having different advantages and shortcomings. Distances can be calculated and windowed on the terminal, offsetting scaling problems.

It has been said at various times that all worthwhile map projections have already been developed. The same was said decades ago of inventions in general. As Robinson and Sale have written, "It may reasonably be asserted that at present cartographers need to devote little time to devising new projections but rather would do better to become more proficient in selecting from the ones available. On the other hand," they added, "if a new and particular use of maps requires a special type of projection, undeveloped as yet, such a projection might well be worth the time and effort spent in devising it." [210] The subject remains at least as alive as it was in the past.

TABLE 4.2.　Map Projections Developed during the Twentieth Century

Names	Figures	Inventor (Date)	Design
Cylindrical			
Gauss-Krüger	—	Gauss (1822) Krüger (1912)	Transverse Mercator. Scale correct along central meridian for ellipsoid; conformal.
Gauss-Boaga	—	Gauss (1822) Boaga (ca. 1950)	Transverse Mercator. Scale correct along central meridian for ellipsoid; conformal.
Gauss-Schreiber	—	Gauss (1822) Schreiber (ca. 1880)	Transverse Mercator. Scale varies along central meridian for ellipsoid; conformal.
Mod. of transverse Mercator	—	Li (1981)	Transverse Mercator modified for two standard meridians; conformal.
Oblique Mercator (ellipsoid)	—	Rosenmund (1903)	Double projection through conformal sphere.
		Cole (1943)	Double projection through conformal sphere.
		Hotine (1946)	Double projection through aposphere.
		Laborde (1926)	Adaptation of transverse Mercator with complex algebra.
Modified Gall cylindrical	—	Kamenetskiy (1929)	Gall stereographic with std. parallels 55° N and S.
	4.2	BSAM (1937)	Gall stereographic with std. parallels 30° N and S.
Behrmann	4.3	Behrmann (1910)	Cylindrical equal-area with std. parallels 30° N and S.
Trystan Edwards	—	T. Edwards (1953)	Cylindrical equal-area with std. parallels 37°24′ N and S.
Kavrayskiy I	—	Kavrayskiy (ca. 1933)	Fusion of Mercator and equirectangular.
Urmayev III	—	Urmayev (1947)	Parallels spaced with series approximations.
Pavlov	—	Pavlov (1956)	Parallels spaced with series approximations.
Kharchenko-Shabanova	—	Kharchenko (1951) Shabanova (1952)	Parallels spaced with series approximations.

(continued)

Note: Map projections are listed in the order discussed in chapter 4.

TABLE 4.2. *Continued*

Names	Figures	Inventor (Date)	Design
Cylindrical			
Miller cylindrical	4.9	O. M. Miller (1942)	Modified Mercator. Also three other proposals.
Arden-Close cylindrical	—	Arden-Close (1947)	Mean of Mercator and cylindrical equal-area.
Poole cylindrical	—	Poole (1935)	Mercator modified for true-scale loxodrome.
B-charts	—	Breckman (1962)	Straight satellite groundtracks.
Satellite-tracking	4.11	Snyder (1977)	Straight satellite groundtracks.
Cassini-hyperbolic	—	McCaw (1917)	Double proj. of oblique equidistant cylindrical through hyperboloid.
Arden-Close spec.	4.12	Arden-Close (1943)	Mean of cylindrical equal-area and its transverse.
Oblique perspective cylindrical	4.13	Solov'ev (1937)	Oblique perspective-cylindrical with secant cylinder. Three variations.
Transverse cylindrical stereographic (ellipsoid)	—	B. Adams (1984)	Transverse perspective-cylindrical with tangent cylinder.
Space oblique Mercator	4.14	Colvocoresses (1974) Snyder (1978)	Conformal strip near groundtrack of Landsat mapping satellite. For ellipsoid.
Pseudocylindrical			
Interrupted sinusoidal and Mollweide	4.5	Goode (1916)	Pointed-polar; interrupted at meridians to emphasize continents (or oceans).
Nell-Hammer	—	Hammer (1900)	Arbitrary; flat-polar; mean of sinusoidal and cylindrical equal-area.
Eckert I	4.16	Eckert (1906)	Arbitrary; flat-polar; equidistant parallels; broken straight-line meridians.
Eckert II	4.17	Eckert (1906)	Equal-area; flat-polar; broken straight-line meridians.
Eckert III	4.18	Eckert (1906)	Arbitrary; flat-polar; equidistant parallels; elliptical meridians.
Eckert IV	4.19	Eckert (1906)	Equal-area; flat-polar; elliptical meridians.
Eckert V	4.20	Eckert (1906)	Arbitrary; flat-polar; equidistant parallels; sinusoidal meridians.

(continued)

TABLE 4.2. *Continued*

Names	Figures	Inventor (Date)	Design
Pseudocylindrical			
Eckert VI	4.21	Eckert (1906)	Equal-areas; flat-polar; sinusoidal meridians.
Winkel I	4.22	Winkel (1914)	Arbitrary; flat-polar; mean of sinusoidal and equirectangular.
Winkel II	—	Winkel (1918)	Arbitrary; flat-polar; mean of modified Mollweide and equirectangular.
Homolosine	4.23	Goode (1923)	Equal-area; pointed-polar; fusion of sinusoidal and Mollweide.
Denoyer semielliptical	—	Denoyer (ca. 1921)	Arbitrary; flat-polar; equidistant parallels.
Rosén sinusoidal	—	Rosén (ca. 1926)	Equal-area; flat-polar; portions of sinusoids for meridians.
Boggs eumorphic	4.24	Boggs (1929)	Equal-area; pointed-polar; mean of sinusoidal and Mollweide.
Craster parabolic	4.25	Craster (1929) Putniņš (P_4) (1934)	Equal-area; pointed-polar; parabolic meridians.
Kavrayskiy V	—	Kavrayskiy (1933)	Equal-area; pointed-polar; sine-function meridians.
Kavrayskiy VII	4.26	Kavrayskiy (1939)	Arbitrary; flat-polar; 120th meridians circular.
Ginzburg VIII (TsNIIGAiK)	—	Ginzburg (1944)	Arbitrary; flat-polar; polynomial series equations.
Wagner I	4.28	Wagner (1932) Kavrayskiy (VI, 1936) Siemon (1937) Werenskiold (1944)	Equal-area; flat-polar; portion of sinusoids for meridians.
Wagner III	—	Wagner (1932)	Arbitrary; flat-polar; portions of sinusoids for meridians.
Wagner IV	4.29	Wagner (1932) Putniņš (P'_2) (1934)	Equal-area; flat polar; portion of elliptic arcs for meridians.
Wagner VI	—	Wagner (1932) Putniņš (P'_1) (1934)	Arbitrary; flat-polar; portion of elliptic arcs for meridians.
Wagner II	4.30	Wagner (1949)	Arbitrary; flat-polar; portion of sinusoids for meridians.
Wagner V	—	Wagner (1949)	Arbitrary; flat-polar; portion of elliptic arcs for meridians.

(continued)

TABLE 4.2. *Continued*

Names	Figures	Inventor (Date)	Design
Pseudocylindrical			
Putniņš P$_1$	—	Putniņš (1934)	Arbitrary; pointed-polar; equidistant parallels; portions of elliptic arcs for meridians.
Putniņš P$_2$	—	Putniņš (1934)	Equal-area; pointed-polar; portion of elliptic arcs for meridians.
Putniņš P$_3$	—	Putniņš (1934)	Arbitrary; pointed-polar; equidistant parallels; parabolic meridians.
Putniņš P$'_3$	—	Putniņš (1934)	Arbitrary; flat-polar; equidistant parallels; parabolic meridians.
Putniņš P$'_4$	4.31	Putniņš (1934) Werenskiold (1944)	Equal-area; flat-polar; parabolic meridians.
Putniņš P$_5$	—	Putniņš (1934)	Arbitrary; pointed-polar; equidistant parallels; hyperbolic meridians.
Putniņš P$'_5$	—	Putniņš (1934)	Arbitrary; flat-polar; equidistant parallels; hyperbolic meridians.
Putniņš P$_6$	—	Putniņš (1934)	Equal-area; pointed-polar; hyperbolic meridians.
Putniņš P$'_6$	—	Putniņš (1934)	Equal-area; flat-polar; hyperbolic meridians.
Loximuthal	4.32	Siemon (1935) Tobler (1966)	Arbitrary; pointed-polar; loxodromes from center straight at true scale.
Quartic authalic	4.33	Siemon (1937) O. S. Adams (1944)	Equal-area; pointed-polar; quartic-curved meridians.
McBryde-Thomas no. 1	—	McBryde, Thomas (1949)	Equal-area; pointed-polar; sine-function meridians.
no. 2	—	McBryde, Thomas (1949)	Equal-area; flat-polar; sine-function meridians.
no. 3—flat-polar sinusoidal	—	McBryde, Thomas (1949)	Equal-area; flat-polar; sinusoidal meridians.
no. 4—flat-polar quartic	4.35	McBryde, Thomas (1949)	Equal-area; flat-polar; quartic-curved meridians.
no. 5—flat-polar parabolic	—	McBryde, Thomas (1949)	Equal-area; flat-polar; parabolic meridians.
Urmayev series	—	Urmayev (1950)	Equal-area or arbitrary; flat-polar; portion of sinusoids for meridians.

(continued)

TABLE 4.2. *Continued*

Names	Figures	Inventor (Date)	Design
Pseudocylindrical			
Modified Gall	—	Bomford (1951)	Arbitrary; flat-polar; curved meridians; spacing of parallels as on Gall stereographic.
The Times	4.36	Moir (1965)	Arbitrary; flat-polar; curved meridians; spacing of parallels as on Gall stereographic.
Fahey modified Gall	—	Fahey (1975)	Arbitrary; pointed-polar; elliptical meridians.
Robinson	4.38	Robinson (1963)	Arbitrary; flat-polar; uses table of coordinates.
Baranyi I–VII	4.39	Baranyi (1968)	Arbitrary; flat- or pointed-polar.
Érdi-Krausz	4.40	Érdi-Krausz (1968)	Equal-area at two scales; flat-polar; fusion of Mollweide and flat-polar sinusoidal.
McBryde Q3	—	McBryde (1977)	Equal-area; flat-polar; fusion of quartic authalic and McBryde-Thomas no. 4.
McBryde P3	—	McBryde (1977)	Equal-area; flat-polar; fusion of Craster parabolic and McBryde-Thomas no. 5.
McBryde S2	—	McBryde (1977)	Equal-area; flat-polar; fusion of sinusoidal and Eckert VI.
McBryde S3	4.41	McBryde (1977)	Equal-area; flat-polar; fusion of sinusoidal and McBryde-Thomas no. 3.
Asymmetric elliptical	—	Hatano (1972)	Equal-area; flat-polar; northern parallels are shorter than southern parallels. Elliptical meridians.
Mayr	—	Mayr (1964) Tobler (1973)	Equal-area; flat-polar; mean of cylindrical equal-area and sinusoidal.
Hyperelliptical	4.42	Tobler (1973)	Equal-area; pointed-polar; bounding curve between ellipse and rectangle.
Hölzel	—	Hölzel (ca. 1960)	Arbitrary; flat-polar; equidistant parallels.
Atlantis	4.44	J. Bartholomew (1948)	Equal-area; transverse oblique Mollweide.
Sinu-Mollweide	4.45	Philbrick (1953)	Equal-area; oblique interrupted fusion of sinusoidal and Mollweide.

Misc. transverse and oblique pseudocylindrical (not itemized)   *(continued)*

TABLE 4.2.   *Continued*

Names	Figures	Inventor (Date)	Design
Azimuthal			
Roussilhe	—	Roussilhe (1922)	Oblique stereographic of ellipsoid, using series.
Lunar	—	Nowicki (1963)	Far-side perspective
Breusing harmonic	—	Young (1920)	Mean of stereographic and Lambert azimuthal equal-area.
Ginzburg I, II	—	Ginzburg (1949)	Arbitrary modifications of Lambert azimuthal equal-area.
Urmayev I, II	—	Urmayev (1950)	Arbitrary modifications of Lambert azimuthal equal-area.
Logarithmic	4.46	Hägerstrand (1957)	Enlarged center.
"Magnifying-glass" series	4.47	Snyder (1987)	Enlarged center with various properties.
Modified azimuthal			
Mecca retro-azimuthal	4.48	J. I. Craig (1909)	Azimuths to center true; meridians parallel.
Hammer retro-azimuthal	4.49	Hammer (1910) Hinks, Reeves (1929)	Azimuths and distances to center true.
Schoy	—	Schoy (1913)	Azimuthal; meridians parallel and straight.
Two-point retro-azimuthal	—	Maurer (1919)	Both points on equator.
Misc. retroazimuthals (not itemized)			
Winkel tripel	4.50	Winkel (1921)	Mean of equirectangular and Aitoff.
Two-point azimuthal	4.51	Maurer (1914) Close (1922)	Azimuths true from two points; compressed gnomonic.
Two-point equidistant	4.52	Maurer (1919) Close (1921)	Distances true from two points.
Two-point azimuthal-equidistant	—	Close (1935)	Distances true from one point, azimuths true from other point.
Chamberlin trimetric	—	Chamberlin (1946)	Distances approx. true from three points.
Donald elliptic	—	Donald (1956)	Two-point equidistant adapted to ellipsoid.
Rosén	—	Rosén (1926)	Modif. of equatorial Lambert azimuthal equal-area.

(continued)

TABLE 4.2. *Continued*

Names	Figures	Inventor (Date)	Design
Modified azimuthal			
Eckert-Greifendorff	4.53	Eckert-Greifendorff (1935)	Modif. of equatorial Lambert azimuthal equal-area.
Wagner VII (Hammer-Wagner)	4.54	Wagner (1941)	Equal-area modif. of Hammer to give curved pole-line.
Wagner VIII	—	Wagner (1949)	Arbitrary modification of Wagner VII.
Wagner IX (Aitoff-Wagner)	4.55	Wagner (1949)	Arbitrary modification of Aitoff.
Spilhaus ocean	—	Spilhaus (1942–89)	Equal-area; three-lobed transverse oblique Hammer; also variations and conformal versions using August.
Nordic; misc.	—	J. Bartholomew (1950) Bomford (1951)	Equal-area; oblique Hammer variations.
Briesemeister	4.56	Briesemeister (1953)	Equal-area; modified oblique Hammer.
Maurer star variations	—	Maurer (1935)	Equal-area or arbitrary; six equal points.
Fawcett	—	Fawcett (1949)	Equal-area; polar or oblique Lambert azimuthal equal-area inner hemis. with lobes for outer continents.
William-Olsson	—	William-Olsson (1968)	Equal-area; polar Lambert azimuthal equal-area with four lobes attached.
Oblated equal-area	4.57	Snyder (1987)	Equal-area; modif. Lambert azimuthal equal-area.
Minimum-error equal-area	—	Dyer, Snyder (1984)	Equal-area; modif. Lambert azimuthal equal-area for irregular regions.
Tetrahedral	—	J. Bartholomew (1942)	Polar azimuthal equidistant with three Werner lobes attached.
Matter-most	—	Gilfillan and Wray (1946)	Oblique azimuthal equidistant with lobes attached.
Davies	—	Davies (1949)	Oblique azimuthal equidistant with lobes attached.
Bomford modif. azimuthal equidistant	—	Bomford (1951)	Approx. equal scale error along oval; pseudoazimuthal.
Oblique TsNIIGAiK with oval isocols	4.58	Ginzburg (1952)	Approx. equal scale error along ovals; pseudoazimuthal.

(continued)

TABLE 4.2. *Continued*

Names	Figures	Inventor (Date)	Design
Modified azimuthal			
Oblated stereographic	4.59	O. M. Miller (1952)	Conformal; constant scale factors along ovals.
New Zealand Map Grid	—	Reilly (1973)	Conformal; constant scale factors along irregular lines.
Modif.-stereographic conformal	—	Snyder (1984, 1986)	Conformal; constant scale factors along polygonal, rectangular, or irregular lines.
Hammond optimal conformal	—	Feigenbaum (1991)	Conformal; constant scale factors along irregular lines.
Conic			
Minimum-error	—	Young (1920)	Arbitrary properties.
Satellite-tracking	—	Snyder (1977)	Straight Landsat groundtracks.
Stereographic conic (ellipsoid)	—	B. Adams (1984)	Perspective onto tangent cone.
Modified polyconic (IMW)	—	Lallemand (ca. 1909)	Straight meridians, two true to scale.
Modified rectangular polyconic	—	McCaw (1921)	Approximately conformal.
Maurer equal-area polyconic	—	Maurer (1935)	Equal-area; central meridian true to scale.
Bumstead equal-area polyconic	—	Bumstead (ca. 1937)	Equal-area; devised from zones.
KhISI modif. polyconic	—	KhISI (1937)	Arbitrary.
Taich modif. polyconic	—	Taich (1937)	Arbitrary.
Ginzburg modif. polyconics (10)	4.60	Ginzburg (1934–66)	Arbitrary; polynomial series used; some asymmetrical.
Salmanova modif. polyconic	—	Salmanova (1951)	Arbitrary; polynomial series used.
Bousfield modif. polyconic	—	Bousfield (ca. 1950)	Arbitrary for Canada north of 80° N.
Transverse polyconic	—	Deetz (1919)	
Bipolar oblique conic conformal	4.61	O. M. Miller, Briesemeister (1941)	Two oblique conics back to back for map of Americas.

(continued)

TABLE 4.2.   *Continued*

Names	Figures	Inventor (Date)	Design
Misc. oblique conics (not itemized)			
Misc. selections of standard parallels: conformal (Vitkovskiy III, 1907; Tsinger I, 1916; Kavrayakiy III, 1934); equal-area (Vitkovskiy II, 1907; Tsinger II, 1916; Krasovskiy I, 1922); equidistant (Everett; Vitkovskiy I, 1907; Kavrayskiy II and IV, 1934; Mikhaylov; Mendeleev, 1907).			
Pseudoconic			
Schjerning II	—	Schjerning (1904)	Arbitrary; mod. azimuthal equidistant with ellipse-like boundary.
Schjerning III	4.63	Schjerning (1904)	Arbitrary; mod. azimuthal equidistant bounded in two circles.
Schjerning VI	4.64	Schjerning (1904)	Equal-area; three-lobed interrupted Werner.
Polar equal-area	—	Goode (1929)	Equal-area; multilobed interrupted Werner.
Kite	—	J. Bartholomew (1942)	Arbitrary; fusion of equidistant conic and interrupted Bonne.
Regional	—	J. Bartholomew (1948)	Arbitrary; fusion of equidistant conic and interrupted Boone.
Lotus	—	J. Bartholomew (1958)	Arbitrary; fusion of equidistant conic and interrupted Bonne.
Watts series of 55	—	Watts (1970)	Equal-area or arbitrary; fusion of various pseudocylindrical and pseudoconic projections.
Solov'ev	—	Solov'ev (1937–38)	Arbitrary; modified Bonne.
Oxford	—	Bomford (1950)	Arbitrary; modif. pseudocylindrical and pseudoconic.
Miscellaneous			
Van der Grinten I	4.65	van der Grinten (1904)	Arbitrary; world in circle; meridians and parallels circular arcs.
Van der Grinten II	4.66	Bludau (1912)	Arbitrary; world in circle; meridians and parallels circular arcs intersecting at right angles.
Van der Grinten III	4.67	Bludau (1912)	Arbitrary; world in circle; meridians circular arcs; parallels are straight lines.
Van der Grinten IV	4.68	van der Grinten (1904)	Arbitrary; world in two intersecting circles; meridians and parallels circular arcs.
Maurer modif. of Van der Grinten II	—	Maurer (1935)	Arbitrary; curved poles.

(continued)

TABLE 4.2.   *Continued*

Names	Figures	Inventor (Date)	Design
Miscellaneous			
Cartograms	4.69, 4.70	Raisz (1934) Tobler (1973) others	Reshapes map to maintain area or distance based on thematic variables.
Cahill butterfly	4.71	Cahill (1909)	Arbitrary or conformal; octants of earth arranged like butterfly.
Pitner	—	Pitner (1943)	Equal-area; octants of globe.
Maurer globular	—	Maurer (1922)	Arbitrary.
Conformal world or hemis. in polygons	4.72	O. S. Adams (1925) L. P. Lee (1965, 1976)	Conformal; bounded by various polygons or mounted on polyhedra.
Raisz armadillo	4.74	Raisz (1943)	Arbitrary; orthographic projection of world on portion of torus.
Raisz orthoapsidal	—	Raisz (1943)	Arbitrary; general series of certain double projections including armadillo.
Polyhedral globes	4.75	Fisher (1943) Fuller (1943) others	Arbitrary; using icosahedron and numerous other types.
Anderson	—	Anderson (1974)	Arbitrary; combination.
Poly-centered oblique orthographic	—	Dent (1987)	Arbitrary; separate orthographic projections.
Optimum continental	—	Macdonald (1968)	Arbitrary; combination.
Polyfocal	—	Shlomi, Kadmon (ca. 1977)	Enlarges various points of interest on same map.
Dinomic	—	Baker (1986)	Arbitrary; modif. of Mercator.
Misc. others not itemized.			

NOTES

CHAPTER ONE: EMERGENCE OF MAP PROJECTIONS

1. See Dilke (1987) and subsequent discussion in this volume, pp. 10–14.

2. Dilke (1987, 177 n.5, 179 n.17) finds the Stevenson translation "inadequate" and did not use it. Neugebauer (1959, 22), regarding Ptolemy's book 7, says Stevenson "did not understand, even remotely, what he was translating." Dilke also points out that it is the only complete English edition, and in effect quotes it almost verbatim except for one clause on page 186, although calling it his own translation. I have used it to provide Ptolemy's flavor and insight more directly.

3. D'Avezac-Macaya (1863, 473–85); Maurer (1935); Lee (1944); Goussinsky (1951); Tobler (1962a); Starostin, Vakhrameyeva, and Bugayevskiy (1981); and Calderera (1910). Numerous other papers directed to map projection classification and nomenclature are listed in Snyder and Steward (1988, 103). See also p. 271, this vol.

4. D'Avezac-Macaya (1863, 474–78), refers to "les *projections perspectives*, les *développements* de surfaces osculatrices ou pénétrantes," that is, tangent or secant surfaces, specifically "les développements cylindriques" and "cóniques," and the catch-all "les *systèmes conventionnels*" (his italics). I did not find an earlier cartographic reference to developable surfaces.

5. These names were originated by d'Avezac-Macaya; see pp. 10, 12, this vol.

6. Woodward (1987) gives a detailed account.

7. Dickinson (1949–50; 1968 reprint, 39, 42); Isaiah 44:13.

8. Dickinson (1949–50; 1968 reprint, 38, 42–43).

9. Keuning (1955, 3); Varenius (1650). There is also reference to a 1672 Latin edition with notes by Isaac Newton in d'Avezac-Macaya (1863, 336–37). Varenius describes the construction of meridians and parallels for some eight projections and aspects then in use, although the projections are given mode numbers, not names.

10. Aujac et al. (1987a); L. A. Brown (1949, 47–70).

11. D'Avezac-Macaya (1863, 476). See also Campbell (1987, 385), stating an origin for the plate carrée of 1450–1500. Mead (1717, 88) used the term "plain Chart."

12. Keuning (1955, 13) supports Eratosthenes, but Aujac et al. (1987b, 154–57) state why this cannot be the case.

13. Nordenskiöld (1889, 85); Dilke (1987, 178–80, 185).

14. Stevenson (1932, 39–40); Keuning (1955, 13).

15. Neugebauer (1975, 2:879–80); Keuning (1955, 13–14). The first Latin manuscript is Rome, Biblioteca Apostolica Vaticana, Vat. Lat. 5698. Keuning states that the maps of the Latin codex are reproduced full size by Joseph Fischer, ed., *Claudii Ptolemaei geographiae codex urbinas graecus 82*, 2 vols. in 4, Codices e Vaticanis selecti quam simillime expressi, vol. 19 (Leiden: Brill; Leipzig: Harrassowitz, 1932), vol. 2, pt. 2.

16. Nordenskiöld (1889, 49), nearly full-size reproduction.

17. Fischer and von Wieser (1903). A small reproduction of *Carta marina* is found in Bagrow (1964/1985), pl. 62. Thorne map is reproduced in Nordenskiöld (1889, pl. 41[1]).

18. *Planisphere* is the term frequently applied to a map as a projection of all or part of a sphere onto a plane surface, sometimes implying a pole-centered aspect.

19. Keuning (1955, 15–16). Stevenson (1903–6, nos. 5–7, 9, 11); Diogo Ribeiro, *Carta uniuersal en que se contiene todo lo que del mundo se ha descubierto fasta agora* (manuscript map, 1529), reproduced in Fite and Freeman (1926, 46–49). Nordenskiöld (1889, pl. 39[1]); Laurentius Frisius's graduated world map of 1522, *Orbis typus universalis iuxta hydrographorum traditionem exactissima depicta* is reproduced. Recent studies of possible portolan projections include Tobler (1966a), Loomer (1986), Campbell (1987, 385), and Lanman (1987, 49–54).

20. Uhden (1937). Durand (1952, pl. I), reference in Keuning (1955, 19); the 1426 map by Conrad is contained in Rome, Biblioteca Apostolica, Palatinus Latinus 1368. Donnus Nicolaus Germanus, *Geography* (Ulm: 1482), dedicatory letter to Pope Paul II, quoted in Nordenskiöld (1889, 14, 86): "Cogitare cepimus quo pacto nos aliquid glorie comparemus. Rati enim nobis oblatam esse occasionem uti aliquid industrie nostre monimentum extaret et ingenii vires ducescere possent statim picturam orbis proper aratione aggressi sumus."

21. Varenius (1650; 1693 trans., 332), 2d mode for particular maps:

First, a transverse *line* is drawn in the extremity of the *Table*, for the Circle of *Latitude*, in which the ends of the Regions respecting the *Æquator*, are to be drawn; in that so many parts are taken equally, through how many *deg.* of *Longitude* that Region is extended from that part. Then from the middle of this *line*, a perpendicular is drawn, which hath so many parts as there are *deg.* of *Longitude* between the bounds of that Region towards the *Æquator*, and the Pole. But how great these parts should be, is known from the proportion of the *deg.* of the first Circle, which is greatest to the *deg.* of Parallel, which is represented from the lowest transverse *line*. Through the term of this perpendicular, another perpendicular, or Parallel to the inferiour *line*, is drawn, in which so many *deg.* of *Longitude* must be taken as are in the lower *line*, and equal to them of the lower *line*; if these *Latitudes* be not much distant from the *Æquator*, or mutual from themselves. . . . After the parts are thus taken for the *deg.* of *Longitude* in the superiour and inferiour *line*, the right *lines* are to be drawn through the beginning and end of the parts of the same number: which right *lines* shall represent the Meridian *lines*. Then through every *deg.* of its perpendicular, which we have ordered to be erected from the middle *point* of the inferiour *line*, *lines* Parallel to that lower *line* must be drawn through the beginnings of every *degree* which shall shew the Parallels of *Latitude*.

22. Keuning (1955, 19); Nordenskiöld (1889, 86, 87).

23. Stevenson (1932, 42); Keuning (1955, 9–10); Dilke (1987, 185–86, 187); Neugebauer (1975, 2:880–82).

24. Keuning (1955, 9–10); Hopfner (1938). Mollweide (1805a, 11:319–40) has a detailed analysis of the geometry of Ptolemy's coniclike projections.

25. See also Dilke (1987, 186–88); Neugebauer (1975, 2:883–85).

26. Keuning (1955, 10); Nordenskiöld (1897, 173; 1889, 85–86, pls. 29, 31 [6], and 33); Sylvano (1511, 1st map); and Münster (1540).

27. Ptolemy as translated in Neugebauer (1959, 23). See also Mollweide (1805a, 11: 504–14); Dilke (1987, 188–89).

28. *Globular* was the name given by Aaron Arrowsmith in 1794 to a specific globular projection (see n. 101), but it has since been applied to the general type described here at the suggestion of d'Avezac-Macaya (1863, 482).

29. Keuning (1955, 20); Nordenskiöld (1889, 93) interprets Bacon differently; Bacon (1265–68); Woodward (1987, 305, 322, 342).

30. Keuning (1955, 20); Osley (1969, 41).

31. Woodward (1987, 353–54); Keuning (1955, 20).

32. Apian (1524, 1551); Keuning (1955, 20); Tissot (1887, 60, 61, 114–16, 119).

33. Reproduced in Keuning (1955, 20). Nordenskiöld (1889, 93–94) calls it the Bacon projection.

34. Ruscelli (1561–74), reference in Keuning (1955, 20); Bodel Nyenhuis Collection at Leyden, *Remarkable Maps of the XVth, XVIth & XVIIth Centuries, Reproduced in Their Original Size* (1894, 1:1–4; and *Periplus of the Erythraean Sea* (New York, 1912), fig. 76.

35. Other types of maps partitioned in different manners are described as zonal in Woodward (1987, 353–55). These include tripartite T-O maps and others that are not divided primarily by lines of latitude.

36. Lilius (1493), reference in Nordenskiöld (1889, 38).

37. d'Ailly (ca. 1483), reference in Nordenskiöld (1889, 37, 38).

38. Eastwood (1489); Macrobius (1483). References in Nordenskiöld (1889, 37, 40, pl. 31[1, 3–5]). Woodward (1987, 300, 354).

39. Neither term was used by d'Avezac-Macaya (1863) in his detailed account. *Zenithal* was used by Gretschel (1873, 234). *Azimuthal* was used by Velten (1898). Both terms were common by 1920.

40. The terms *polar, equatorial,* and *horizontal,* as applied to aspects of azimuthal projections, were used by d'Avezac-Macaya (1863, 337). *Meridian* was used by Thomas Craig (1882, 14) because "the plane of projection will pass through a meridian." Using consistent reasoning, he used *equatorial* for what d'Avezac-Macaya and present convention call the polar aspect (T. Craig 1882, 14) because "the plane of projection will coincide with the equator." He had no term corresponding to *oblique.* As applied to azimuthal projections, *oblique* did not become popular until the twentieth century, but d'Avezac-Macaya (1863, 478) used it, *direct,* and *transverse* for the aspects of cylindrical projections.

41. Von Braunmühl (1900, 1:101); Vitruvius Pollio (n.d., bk. 10:9.7); references in Keuning (1955, 6).

42. Keuning (1955, 6). D'Aiguillon (1613, bk. 6:498), quoted in d'Avezac-Macaya (1863, 332–33): "Tria itaque esse dicimus projectionum genera absolutè prolata. Primum quod et *orthographice* nuncupatur, ex infinita oculi distantia. Secundum ex contactu, quod et *stereographice* non incongruè poterit appellari, quare ut ea vox in usum venire liberè possit dum alia mèlior non occurrit, Lector, veniam dabis. Tertium quod *scenographice* vulgo nominatur, ex justo oculi intervallo."

43. Vitruvius Pollio (n.d., bk. 1, chap. 2), reference in d'Avezac-Macaya (1863, 332, 475).

44. *Globus mundi* (Strasbourg, 1509), reference in Nordenskiöld (1889, 40, 41). Schöner (1533), reference in Nordenskiöld (1889, 80, 83). Schöner (1551), reference in Keuning (1955, 6). Apian (1524, 1551), reference in Nordenskiöld (1889, 93, 102, pl. 44[3]).

45. Bagrow (1964, 127, 128). Bagrow states that this projection was proposed by Werner in 1514 as new, but Bagrow is apparently confusing this with the oblique stereographic (see n. 58). Varenius (1650; 1693 trans., 329) described the construction of the equatorial aspect of the orthographic projection as follows (no illustration included by Varenius or Blome): With the eye

supposed to be removed by an infinite space from the Earth, . . . take any *point* in the *plain* E, and from that as from a *Center* let the *Periphery* ABCD be described, let the *Quadrants* be AB, BC, CD, DA; let every one be divided into 90 *degrees,* beginning from AC, towards B and A [D(?)], BAD, shall be the first Meridian, BCD the opposite, in the right *line* BD, the middle between these is the 90 from the first BAD. Let them be drawn to AC, which sheweth the *Semiperiphery* of the *Æquator;* Right *lines* Parallel through every *degree* of the *Quadrants,* or *quarters,* they shew the Parallels of the *Æquator* or the *Circles* of *Latitude,* and the *Tropicks* and *Polary Circles* shall also be found out. The parts into which EB, ED is divided, through these that are drawn, are the Meridian *degrees* BD, which are noted, 1, 2, 3, and so on. The same are taken in the *Quadrant* EA, of the *Æquator,* and the *Quadrant* EC, and the number 1, 2, 3, are ascribed, even to 180, beginning from the first *point,* or next to the Meridian BAD. So the parts AEC shew the *degrees,* into which the *Semiperiphery* of the *Æquator* is divided, through which the Poles BD, the *Semiellipsis* must be drawn for the Meridians. Because through BD, is the greater *Axis* of *Ellipsis* which are to be drawn, but the *Semissis* EB, or ED: but the *Axis* of the lesser *Semissis* is various in divers, *viz.* part of EA, intercepted between E, and the *degree* of *Longitude,* and therefore from those given it is easy by an apt Instrument, to describe these *Ellipses,* which *Instrument* is vulgar at this day, neither is it difficult to make it. Yet the *points* of every one of the *Ellipsis* may be easily found, through which they must be drawn with a free hand: but it is better to delineate them with an *Instrument.*

46. D'Avezac-Macaya (1863, 333, 475); Nordenskiöld (1889, 93); Nikolaus Visscher, *Planisphaerium terrestre cum utroque coelisti hemisphaerio,* reproduced in Portinaro and Knirsch (1987, 182–83).

47. Keuning (1955, 6–7); d'Avezac-Macaya (1863, 265–66, 275–77).

48. Published in Ritter (1610), referenced and reproduced in Woodward (1979). See also Shirley (1983, 290–91).

49. Kepler (1606, opp. 76); Grienberger (1612); Grassi (1619). References in Warner (1979, 135, 100, and 99, respectively).

50. Keuning (1955, 7); d'Aiguillon (1613); d'Avezac-Macaya (1863, 275–77). See also n. 42.

51. Halley (1695–96; 1809 reprint, 69–70). See also Neugebauer (1975, 2:859).

52. Riel (1875, 239, 294); and Lagrange (1779; 1894 German ed.), 5, 87). References in Keuning (1955, 7). Dürer maps are described and reproduced in Warner (1979, 71–75).

53. Bound in some copies of Petri (1541), reference in Warner (1979, 123–25).

54. Lud (1507); reference in Keuning (1955, 7–8).

55. Reisch (1512); Apian (1524, 1551). Reference in Nordenskiöld (1889, 92).

56. Keuning (1955, 8); Dahlberg (1962, 42, fig. 3); Warner (1979, 32).

57. Keuning (1955, 8). Keuning states that, while Fiorini (1881, 117) said the Arab astronomer az-Zarqalī (Arzachel) (at Toledo in 1080) invented the oblique stereographic projection, "an Arabic table of contents from the lost treatise of" Theon proves its earlier use.

58. Werner (1514), containing "Libellus de quatuor terrarum orbis in plano figurationibus ab eodem Ioanne Vernero nouissime compertis & enarratis"; reference in Keuning (1955, 11), and Nordenskiöld (1889, 88, 92).

59. Werner (1514): The original Latin is "Talis profecto terrarum orbis figuratio, plurimum honestatis atque ingens ornamentum viro adiiciet philosopho, si super ipsius mensae plano depicta fuerit," quoted in Nordenskiöld (1889, 92).

60. De Vaulx (n.d.), referenced in Keuning (1955, 8). John Blagrave, *Astrolabium uranicum generale* (London, 1596), referenced and reproduced in Warner (1979, 32–33). John Speed, *A Prospect of the Most Famous Parts of the World* (London, 1627).

61. Keuning (1955, 8). The use of the stereographic projection in all aspects on astrolabes was extensive during the several centuries of astrolabe use. For numerous illustrations and descriptions see Gunther (1932).

62. De Vaulx (n.d.). Reference in Keuning (1955, 9).

63. Keuning (1955, 9). He quotes Mercator's reason for use of the projection here: "Sciet lector nos eam complanandae sphaerae rationem secutos esse, quam Gemma Frisius in suo Planisphaerio adiuvenit, quae omnium longe optima est."

64. Jodocus Hondius, *Vera totius expeditionis nauticae descriptio D. Franc. Draci*, reproduced in Fite and Freeman (1926, 96–99).

65. *Noua orbis terrarum delineatio singulari ratione accommodata meridiano tabb. Rudolphi astronomicarum*, reproduced in Tooley and Bricker (1968, 252–53).

66. Keuning (1955, 4); Uhden (1935). See further discussion of the projections for portolans in Campbell (1987, 385–86).

67. Uhden (1937). Reinganum (1839, 75); Herodotus II, 109; Dürer's map was based on instructions by Heinfogel (ca. 1515, 195); references in Keuning (1955, 4).

68. Keuning (1955, 5); Nordenskiöld (1897, pl. 47).

69. Fite and Freeman (1926, 76); Tooley and Bricker (1968, 120).

70. Keuning (1955, 5). D'Avezac-Macaya (1863, 320–22, 336, 470) considered Postel's use of 1581 to be first. Maling (1960, 203); Shirley (1983, 166–67).

71. Untitled, but dedicated by Lok to Sir Philip Sidney and contained in Hakluyt (1582), reproduced in Fite and Freeman (1926, 90–91).

72. Keuning (1955, 5). The tally of celestial maps is based on entries in Warner (1979).

73. Shirley (1983, 254). Louis de Mayerne Turquet, *La nouvelle maniere de representer le globe terrestre . . . inventée par Louis de Mayerne Turquet geographe*, in his *Discours sur la Carte universelle* (Paris, 1648), reproduced and described with successors in Shirley (1983, 397, 398, 531, 532, 573–75, 577–78).

74. *Geographia Ptholomei ad g͂o 180. Cum additione ulterius emisperij co ordine enam in planū g͂o 180 et sig͂o cū gradibus*, reproduced in Fite and Freeman (1926, 18–20), Cumming et al. (1972, 54–55). See also Keuning (1955, 10).

75. *Universalior cogniti orbis tabula*, reproduced in Fite and Freeman (1926, 28–31), and Nordenskiöld (1889, 85–86, pl. 32).

76. Schickard (1623), referenced and reproduced in Warner (1979, 224–28); see also Applebaum (1970–80). Uses of the conic projection for star maps after the Renaissance are mentioned on p. 68, this vol.

77. Rumold Mercator, *Europa, ad magnae Europae Gerardi Mercatoris P. imitationem, Rumoldi Mercatoris F. cura edita. . . . Medius meridianus 50. reliqui ad hunc inclinantur pro ratione 60. & 40. parallelorum*, in Gerardus Mercator, *Atlas sive cosmographicae meditationes de fabrica mundi et fabricati figure* (Duisberg, 1595); Willem Janszoon Blaeu, *Aethopia [Africa] inferior vel exterior* (1642); Henry Briggs, *The North Part of America* (1625); all reproduced in Tooley and Bricker (1968, 66, 167, 215, respectively). Briggs's map is also discussed and reproduced in Fite and Freeman (1926, 128–31).

78. Huxley (1970–80); Glaisher (1911, 869–74); Struik (1948, 121).

79. Mercator, *Europa* (n. 77), and *Russia cum confinijs* from the same 1595 atlas. John Speed, *A Newe Mape of Tartary* (date not given), reproduced in Gohm (1972, 67). *Mercator-Hondius-Janssonius Atlas* (1636), *Tartaria* (between 413 and 414), *Turcici imperii imago* (between 403 and 404). This projection is discussed further on pp. 68–70, this vol. Nordenskiöld (1889, 7, 86) shows or lists other examples of early true and untrue conics.

80. This map is reproduced in Vietor (1963). Dilke (1987, 187) states that this map has all parallels at true scale, instead of using circular arcs for meridians. Henricus Martellus Germanus, however, used three circular arcs for each meridian instead of Ptolemy's single arc; therefore, only seven parallels can be technically true to scale, and the rest of them are (very slightly) off scale. It is necessary to use special noncircular curves so that all parallels are true to scale (and thus the projection is correct in area scale).

81. Waldseemüller (1507); Keuning (1955, 11); Fischer and Wieser (1903); Fite and Freeman (1926, 24–27).

82. Peter Apian, *Tipus orbis universalis iuxta Ptolomei cosmographi traditionem et Americi Vespucii aliorque lustrationes a Petro Apiano Leysnico elucbratan.do MD.XX*, from *Joannis Camertis Minoritani in C. Julii Solini . . . enarrationes* (Vienna, 1520); Peter Apian, *Charta cosmographica, cum ventorum propria natura et operatione*, from his *Cosmographia* (Paris, 1551); reproduced in Nordenskiöld (1889, pls. 38[2] and 44[4], respectively). Johannes Honter, *De cosmographiae rudimentis* (Basel, 1561); map reproduced in Nordenskiöld (1889, 119).

83. Keuning (1955, 11); Nordenskiöld (1889, pl. 34).

84. Sylvano (1511), last map, reproduced in Nordenskiöld (1889, pl. 33).

85. Keuning (1955, 11). See also n. 58.

86. Werner (1514, propositions 4, 5 and 6 [unpaginated]). Werner used 22/7 (3.14286) or 180°/57°16′ (3.14319) as approximations for π (3.14159); "ratio gra.lvii.m[inutiae primae].xvi. ad.ccclx.gradus totius perimetri circuli est sicut ratio eius quae ex centro ad totam circuferetia circuli.id est.sicut septem ad.xliv" (the ratio 57°16′ to the 360° of the total perimeter of the circle is as the ratio from the center to the total circumference of the circle, that is, as 7 to 44; trans. by author) under proposition 5. Keuning (1955, 11–13); Nordenskiöld (1889, 88–90); Bagrow (1964, 127); Fiorini (1889); Kish (1965).

87. Keuning (1955, 12). Nordenskiöld (1889, 88–90) shows the Werner graticules, but states that Finé's map is based on the second projection instead of the third. Inspection shows this cannot be the case. Keuning is correct. See also d'Avezac-Macaya (1866) and Kish (1957). David Woodward (pers. com., 1985, 1987).

88. Keuning (1955, 12); Nordenskiöld (1897, pl. 44[2]).

89. Keuning (1955, 12). Finé's 1531 map, *Nova, et Integra universi orbis descriptio,* was apparently first printed as a separate sheet, but was included in Grynaeus (1532). Nordenskiöld (1889, 90, pl. 41[2]).

90. Untitled but dedicated to John Drosius. Keuning (1955, 12); Nordenskiöld (1889, 90, pl. 43).

91. Christian Sgrooten, *Praeseti tabula totius terrae hemisphaerium arcticum,* 1592 manuscript atlas prepared for Philip II, king of Spain, in Biblioteca Nacional, Madrid, folio 2.B.1. Referenced and reproduced in Cumming, Skelton, and Quinn (1972, 109–11, 220–21). See also n. 82 and 84.

92. Paris, Bibliothèque Nationale, reproduced in Cumming, Skelton, and Quinn (1972, 18–19). Keuning (1955, 23).

93. Paris, Bibliothèque Nationale, reference in Keuning (1955, 23); Anthiaume (1911).

94. David Woodward (pers. com., 1986). Portinaro and Knirsch (1987, 44).

95. Ortelius map is reproduced in Cumming, Skelton, and Quinn (1972, 149). Agnese's maps, untitled and undated but estimated 1538–48, are reproduced in Fite and Freeman (1926, frontispiece, 58).

96. *America* in *Theatrum,* Ortelius's 1570 atlas.

97. Without title, date, or name of author, described and reproduced in Fite and Freeman (1926, 60–63).

98. Keuning (1955, 22–23); Nordenskiöld (1889, 77, 94); Fiorini (1894). See also Hennig (1948, 36); True (1954, 80).

99. Nordenskiöld (1889, 94); Keuning (1955, 23; 1954).

100. Fournier (1643, bk. 14, 524); reference in d'Avezac-Macaya (1863, 334–35). Steers (1970, 160, 240–41); Keuning (1955, 20).

101. Steers (1970, 160, 241); Keuning (1955, 20). Nicolosi (1660). Aaron Arrowsmith, *Map of the World on a Globular Projection*, and *A Companion to a Map of the World* (London, 1794); the projection was also called the English and the Arrowsmith by some writers. References in d'Avezac-Macaya (1863, 342, 359; see also 359–61, 482).

102. The Floriano map, undated, is in Antonio Lafreri's *Atlas*, referenced and reproduced in Nordenskiöld (1889, 81; see also 58, 73, 94). The Santa Cruz map is reproduced in Bagrow (1964, pl. 65), and Nordenskiöld (1897, pl. 50).

103. From a manuscript atlas, author unknown, referenced and reproduced in Dahlberg (1962, 37, 39, fig. 2). Stevenson (1921, 2:205–7).

104. Kish (1970–80a). Osley (1969, 19–25, 185–194, 196). See also Averdunk and Müller-Reinhard (1914).

105. The original Latin is clearly reproduced in Osley (1969, 73).

106. See also Heinrich Wagner (1915) for another detailed account.

107. E. Wright, (1610); Halley (1695–96; 1809 reprint, 68). Halley said, "though it generally be called Mercator's, was yet undoubtedly Mr. Wright's invention." D'Avezac-Macaya (1863, 316–17) quotes Halley and several others in the same vein, including Mead (1717, 94): "the truth is, Mr. Wright having found out his method, communicated it too freely to Mercator . . . who publish'd a chart thereupon, and Batillus like, took the invention of it to himself."

108. Wallis (1970–80); Glaisher (1911, 869).

109. Bagrow (1964, 148, 150); Drecker (1917); Hammer (1917). Ronan (1983, 138, 162–63) states that a Chinese star map of A.D. 940 is based on a "Mercator" projection, but his illustration shows only near-equatorial constellations which can only support the probable use of a cylindrical (equirectangular?) projection.

110. Keuning (1955, 18). E. Wright (1610, 12 of unpaginated preface) was angered by Hondius's use of Wright's sea chart design as if it were Hondius's own invention.

111. Hakluyt (1598–1600, vol. 1, pl. 5); reference in Keuning (1955, 18). Comments and reproduction in Nordenskiöld (1889, 96, pl. 50). He notes that Coote (1878) concludes that this is the map referred to by William Shakespeare in *Twelfth Night,* act 3, scene 2, when Maria says, "he does smile his face into more lines than is in the new map with the augmentation of the Indies." The map is bedecked with thirty-two straight lines extended from each of numerous compass roses.

112. Dudley (1646–47), reference in J. Potter (1989, 60, 61).

113. The history of algebraic notation is long and diffuse, but many common symbols $(+, -, \sqrt{}, \text{ and } =)$ were introduced in the sixteenth century. In the seventeenth century appeared $<$, $>$, and $\times$ (times). See Everitt (1911, esp. 619).

114. Jehan Cossin, "Carte cosmografique ou universelle description, du monde avec le vrai traict des vens, faict en Dieppe par Jehan Cossin, marinnier en l'an 1570," ms. world map, Bibliothèque Nationale, Département des Cartes et Plans, Ge D 7896. References

from David Woodward and Richard T. Porter. See also Keuning (1955, 24). Keuning mistakenly says the projection is conformal instead of equal-area. He refers to Anthiaume (1916).

115. For example, some gore maps with gores touching along the equator, including the map with large gores references in n. 103. Another example is Jodocus Hondius, *Europa* (1602), reproduced in Tooley and Bricker (1968, 74–75).

116. Keuning (1955, 24); Flamsteed (1729). Quotation by Flamsteed given in Baily (1835, 131): letter from Flamsteed to Newton dated Feb. 24, 1691/92. Sanson used, for example, the sinusoidal projection for maps of North America, South America, Europe, Asia, and Africa in an atlas (Sanson 1675) and a geography (Sanson 1683).

CHAPTER TWO: MAP PROJECTIONS, 1670–1799

1. Nordenskiöld (1889); Keuning (1955); d'Avezac-Macaya (1863).

2. Matthew Seuter, *Siciliae Regnum cum adjacente insula Sardinia, Marchionat us Brandburgens is Ducatus Pomeraniae et Ducatus Mecklenburgicus* (1730), reproduced in Gohm (1972, 55, 57).

3. Louis Renard, *Daniae, Frisiae, Groningae et orientalis Frisiae* (from *Atlas de la navigation et du commerce*, Amsterdam, 1715), reproduced in Tooley (1949, facing 44). Joshua Ottens, *Barbariae et Guineae Maritimi à Freto Gibraltar ad Fluvium Gambiae* (from *Atlas van Zeevaert en Koophandel door de Geheele Weereldt*, 1745, reproduced in Tooley (1949, facing 99).

4. *English Pilot* (1689). For example, John Thornton, *A Chart of ye North part of America; A General Chart of the West India; A Generall Chart of the West Indies;* all these extensive in latitude range.

5. For example, zodiacal maps in Royer (1679), reproduced in Warner (1979, 215); *Zodiacus stellatus cujus limitibus planetarum omnium visibiles viae comprehenduntur* (from Seller 1679), reproduced in Warner (1979, 235).

6. John Senex, *Zodiacus stellatus fixas omnes hactenus cognitas, ad quas lunae appulsus ullib terrarum telescopio observari poterunt, complexus* (1718), reproduced in Warner (1979, 240–41).

7. Examples are maps by Alexis Hubert Jaillot (*Carte particuliere des postes de France*, 1695), Jean Baptiste d'Anville (*Royaume de Corée*, 1737), and John Mitchell (*A Map of the British and French Dominions in North America*, 1755), reproduced in Tooley and Bricker (1968, 88, 129 and 241, respectively). See Fite and Freeman (1926, 180) for a better reproduction of the Mitchell map. Other maps on the trapezoidal projection reproduced in Fite and Freeman (1926) are on pp. 172 (Hennepin, 1683), 186 (Kitchin, 1755), 218 (anon., 1763); in D. Gohm (1972) on pp. 9 and 56 (Seutter, 1730), 42 (Bellin, 1753), 52 (Visscher, 1690), 64 (Seutter, ca. 1740), 94 (Jefferys, 1758), 109 (Nicholls, ca. 1700).

8. Hevelius (1687), 54 maps of constellations, reference in Warner (1979, 113). Charles Messier ("les dernières observations faites en Mai et Juin" of comet of 1759), reference in Warner (1979, 178).

9. Reproduced in Tooley and Bricker (1968, 39). See also J. R. Smith (1986, 57–66).

10. Reproduced in Fite and Freeman (1926, 290–91); see 180 for first edition. Also see Ristow (1972, 102–13).

11. See n. 7; also Sanson (1675); G. De l'Isle (1730), maps of *Guienne et Gascogne* and other provinces and small countries; Homann (1750), *Magna Britannia, Scandinavia*, etc.

12. Pietro Todeshi, *Nova et acurata totius Americae tabula* (1673), reproduced in Ristow (1972, 66–69).

13. Willem Janszoon Blaeu, *Nova totius terrarum orbis geographica ac hydrographica tabula* (Amsterdam, 1606 through much of the seventeenth century), reproduced in Shirley (1983, 272).

14. G. De l'Isle (1730), *Hemisphere septentrional pour voir plus distinctement les Terres Arctiques* and ditto but *Hemisphere meridional . . . Australes,* marked 1714. De l'Isle and Buache (1700–1763), same 1714 maps as those in G. De l'Isle (1730). For polar stereographic celestial hemispheres, see Warner (1979, 45 [Brunacci, 1687], 50 [G. D. Cassini, 1681], 67 [Doppelmayr, 1730], 77 [Eimmart, 1705], 108 [Halley, 1678]) for just the first few illustrations in her alphabetically listed book. Several others are illustrated, or are listed in the text. She lists about forty maps between 1670 and 1780.

15. Sanson (1675). Edward Wells, *A New Map of the Terraqueous Globe* (1701), reproduced as endmaps in Lewis and Campbell (1951). John Senex, *A New Map of the World* (1721), reproduced in van de Gohm (1972, 88). J. G. Doppelmayr, *Basis Geographiae Recentioris Astronomica* (Nuremberg, 1733), reproduced in Bagrow (1964, pl. 111). G. De l'Isle (1730), *Mappe-Monde.*

16. De l'Isle and Bauche (1700–1763), *Carte d'Afrique* and *Carte d'Amerique.* Herman Moll, *Map of North America* (1720), reproduced in Schwartz and Ehrenberg (1980, 140). Other examples of continental maps on the equatorial stereographic are by John Senex (1720) and Thomas Kitchin (1765), reproduced in Gohm (1972, 91, 92), and by Thomas Jefferys (1758), reproduced in Gohm (1972, 110).

17. Homann (1750), *Europa, Asiae . . . , Africae . . . , totius Americae septentrionalis et meridionalis.*

18. Matthias Seutter, *Accurata utopiae tabulae, Das ist . . . des . . . Schlaraffenlandes Neu-erfundene lächeriche Land-Tabell* (Augsburg, ca. 1730), reproduced in Hill (1978, 57, 58).

19. Homann (1750), *Représentation de l'Eclipse Partiale de la Lune,* Aug. 8–9, 1748.

20. Noel André, *Planisphere céleste boréal projeté sur le plan de l'équateur* (Paris, 1778), reproduced in Warner (1979, 5). Warner calls it polar stereographic, and it is centered at the point over the earth's North Pole, but the principal graticule is relative to the ecliptic, 23½° oblique.

21. Thomas Jefferys, *The Royal Geographical Pastime: Exhibiting a Complete Tour Round the World* (London, 1770), reproduced in Hill (1978, 9).

22. Mead (1717, 64–69); Lambert (1772; 1972 trans., 66–71).

23. Based on a count of the listings in Warner (1979).

24. L. A. Brown (1949, 219). See also p. 29, this vol.

25. Philippe Buache, *Carte des terres australes* (Paris, 1739, 1754), reproduced in Hill (1978, 30–31).

26. Mead (1717); see chap. 1 for various quotations from Varenius/Blome.

27. Homann (1750), *Sphaerarum artificialium typica repraesentatio.*

28. Ignace-Gaston Pardies, *Globi coelestis in tabulas planas redacti descriptio* (Paris, 1674), description and one reproduced in Warner (1979, 196–97). See also Fisher and Miller (1944, 92–97).

29. Warner (1979, 191, 102, 66, 140–41, respectively).

30. See chap. 1, n. 45, for a translation of Varenius's 1650 description of construction of the equatorial orthographic projection.

31. Reproduced and discussed in Robinson (1982, 69–71), and Thrower (1972, 65, 66, 68), respectively.

32. In Dalrymple (n.d.), reference and map reproduced in Tooley and Bricker (1968, 249).

33. Schwartz and Ehrenberg (1980, 204) shows a detail of the map.

34. Tooley and Bricker (1968, 132); Schwartz and Ehrenberg (1980, 152).

35. Guillaume De l'Isle, *L'Amerique septentrionale* (1700), reproduced in Lunny (1961, 33). G. De l'Isle, *Carte du Mexique et de la Floride* (1703), reproduced in Schwartz and Ehrenberg (1980, 143). G. De l'Isle (1730), *L'Europe, l'Asie, l'Amerique Septentrionale, Canada, America Meridionalis, Mexique et de la Floride.* De l'Isle and Buache (1700–1763), sectional maps of Africa and South America, *Mexique et de la Floride, Canada.*

36. Edward Wells, *A New Map of Africk* (1700), detail reproduced in Tooley and Bricker (1968, 172). For celestial map references to the use of the sinusoidal, see Warner (1979, 178–85 [Messier, 1762–90], 80–81 [Flamsteed 1729], 35 [Bode, 1782]). For Flamsteed's rationale in using the sinusoidal, see quotation in this volume, p. 50.

37. Warner (1979, 35 [Bode, 1782], 80 [Flamsteed 1729], 84 [Fortin, 1776]).

38. Guillaume De l'Isle, map of Africa (detail; title not given), reproduced in Tooley and Bricker (1968, 171). Vincenzo Maria Coronelli, map of Australia (Niew Hollandt), reproduced in Tooley (1949, 119).

39. De l'Isle and Buache (1700–1763), *Carte d'Europe, Carte d'Asie.*

40. For various descriptions of the survey of France and of the expeditions to Lapland and Peru, see L. Brown (1949, 241–55); Clarke and Helmert (1911, 802); J. R. Smith (1986, 69–193); Konvitz (1987, 1–31).

41. Glaisher (1911, 869–70); Struik (1948, 119–21).

42. D'Alembert (1752, 60ff.). Jim Cross, University of Melbourne, pers. com., 1983. Struik (1948, 182–83, 219–22).

43. Maurer (1935, [S8–S11, S13], 19, tab. pl. 1; 1968 trans., 27–28, 212, quotation from p. 212). Parent (1704); reference in d'Avezac-Macaya (1863, 344–45).

44. Philippe Buache, *Planisphere physique* (Paris, ca. 1760), reproduced in J. Potter (1989, 178–79).

45. For the Fournier and Nicolosi globular projections, see pp. 40–42.

46. Mead (1717, 73–75): "Another Way to project Circular Maps."

47. Mead (1717, 77–78): "To draw a Map of a mix'd Kind, whose Meridians are strait lines, and Parallels Arches of Circles."

48. J. N. De l'Isle (1745). See also Bagrow (1964, 175); Fiala (1952; 1957 German trans., 307).

49. Steers (1970, 239). Gesellschaft (1891, pls. 1–15).

50. Herman Moll, *A New and Exact Map of the Dominions of the King of Great Britain and ye Continent of North America* (1715), reproduced in Schwartz and Ehrenberg (1980, 138). Emanuel Bowen, *A New & Accurate Map of the Southern Parts of Africa*, in his *Complete System of Geography of the Known World* (1744–47), reference and map reproduced in Tooley and Bricker (1968, 172).

51. G. De l'Isle (ca. 1730), *Italie*, map reproduced in Tooley and Bricker (1968, 86–87). De l'Isle and Bauche (1700–1763). Lewis Evans, *A Map of Pensilvania, New-Jersey, New-York, and the Three Delaware Counties* (1749), reproduced in Schwartz and Ehrenberg (1980, 154). Evans, *A General Map of the Middle Colonies, in America* (1755), reproduced in Schwartz and Ehrenberg (1980, 165); as copied by Thomas Jefferys (1758), reproduced in Lunny (1961, 37). William Faden, *The British Colonies in North America* (London, 1777),

reproduced in Fite and Freeman (1926, 232). Charles Messier, *Route de la comete observée à Paris en . . . 1758*, in *Académie Royale des Sciences, Mémoires* 1759 (pub. 1765), opp. 188, reproduced in Warner (1979, 177).

52. Schickard (1623, 1687 ed.), star maps, reference, and reproduction in Warner (1979, 224–28).

53. Warner (1979, 268, 55, respectively). Lambert (1772; 1972 trans., 37, 38) states that Zimmermann's maps in effect use a cone constant of 5/6.

54. Murdoch (1758); Schott (1882, 291); Steers (1970, 111–15); Hinks (1941b).

55. Büsching (1754ff.), introduction by Murdoch containing description of his third projection. Reference and quotation of geometric description in Maling (1983, 110, 115, 116).

56. Euler (1777); Biernacki (1949; 1965 trans., 58, 209, 301–3).

57. Colles (1789; 1961 reprint, 3–116, biography of Colles).

58. Colles (1794, iii–viii). See also Ristow (1979, 328–33).

59. L. A. Brown (1949, 51–52, 254). T. Craig (1882, 80).

60. Soldner developed a coordinate conversion for the ellipsoid that was comparable in intent to that of Cassini. See p. 97, this vol.; Jordan and Eggert (1939, 309, 327, 385; 1962 trans., 115–16, 134, 191). Reignier (1957, 98–99). Steers (1970, 229, 233).

61. See also Lagrange (1779, 1869 *Oeuvres*, 4:692), using Lambert, and Close and Clarke (1911, 663), discounting the transverse Mercator.

62. Christlieb Benedict Funke, *Coniglobium boreale* and *coniglobium australe*, in *Anweisung zur Kenntniss der Gestirne auf zwei Planiglobien und zwei Sterntegeln, nach Bayern und Vaugondy* (Leipzig, ca. 1777), reference and southern conic map reproduced in Warner (1979, 86–87).

63. The Zimmermann "star cones" are the equidistant conic maps mentioned on p. 68 in this volume and listed in Warner (1979, 268).

64. Lagrange (1779). For a thorough description of the derivations, see O. S. Adams (1919a, 80–95, 111–14).

65. Lagrange (1779; 1869 *Oeuvres*, 4:692). O. S. Adams (1919a, 111–14).

66. Germain (1866, 101), adding a footnote that the Greek *stenoteros* means "more right"; O. S. Adams (1919a, 116; 1945, 34–35). Steers (1970, 103) says the projection "is often called Lambert's conical equal-area projection."

CHAPTER THREE. MAP PROJECTIONS OF THE NINETEENTH CENTURY

1. Lambert's projections that are now called the Lambert azimuthal equal-area projection, the transverse Mercator (as adapted by Carl Friedrich Gauss to the ellipsoidal earth), and the Lambert conformal conic were used during the nineteenth century, but frequently as independently derived projections. The other Lambert developments, the cylindrical equal-area, its transverse form, the conical equal-area, and the "Lagrange," have remained relatively dormant except among projection scholars, aside from recurrent attempts in the twentieth century to promote reinvented cylindrical equal-area projections with different parallels of no distortion (see pp. 164–66).

2. A Mercator world map is the frontispiece of Morse (1814). Both Mercator and globular world maps are shown in Warner and Carey (1820, maps 1 and 2); J. G. Bartholomew (1890, pls. 2 and 7); and Rand McNally (1896, 33–38). Both Mercator and equatorial stereographic world maps appear in Stieler (1855, maps 6–9); and Fay (1869, pls. II–IV).

3. William Croswell, *A Mercator Map of the Starry Heavens* (Boston, 1810), reproduced in Boston (1985, 44–45, 56).

4. For example, Stieler (1855, Australia and Polynesia, map 50a); J. G. Bartholomew (1893?, Oceania and Pacific, 120); Rand McNally (1896, Australia and Oceania, 304–5).

5. For more comments on the geographic use of the Mercator projection, see pp. 156–57.

6. Warner and Carey (1820, map 51); Stieler (1855, map 44c).

7. See pp. 59–60, this vol.; Robinson (1982, 104, 105, 107, 115, 180, 204). See also examples in Debes (1895, map pls. 2–5, 6–9); Berghaus (1845, vol. 1, group 1, pls. 6, 7, 9; group 2, pl. 1; group 4, pls. 1, 3, 5; group 5, pls. 1, 2.

8. United States maps are shown in Cornell (1859). Abraham Bradley, Jr., *Map of the United States, Exhibiting the Post Roads* (1796), reproduced in Schwartz and Ehrenberg (1980, 210); also in Ristow (1985, 70). Henry S. Tanner, *Travellers Guide or Map of the Roads, Canals, and Rail Roads of the United States* (1834), reproduced in Ristow (1985, 202, 205). Map of the Mediterranean Sea in Bradford (1835, 116).

9. Debes (1895, pls. 32 [Russia] and 48 [the Nile land]).

10. Jordan and Eggert (1939; 1962 trans., 115–16, 134). On p. 134, Jordan and Eggert state that "Soldner's developments, originating from 1810, were not published until 1873 in *Bayerische Landesvermessung in ihrer wissenschaftlichen Grundlage,* München, 1873, pp. 263–281, and are newly edited by J. Frischauf: 'Theorie des Landesvermessung von Johann Soldner,' *Ostwalds Klassiker der exakten Wissenschaften,* No. 184, Leipzig, 1911."

11. Steers (1970, 208–10, 229, 242). Maling (1973, 207–8, 216–17).

12. Biernacki (1949; 1965 trans., 211); Lambert (1772; 1972 trans., xiii–xiv).

13. Gauss (1825). See also Lee (1976, 100–101). Schreiber (1866), reference in Jordan and Eggert (1941; 1962 trans., 164).

14. Schreiber (1897, 1899–1900). References and discussion in Jordan and Eggert (1941; 1962 trans., 241, 263, 266, 283); Biernacki (1949; 1965 trans., 190–92, 194, 212). O. S. Adams (1921, 8–10, 16–60) derived the relationship between conformal (he calls it isometric) and geographic latitude in great detail without mentioning the earlier works.

15. Bell (1937, 218–69); Struik (1948, 203–9).

16. For eastern and western hemispheres, see De l'Isle and Buache (1833, *Mappemonde,* no pagination); Fay (1869, pls. II–III); Stieler (1855, maps 6–8). For northern and southern hemispheres, see Bradford (1835, 144); Stieler (1855, maps 6–8). Oblique stereographic hemispheres appear in Fay (1869, pl. I); Stieler (1855, maps 6–8); B. E. Smith (1897, map 2). In Berghaus (1845), there is considerable use of the polar, oblique, and equatorial stereographic (polar: group 1, pl. 2; group 4, pl. 2; oblique: group 3, pl. 1; group 4, pl. 4; equatorial: group 2, pls. 3–5; group 3, pl. 9). By the third edition of Berghaus (Justus Perthes, 1892), there are no stereographics, but there are seven Lambert azimuthal equal-area projections (see n. 20). The oblique stereographic regional maps are in Debes (1895, pls. 47 [Kapland, or south Africa] and 50 [equatorial Africa]).

17. J. G. Bartholomew (1893?, 1); Rand McNally (1896, 37, 39).

18. A north polar hemisphere of 1814 by John Thomson is reproduced in Gohm (1972, 32); another is shown in Cornell (1859, no pagination). North polar regions are shown in Mathew Carey (1818, map 57); Rand McNally (1896, 282–83); B. E. Smith (1897, maps 2 and 3).

19. George Buchanan, *Northern Hemisphere Projected on the Plane of the Horizon of*

London, in Thomson (1816), reference and reproduced in Gohm (1972, 29); Debes (1895, pls. 38 [Asia, centered at 40° N, 90° E] and 45 [Africa, centered on the equator at 15° E]).

20. North and south polar aspects were used by J. G. Bartholomew (1890, pl. 8); Justus Perthes (1892, maps 4, 5, 44). Equatorial aspects (eastern and western hemispheres) are in Justus Perthes (1892, maps 1, 4, 62). Oblique aspects (hemispheres centered on Berlin and its antipode) are in Justus Perthes (1892, map 16). Debes (1895) did not use this projection among his generally wide variety.

21. Lorgna (1789, chaps. 9–11, pp. 68–94); reference and discussion in d'Avezac-Macaya (1863, 443–45). D'Avezac-Macaya quotes Lorgna's colleague Cagnoli (1799), who treated Lorgna as the originator. Also see Schott (1882, 290).

22. See maps of Russia in De l'Isle and Buache (1833), of North America in Carey (1818, map 3) and J. G. Bartholomew (1890, pl. 59).

23. Debes (1895, pls. 11–12, 39, 40, 42, 44, 46, 52, 55, 59).

24. Debes (1895, pl. 49). The pole of the projection was made 10° S, 5° W.

25. All continents (except Antarctica) were plotted on the Bonne or sinusoidal projection in Stieler (1855), J. G. Bartholomew (1890), and Justus Perthes (1892). South America is on the sinusoidal only in Bartholomew of these three, but they are not representative of the atlases checked by this writer (see table 3.1).

26. This was encouraged by Bonne (1788) in his own atlases, where the Bonne projection was used for most maps.

27. Biernacki (1949; 1965 trans., 231). Steers (1970, 242) gives a date of 1803 for the change to the Bonne, but Biernacki is more detailed in giving the dates used herein. Hinks (1912, 53).

28. Reproduced at 83% of original size (1859 ed., corrected under Ludwig von Bucholtz) in Sanchez-Saavedra (1975, 55–68, and separate map no. 6 in nine parts).

29. The maps of Asia appear to be constructed on the globular projection in Morse (1814, facing 516), and Carey (1818, map 51). North and South America are on the globular in a map by John Arrowsmith (ca. 1835), reproduced in Gohm (1972, 75), and South America seems to be on the globular in Carey (1818, map 32). For a discussion of atlas sheets of India, see Edney (1991).

30. Maurer (1935, [S108], 38; 1968 trans., 68); Tissot (1887, 61, 119).

31. There can be some special properties, such as that of the cylindrical form of the satellite-tracking projections, with groundtracks of artificial mapping satellites shown as straight lines, described in Snyder (1981a).

32. For example, see Deetz and Adams (1934, 30–35). It is mentioned only in passing as the normal form of the Wetch projection in Gretschel (1873, 131).

33. Maurer (1935, [S93], 35; 1968 trans., 61). Also see Herz (1885, 124); Germain (1866, 213). Germain refers without date to *Projection of the Globe on the cylinder of a meridian, by J. Wetch,* in Bibliothèque Impériale, Kl. 1112.

34. Debes (1895, pls. 43 [Southeast Asia, pole of projection 45° N, 207°30′ E] and 57 [Central America, pole at 45° N, 25° E]). Also see Zöppritz and Bludau (1899, 146). Zöppritz died in 1885.

35. Used for several world maps in J. G. Bartholomew (1889, pls. 2–12; 1890, pls. 3–5).

36. Braun (1867, 272). The quotation is translated by the author from Braun's statement "so erhält man ziemlich richtig die Abstände der Parallele vom Aequator für die Mercator's Projection."

37. Mollweide (1805b). The title refers to Schmidt (1801–3), in which Schmidt presents his globular projection. See also d'Avezac-Macaya (1863, 449, 451).

38. Freiesleben (1970–80); Snyder and Steward (1988, 66).

39. Germain (1866, 319–22); Deetz and Adams (1934, 155–56).

40. Babinet (1859). Babinet (1861, "Planisphère Babinet, physique et politique. Projection homalographique de M. Babinet").

41. Justus Perthes (1892, maps 3, 37–38, 45–46, 51, 61, 63, 64–65, 74). Also used by J. G. Bartholomew (1889, pl. 13).

42. See also Gretschel (1873, 183, fig. XXVIII); Herz (1885, 170); O. S. Adams (1919b, 17).

43. Albers (1805a, 457–59). The "small individual part" is on p. 457, where Albers says: "3. Der *Flächeninhalt* stimmt mit dem der Kugel auf das genaueste überein, und zwar nicht bloss der Inhalt der ganzen Zone, sondern auch der eines jeden noch so kleinen einzelnen Theils derselben."

44. Germain (1866, 104–6); d'Avezac-Macaya (1863, 355, 479); Albers (1805b); T. Craig (1882, 113–14).

45. Bonacker and Anliker (1930, 239); d'Avezac-Macaya (1863, 355); Germain (1866, 104, 190).

46. For the generic use of *polyconic,* see O. S. Adams (1919a, six categories on p. 13). For a biography of Hassler, see Wraight and Roberts (1957, 4–14). For the naming of Hassler's polyconic, see Hunt and Schott (1854, 99).

47. Wraight and Roberts (1957, 5). The act was dated Feb. 10, 1807.

48. On p. 232, it is noted that this is one of his papers communicated in 1820.

49. Samples are reproduced in Guthorn (1984, 30–207 *passim*).

50. For maps of the United States, see Emil Mahlo, "Political Map of the United States Showing by Congressional Districts the Geographical Distribution of the Political Parties", 1877, reproduced in Schwartz and Ehrenberg (1980, 305); B. E. Smith (1897, map 5). A map of Canada is in B. E. Smith (1897, map 59). For maps of North America, see U.S. Coast Survey (1867), with an outline map on p. 177 and a table of coordinates on pp. 177–86; John Wesley Powell, "Linguistic Stocks of American Indians North of Mexico," 1891, reproduced in Schwartz and Ehrenberg (1980, 316); B. E. Smith (1897, map 4). For a map of Asia, see Appleton (1872, facing 81). For a map of Oceania, see Mitchell (1866a, map 43); but Mitchell (1866b, map 88 of Oceania) is based on the Mercator projection.

51. Hunt and Schott (1854, 99, 100). See also Shalowitz (1964, 138–39).

52. H. James (1860). Also see O. S. Adams (1919a, 13–23, esp. 18); Steers (1970, 118–19).

53. Hunt and Schott (1854, 100); T. Craig (1882, 208–9); Shalowitz (1964, 139–40).

54. Harding (1808–22, pls. 19–26). See also d'Avezac-Macaya (1863, 355–56, 479); Herz (1885, 254).

55. Tissot (1881, 139; 1887, 92). See also Herz (1885, 113, 254); Khanikof (1862).

56. Biography in Evans (1970–80). T. Craig (1882, 49, 55–59, 242) discusses Herschel's and Lambert's work on the conformal conic a few pages apart without noting the commonality.

57. T. Craig (1882, 59–66, 242). Polar coordinates given in the "Boole's Projection" and "John Herschel's Projection" tables on p. 242 agree with those for the Lambert conformal conic using the same cone constant and sphere or ellipsoid.

58. "Uebersicht der Literatur" (1877); Tissot (1887, vi, 64, 133). Maurer (1935, [S22], 19–20; 1968 trans., 29–30). Royal Society (1966, 54).

59. Tissot (1881, 226; 1887, 143–44). Also see Maling (1960, 260 and table facing 256).

60. Tissot (1881, 226). Kavrayskiy (1934, 104–6); Maling (1960, 260).

61. See also Maurer (1935, [S7], tab. pl. 1; 1968 trans., 212).

62. Tissot (1887, 65, 75, 127, 128); Nell (1885–86); Maurer (1935, [S16], 19, tab. pl. 1; 1968 trans., 28, 212).

63. Murchison (1857). Also see James and Clarke (1862, 311).

64. Also see Maurer (1935, [S6], 18, tab. pl. 1; 1968 trans., 27, 212).

65. Herz (1885, 64–77); Tissot (1881, 98, 204).

66. Bell (1937, 259). Scarborough (1930, 304).

67. T. Craig (1882, 91–97); Young (1920, 2–7); Andrews (1938b); Daskalova (1982); Hojovec (1977); Tolstova (1969).

68. James and Clarke (1862, 309–11). See also T. Craig (1882, 95–97); Close and Clarke (1911, 655–56); Snyder (1985a, 65–67).

69. Clarke (1879, 204–5). Close and Clarke (1911, 656), but the illustration (fig. 12) in this reference is incorrect, since the height used is apparently 3.236, based on a horizon of 108°, rather than Clarke's 1.4 for a minimum-error range of 108° (with a horizon of 135°35′).

70. Breusing (1892). The year 1892 has been given as the date of introduction, e.g., in Maurer (1935, [S25], tab. pl. 1; 1968 trans., 213); it is described in the earlier Tissot (1887, 63). See also Young (1920, 7).

71. Debes (1895, pls. 54, 58). These were oblique aspects, centered at 45° N, 100° W, and 20° S, 59° W, respectively.

72. Tissot (1887, 127–28); Maurer (1935, [S18], 19, tab. pl. 1; 1968 trans., 28, 212). See Hermann Wagner (1925) for biographical notes about Hammer.

73. Arden-Close and George (1952) reminds the reader of the Wiechel projection.

74. A popular user of the name *Aitoff equal-area projection* was Deetz and Adams (1934, 150–53), although a footnote on p. 150 credits Hammer and says the "projection may justly be termed the Hammer-Aitoff projection." O. S. Adams (1925, 106) also apologized separately.

75. Also see Dahlberg (1962, 46, 47).

76. Also note subtle differences from Mead's approach; see p. 67.

77. Maurer (1935, [S174], 55, tab. pl. 5; 1968 trans., 107, 220). Hinks (1912, 64).

78. Jackson (1968, 328). Maurer (1935, 65; 1968 trans., 132, 133). Weir's *Azimuth Diagram* was published in London (1890), and as *Time Azimuth Diagram* by the U.S. Navy Hydrographic Office (1891), reference in Bowditch (1966, 572). Maurer (1905) proved its identity with the Littrow projection.

79. If n is made 2.0 and $\alpha = 0$ in those formulas (see p. 82), the Littrow is a transverse Lagrange with the North Pole 90° east or west of the central meridian of the normal aspect of the Lagrange.

80. See a detailed mathematical analysis in O. S. Adams (1925, ii, 67, 89–90, 94–98).

81. Boerner (1970–80); Struik (1948, 232–36). See Bell (1937, 406–32, 484–509) for biographies of Weierstrass and Riemann, respectively.

82. For example, Pierpont (1896); Frischauf (1897); Eisele (1963). The U.S. Coast and Geodetic Survey Chart (no. 3092) is the subject of Stanley (1946).

83. Eisele (1970–80a, 480; 1970–80b).

84. For formulas with programmability described, see Snyder and Voxland (1989, 235–36).

85. Jäger (1865a, 1865b). See Jäger (1865b, 67–68) for Petermann's editorial comment. See also Maurer (1935, [S232], 77, pl. 7, tab. pl. 7; 1968 trans., 165, 206, 224); Germain (1866, 369–70).

86. Maurer (1935, [S229], 76, tab. pl. 7; 1968 trans., 162–63, 223).

87. Tissot (1887, 190); Maurer (1935, [S230], 76, tab. pl. 7; 1968 trans., 163, 223).

88. For equations for the Berghaus star projection, see Snyder and Voxland (1989, 231–32).

89. For a biography of Chebyshev, see Youschkevitch (1970–80).

90. For a biography of Grave, see Volodarsky (1970–80). For a listing of eight papers with a Chebyshev projection in the title, see Snyder and Steward (1988, 103).

91. For equations in real (not complex) form see Snyder and Voxland (1989, 235).

92. The August projection is displayed and described in Deetz and Adams (1934, 160, 192, pl. VIII); used in transverse form in Spilhaus (1942); reproduced from Spilhaus in Greenhood (1964, 116); and the subject of Schmid (1974).

93. See also Hinks (1912, 65–66); Zöppritz and Bludau (1899, 127–28).

94. Germain (1866, 133). No original reference is given by Germain.

95. Schott (1882, pl. VI), with one face of the cube centered at 30° N and 100° W, used as the basis of a sketch in Hinks (1912, 45), as pointed out in Hinks (1921, 454).

96. The use of a patent (and occasionally a copyright) to protect the inventor against indiscriminate use of a map projection is generally of mixed value. Because there are so many freely available projections equal to or better than those patented, as viewed by the more objective or naïve user, the protection sometimes insures the dormancy of the proposal, contrary to the inventor's dreams.

97. Fisher (1943, 606). Date of patent Jan. 31, 1851.

98. Tissot (1881, 41–72). See also Tobler (1974b). Earlier related work was Tissot (1860).

99. Hinks (1912; 1921 ed., 149–50); Reignier (1957, 273).

100. Fiorini (1881, 216–18, 526–28). Although Maurer (1935, 65–66, 87; 1968 trans., 134, 196) references p. 526 of Fiorini as the basis of his second "new" projection, the discussion on pp. 526–28 seems to concern the Littrow projection, not a modification.

101. Maurer (1935, [S15], tab. pl. 1; 1968 trans., 212). James and Clarke (1862). Origin of the term *scenographic* is discussed in this volume on p. 18.

102. Tissot (1887, 51, 204); Maurer (1935, [S190], 65–66; 1968 trans., 134). See also n. 100.

103. Quoted in Modelski (1984, xix); see his map reproductions (1984, 90–93, 104–5, 110–13) for examples of maps distorting sizes of states to emphasize particular railroad routes.

104. Robinson (1982, 96–97, 124); counts are from Snyder and Steward (1988).

105. Tissot (1881, 14–40, esp. 14; 1887, 8–24, esp. 8–9).

106. Tissot (1881, 40) refers to extraction from Tissot (1878). Also see Tissot (1859).

107. Tissot (1881, 14–25, 43; 1887, 8–15, 26).

108. Tissot (1881, 15–20, 43; 1887, 9–12, 26).

CHAPTER FOUR: MAP PROJECTIONS OF THE TWENTIETH CENTURY

1. See, for example, Maling (1974a), focused on variations of the cylindrical equal-area projection by Behrmann, Edwards, and Peters.

2. Campbell (1987); Thrower (1972, 24); this vol., pp. 97–99.

3. J. Bartholomew (1948, 1 [Atlantis projection], 2–3 [Regional projection], 8 [azimuthal equidistant projection centered on London]). Revision as J. Bartholomew (1953) reuses these projections. J. Bartholomew (1942, 2–3 [Bartholomew's Tetrahedral and Kite projections], 12–15 [interrupted sinusoidal projection]; 1950, 22–23 [Nordic projection]). In a recent atlas, J. C. Bartholomew (1982), the Atlantis (p. 11), Regional (pp. 14–15), and Nordic (pp. 22–23) projections were still in use.

4. See National Geographic (1990); Rand McNally (1983); Times Books (1985).

5. *New York Times,* "Airplanes and Maps," editorial, Feb. 21, 1943, p. E9. Responding letters by Richard Edes Harrison, p. 18, of Feb. 26 issue, and Logan Cresap, p. E13, of Feb. 28 issue.

6. In sample Rand McNally world atlases (1916, 4–5;1932, 12–13; and 1949b, 14–15), the sole or principal world map is a Mercator, so labeled. The 1932 atlas also has eastern and western hemispheres (pp. 10–11) on the Lambert azimuthal equal-area projection, and the 1949 atlas has a north polar azimuthal equidistant projection (pp. 12–13) extending to latitude 60° S, with a south polar inset. Rand McNally (1983), however, instead of a Mercator map, uses the Robinson projection (pp. 2–5) for the principal world map, and the Goode homolosine and Briesemeister projections for most secondary world maps (pp. 292–312, 315–19). There is also a Mercator time-zone map (p. 313) and a Miller cylindrical ocean-current map (p. 320). These projections are discussed later in this chapter. On the other hand, the listing of different world sheet maps available in 1988 in The Map Store Inc., in Washington, D.C., shows 17 Mercator, 6 Van der Grinten, and 5 other projections, from 15 companies and government agencies.

7. Times Books (1985, pls. 15, 18, 19, 25) for New Guinea, Indonesia, and Indochina. Rand McNally (1983, 112–15) for Indonesia. Equatorial regions of Africa or South America in the three atlases cited in n. 4 are shown on the same non-Mercator projection as other regions of the same continent.

8. Steers (1970, 208–12); Harley (1975, 18–28); Maling (1973, 20–23, 216–17).

9. Mitchell and Simmons (1945); Claire (1968); Snyder (1987b, 51–57, 105–6).

10. O'Keefe (1948, 1952); U.S., Department of the Army (1973); Snyder (1987b, 57–58, 59, 60–64). For other references see the index in Snyder and Steward (1988).

11. Mugnier (1985). Maling (1973; 1992, 2d ed., 311). Graur (1956, 99–105) describes 6°-wide zones using the "Gauss" projection, with use of the Gauss for large-scale maps beginning in 1928.

12. Moccia and Vetrella (1986, 355). This paper includes formulas. Boaga (1948, 1: 556–68). Five other papers, in Italian, and dating from 1947, are indexed under "Gauss-Boaga" in Snyder and Steward (1988).

13. Jordan and Eggert (1941, 222, 244, 246; 1962 trans., 241, 263, 266). Schreiber (1897, 1899–1900).

14. Lagrange (1779; 1869 *Oeuvres,* 660–64). Adams uses the term *isometric* latitude for what is now regularly called conformal latitude.

15. Lee (1976, 92–101); Snyder (1979, 73); Dozier (1980); NASA (1991), program developed by David Wallis.

16. Such a map, but apparently based on the spherical (not ellipsoidal) transverse Merca-

tor, is *Global Navigation and Planning Chart* (St. Louis, Mo.: Defense Mapping Agency Aerospace Center, 1984), chart GNC-1, scale 1:5,000,000, edition 7, with central meridians long. 90° E and W.

17. National Geographic (1990, pls. 73, 78, 92–94); Times Books (1985, pl. 9; also pl. 9 of 1st ed., 1958, vol. 1); Lewis and Campbell (1951, 64–65, 70–71, 76–77).

18. National Geographic (1990, pls. 30, 37, 80, 81, 100); Lewis and Campbell (1951, A2–A3).

19. Laborde (1928); Driencourt and Laborde (1932, 4:360–89); Reignier (1957, 130–31). Reignier omits some special formulas and has errors in those provided for the Laborde projection.

20. Hotine (1946–47, 66–67). Also see Thomas (1952, 7–9, 70–74); Snyder (1987b, 66, 70–75).

21. Mugnier (1985); Claire (1968, 25–31, 51–52).

22. Berry and Bormanis (1970); Snyder (1987b, 56, 68).

23. P. 110. For twentieth-century usage, see J. Bartholomew (1942, 4, 5, 8–11); J. C. Bartholomew (1982, 16–19).

24. Maling (1960, 213); Kavrayskiy (1934, 162).

25. *Bol'shoy sovetskiy* (1937, title pages for pls. 6–7 and 36 other pls.): "Tsilindricheskaya stereograficheskaya proyetsiya Golla c parallelyami secheniya 30°."

26. Graur (1956, 110, 268); O. M. Miller (1942).

27. Zetterstrand and Rosén (1926, 39 of 2d page numbering, and pls. 36, 45).

28. Pye (1955); Maling (1966); Steers (1970, 191–92).

29. Fifty papers and books on the "Peters" projection (some by Peters or his supporters) are listed in the index of Snyder and Steward (1988). In addition, numerous newspaper articles have been written, some sympathetic to Peters and some balanced. Of special significance are Maling (1974b); Deutsche Gesellschaft (1982); Arno Peters (1983); and a review of the latter by Robinson (1985).

30. Eight world map projections are shown in Arno Peters (1983, 66). He says, "All these maps have fidelity of area, but they bought this quality at the price of abandoning important qualities of Mercator's map and could therefore not supplant it." The existence of the equal-area property of these and many other projections is not mentioned in the more popularly oriented literature by promoters of the Peters approach.

31. The name *Gall-Peters* is introduced in Robinson (1986, 12).

32. Germany, Press and Information Office (1977). The reprint of this paper in the *ACSM Bulletin* was then reprinted, lacking only the illustrations, last sentence, *and credit line,* as an endorsement by the American Congress on Surveying and Mapping (ACSM), in Universum-Verlag (ca. 1980, 17). The ACSM at no time endorsed the projection.

33. *American Cartographer* (1989); Robinson (1990). Six endorsements are noted therein: the American Geographical Society, the Association of American Geographers, the Canadian Cartographic Association, the National Council for Geographic Education, the National Geographic Society, and the Geography and Map Division of the Special Libraries Association. The Canadian Association of Geographers subsequently approved the resolution.

34. Rand McNally (1916, 84 [South America], 112 [Africa is polyconic here]; 1932, 96 [South America], 128 [Africa]; 1971, 42–49 [Africa], 53 [South America]).

35. Mugnier (1985). For formulas, see Snyder (1987b, 248).

36. For example, J. Bartholomew (1942, 12–15), nonsymmetrically about the equator,

and symmetrically in Zetterstrand and Rosén (1926, pls. 38, 39, 44, 46). The interruptions in the latter are at longitudes 20° W, 160° W, and 60°E, with central meridians midway between the interruptions. Use of this interruption in 1978 by the U.S. Geological Survey is described in Snyder (1987b, 246, 247).

37. See Goode (1917) for the interrupted Mollweide projection and (1925) for the homolosine. See also Dahlberg (1962, 48–49) for a detailed account of the development.

38. Wong (1965, 85). The interrupted Mollweide is also used for five thematic world maps in National Geographic (1990, pls. 7–11).

39. The regular Mollweide was used for a map of the Atlantic Ocean in *Bol'shoy sovetskiy* (1937, pls. 14–15); for a world map in Geographical Survey, Japan (1977, 206); and for thematic world maps in National Geographic (1990, pls. 2, 3, 6).

40. Snyder (1987b, 157); United Nations (1963, 10).

41. For example, the polar azimuthal equidistant is used in National Geographic (1990, pls. 102, 107, 108, 113, 117–18 [celestial]); Times Books (1985, pl. 48); and Lewis and Campbell (1951, 9–11). The polar Lambert azimuthal equal-area is used in Rand McNally (1983, 9; 1986, 93, 232; and 1932, 136). The polar stereographic is used in Times Books (1985, pl. 123) and for Arctic and Antarctic regions in Morskogo (1950, 1: pls. 2, 7).

42. For example, *American Ephemeris* (1941, 327, 329); *Astronomical Almanac* (1982, A-80, A-81).

43. Thomas (1952, 136–39). Snyder (1987b, 15, 160–62). Approaches taken by Krüger (1922); Hungary and the Netherlands are referenced in Jordan and Eggert (1941, 221, 246, 247; 1962 trans., 240, 265–66, 266–67).

44. See this vol., pp. 98, 161, for references on the conformal sphere concept.

45. Roussilhe (1922), reference in Thomas (1952, 136). Letoval'tsev (1968). See also Jordan and Eggert (1941, 218–21; 1962 trans. 236–40).

46. Simpler perspective views also appeared in Encyclopaedia Britannica (1951, pls. 4–7).

47. Dumitrescu (1966, 1968); nine other such papers by Dumitrescu are cited in Snyder and Steward (1988, 24). Schmid (1962); Snyder (1981b; 1987b, 169–81); Deakin (1990).

48. Hammond (1966, 1). U.S. Geological Survey (1970, 329). Rand McNally (1971, 1). The north and south polar hemispheres appear in Union of Soviet Socialist Republics (1967, 3).

49. For polar examples see Snyder 1987b, 187–90. For oblique aspects see Lewis and Campbell (1951, A4–A7, 20–21, 52–53, 60–61, 66–67); National Geographic (1990, pl. 29).

50. Claire (1968, 52–53); Snyder (1987b, 194, 198–202).

51. Rand McNally (1986, 76–77, 93, 94, 138–40, 186–91, 212–15, 220–22, 232; 1983, 9–19). Times Books (1985, pls. 16, 26, 85, 87–92, 94, 96, 114, 117–22). Hammond (1966, 6, 54, 87, 102, 120, 146).

52. Rand McNally (1932, 10–11); Union of Soviet Socialist Republics (1967, 1–2); Graur (1956, 267).

53. Rand McNally (1916, 6–7, 105, 117; 1932, 16–17, 134; none in 1949b, 1983, 1986). Times Books (1985, pls. 10, 17, 27, 49).

54. Beaman (1928, 163); Snyder (1987b, 2, 126–27).

55. Rand McNally (1986, 108–9; also corresponding maps in 1932 ed. of the same atlas); Hammond (1966, 188–89).

56. Rand McNally (1916, 112 [Africa]; 1986, 128–29 [Gulf of Mexico and Caribbean],

192–93 [Middle East], 198–99 [China and Japan], 206–7 [Indonesia]; also corresponding maps in 1932 ed.).

57. Bowie and Adams (1919, 2); Deetz and Adams (1934, 87–90).

58. Raisz (1938; 1948 2d ed., 206–7, 228–29). See this vol., pp. 159–60, for the British and UTM grids.

59. P. 143, this vol.; Hinks (1912; 1921 2d ed., 136–37, 149–50). Maling (1973; 1992, 2d ed., 311). Adjusting Maling's figures after comparing his source, the relative projection usage is transverse Mercator 83%, polyconic 9%, Lambert conformal conic 5%, all others 3%. These are by areas, with overlapping treated as additional area.

60. O. S. Adams (1918, 6–9, 34); Snyder (1987b, 15, 17–18, 108–9).

61. Lewis and Campbell (1951, A8–A21, 22–45, 48–51, 54–59, 62–63, 68–69, 74–75, 80). Rand McNally (1986, 194–97; 1983, 24–25, 32–33, 40–103, 108–11, 116–29, 132–33, 142–43, 158–73, 175–77, 180–87, 206–29, 232–41). In Canada (1974), nearly all maps of Canada have standard parallels 49° and 77° N, although the region north of latitude 80° is generally on the Bousfield modified polyconic projection (not stated) with curved meridians. Geographical, Japan (1977, 6th numbered page from beginning). Times Books (1985, pls. 30, 104–12). National Geographic (1990, pl. 18).

62. Maling (1960, 259–60, 301), referencing Vitkovskiy (1907).

63. Snyder (1985a, 71–74). Reference to Tsinger (1916) and analysis in Graur (1956, 170–71).

64. Snyder (1987b, 98); National Geographic (1990, pls. 19–28); Rand McNally (1983, 178–79).

65. Hammond (1966, 106, 110–11); National Geographic (1990, pls. 32–36, 57, 59, 61–63, 83).

66. For Tissot's approach, see this vol., pp. 124–25. See also Maling (1960, 259, 265); Graur (1956, 177–78); Snyder (1985a, 74–75).

67. Rand McNally (1916, 1932, 1986); Times Books (1985).

68. Everett (1904); Close and Clarke (1911, 657, 658; Murdoch is not mentioned); Young (1920, 40; 1923).

69. Maling (1960, 259–60). Reference to Mikhaylov (1911–12) in Kavrayskiy (1934, 122). Snyder (1978b, 374–76; 1985a, 75–76).

70. Maling (1960, 265); Graur (1956, 156–63, 267–68); *Bol'shoy sovetskiy* (1937); U.S. Central Intelligence Agency (1973, 5, 11).

71. For the globular projection in twentieth-century atlases, see Cram (1941, 6–7); Encyclopaedia Britannica (1942, 1), but not in (1951). These were the only atlas examples found by Wong (1965, 74–75), for the period 1940–60. I found no others.

72. Wong (1965, 102–3). Finch and Trewartha (1936, 611).

73. Graur (1956, 120–21), with no date or reference for Kavrayskiy.

74. Urmayev (1947), reference in Maling (1960, 210). See also Graur (1956, 116, 268). Series formulas in these references are given for latitude in degrees, but are converted here.

75. Kharchenko (1951); Shabanova (1952); references in Maling (1960, 212).

76. Goodman (1979); American Geographical Society (1962–63).

77. Greenhood (1964, 136); Snyder (1987b, 86).

78. Wong (1965, 71); Rand McNally (1949a, xiv–xv); Encyclopaedia Britannica (1951, pl. 23, p. 2–3 of maps).

79. For a biography of Arden-Close, see an obituary by MacLeod? (1953).

80. Graur (1956, 115, 267–68). Solov'ev (1937; 1946 2d ed., 251–57; 1969, 148–52).

81. For a compilation of most known pseudocylindrical projections, see Snyder (1977).

82. For iteration techniques, see Snyder (1987b, 256–57).

83. "Projection d'Ortelius ou d'Eckert III," stated in Reignier (1957, 265). Maling (1973, 242) repeats the error.

84. Wong (1965, 83–84, 96, 98). Encyclopaedia Britannica (1945, pl. 3-I [air pressure and winds maps]). Hammond (1952, economic and other thematic maps). Eckert IV also appears, e.g., in Hammond (1946, inside back cover; 1957, pp. H-45, H-46 and both sides of the back cover). Geographical, Japan, (1977, 234, 242, 244–45, 248–49, 250) uses the Eckert IV projection for the extent of communications and production. Eckert IV insets on *The World* (Washington, National Geographic Society, wall map with the Van der Grinten as the principal map projection, at least 1965–85), and in National Geographic (1990, pls. 7–11).

85. Wong (1965, 84). Williams (1958, 2–3, 6–7; 2d ed. 1964, 6–7). *Bol'shoy sovetskiy* (1937). See this vol., pp. 163–64.

86. Graur (1956, 268). *Bol'shoy sovetskiy* (1937, pl. "11–12–13," scale 1:30,000,000).

87. R. Miller (1948), proposed the 50°28′ version independently, received an objection from Winkel, and apologized in R. Miller (1948, but 1951 reference).

88. Winkel (1921, 251). Snyder (1977, 77) erroneously claims that Winkel erred; the equal-area Mollweide is not involved.

89. Rand McNally (1986, 2–5, 14–15); Wong (1965, 81).

90. Rand McNally 1959 catalog, p. 17. Courtesy of Richard E. Dahlberg.

91. Richard E. Dahlberg, pers. comm., 1989; Dahlberg (1961).

92. *Newsweek,* July 11, 1988, 24. American Congress on Surveying and Mapping (1989).

93. Dean P. Westmeyer, Denoyer-Geppert Co., pers. com., 1976. For a computer approximation with formulas, see Snyder and Voxland (1989, 80–81, 222).

94. Zetterstrand and Rosén (1926, 40 of 2d page numbering, pls. 40–42, 44, 47).

95. J. K. Wright (1955); Deetz and Adams (1934, 162).

96. Maling (1960, 298). Kavrayskiy (1933, 53–59), reference in Graur (1956, 235), who says true scale is at latitudes ±30°, but constants *a* and *b* require ±35°.

97. Maling (1960, 297–98); Graur (1956, 71, 235–36); Solov'ev (1969, 161–62); Morskogo (1950, 2: pl. 73).

98. Maling (1960, 299–300); Graur (1956, 237–39); Solov'ev (1969, 170–73). Use for published maps is not mentioned.

99. Maling (1960, 299); Graur (1956, 239–40), referencing Ginzburg (1949b).

100. K.-H. Wagner (1932, 17, 23). Maurer (1935, [S141], 44; 1968 transl., 79–80). K.-H. Wagner (1949, 179–84).

101. K.-H. Wagner (1932, 17). Maurer (1935, [S128], 40; 1968 trans., 72). K.-H. Wagner (1949, 189–90).

102. The origin of this projection and P_1' was mistakenly credited to Putniņš in Snyder (1977, 66, 67), and Snyder and Voxland (1989, 62).

103. Siemon (1937, 92–93). His initial formulas have a reduced scale, but he then enlarges the projection to agree with the form Wagner used.

104. O. S. Adams (1945, 23, 46–50). See also Deetz and Adams (1934, but 1944 5th ed., 207–8). For a description of the Foucaut projection, see pp. 113–14, this vol.

105. McBryde and Thomas (1949, 14). The authors credit Baar (1947, 114–16) with the development of the "sine series" and a related tangent series.

106. Robinson (1953b; 1953a, 108–10, through 1984 5th ed., 98–99); Wong (1965, 107); Finch and Trewartha (1936, 1957 4th ed.).

107. McBryde and Thomas (1949, 12); Baar (1947, 116); Germain (1866, 88–89); Graur (1956, 224). For a ludicrous example derived from these formulas, see this writer's "hourglass projection" in Monmonier (1990, 98).

108. Maling (1960, 297–98). Urmayev (1950); reference in Graur (1956, 236–37).

109. Maling (1960, 298–99); Graur (1956, 240–42).

110. Graur (1956, 237, 241–42, 268, 269). Union of Soviet Socialist Republics (1954, 275–76; 1967 2d ed., 242–43).

111. Royal Society (1966, 63). John C. Bartholomew (pers. com., 1992) providing all technical information.

112. Times Books (1985, xlv, pls. 7, 8).

113. American Congress (1976). Flyer by Meridian Maps, Dunn Loring, Va., 1975.

114. Robinson (1974, 154). See area corrections in Richardson (1989).

115. See scalar adjustment in Robinson (1974, 151); Snyder and Voxland (1989, 222–23).

116. Rand McNally (1986, 50–51; 1974 14th ed., 2–3, 22–23, 52–53; 1983, 2–5).

117. *Rand McNally World Portrait Map, Based on the Robinson Projection, 1989 Edition.* National Geographic (1990, pls. 12, 13).

118. Baranyi and Györffy (1989); Martinovich (1989, 30); Cartographia (1987, 15, 146–47, 154–55).

119. Baranyi and Györffy (1989). Martinovich (1989). Cartographia (1967, 40; 1987, 14, 16–19).

120. See Snyder (1977, 66–67) for general formulas for truncated elliptical pseudocylindrical projections.

121. For example, see Rand McNally (1986, 236–40 [ocean floor maps]); Odyssey Books (1966, 168). Williams (1958, 4–5; 1–5, 8–13, of 1964 2d ed.) and Quilette/Hachette (1984, 156–57) use the Hölzel (Hoelzel) projection with equidistant straight parallels and curved meridians for several world maps. Philip M. Voxland (pers. com., 1990) determined that the latter is essentially the same as Eckert V between 80° N and S, with circular meridians beyond there and poles as lines 0.37 the length of the equator. Also see Maurer (1935, [S98 to S157], 36–47, tab. pls. 3–4; 1968 trans., 64–86, 216–19).

122. Hinks (1912, frontispiece, 61); Close (1927).

123. K.-H. Wagner (1932, figs. 25, 26, charts I, II).

124. Maling (1960, 207–8); Solov'ev (1969, 114–18); Graur (1956, 199–200); see also pp. 161–62, this vol.

125. Maling (1960, 206); Graur (1956, 198, 200).

126. Maling (1960, 208); Graur (1956, 198, 200).

127. See Snyder and Steward (1988, index under "Magnified centers on projections").

128. For one mathematical approach, see Snyder (1987a, 61).

129. Schoy (1913a, 1913b, 1915). See also Maurer (1935, [S199], 72, tab. pl. 6; 1968 trans., 151, 222).

130. Maurer (1919b, 20–22). Maurer (1925, [S196], 75, pl. 7; 1968 trans., 158–59, 206). For biographical notes and bibliography of Maurer, see Gabler (1943).

131. Times Books (1985, xliv, pl. 1-6); J. Bartholomew (1958, pl. 3-5). Steers (1970, 185) makes the same error.

132. Wong (1965, 127–28). James and Davis (1959), reference in Wong (1965, 128).

133. See also Behrmann (1919); Immler (1919); Maurer (1919a); Thorade (1919); Close (1922).

134. Maurer (1935, 31; 1968 trans., 54). Uses for seismology, meteorology, location of aurorae or meteors, and oceanography are suggested by Thomas (1970, 103).

135. Chamberlin (1947; 1950 ed., 104). Nicolls (1986). National Geographic (1990, back endsheet).

136. Nicolls (1986, 6); National Geographic (1990, pls. 16, 17, 31, 44–48, 52, 53, 90, 91, 98, 99).

137. Nicolls (1986, 5); Times Books (1985, 97, 102).

138. Zetterstrand and Rosén (1926, 39 of 2d numbering, pls. 2, 3). Arden-Close (1942, 70–71).

139. K.-H. Wagner (1941; 1949, 205–8); Maling (1973, 244).

140. The other parameters shown are 61° with 65/180, 67°30′ with 75/180, and 60° with 80/180.

141. U.S. Environmental (1969, 2, 3, 4, 26); World Book (1982, maps 2, 3); Prentice-Hall (1983, maps 10–11); Peter Heiler [before 1988], without projection identification or response to inquiries.

142. Spinnaker Press (1983), reproduced in Monmonier and Schnell (1988, 224).

143. J. Bartholomew (1950, 22–23); J. C. Bartholomew (1982, 22–23).

144. Hall (1948); American Geographical Society (1950–55).

145. Wong (1965, 107–8, 169). Hammond (1966, 2–3) uses the Mercator, but the 1970 ed. (pp. 2–3) uses the Briesemeister. Hammond's other atlases were also changed by then.

146. Solov'ev (1946, 379–82; 1969, 121, 123–24).

147. Maurer (1935, [S231], 76–77, pl. 8, tab. pl. 7; 1968 trans., 163–64, 208, 223). See this vol., pp. 138–40.

148. Maurer (1935, [S233], 77, pl. 8, tab. pl. 7; 1968 trans., 165, 208, 224). See p. 138, this vol., for Jäger projection.

149. Maling (1960, 209–10); Graur (1956, 65–66, 210, 266).

150. The term *isocol* apparently originated in Russia in the nineteenth century, but its provenance is unclear (pers. com. to A. H. Robinson from A. V. Postnikov, 1987). Union of Soviet Socialist Republics (1954, 278–79; 2d 1967 (English) ed., 245–46; 1985 ed., 200–201).

151. Graur (1956, 210). For constants for other forms, see Maling (1960, 210). Ginzburg and Salmanova (1957).

152. Times Books (1985, pl. 84); Rand McNally (1983, 22–23, 134–39).

153. See also Reilly and Bibby (1975–76); Snyder (1985a, 79–92, 147–51; 1987b, 203–12).

154. See also Snyder (1985a, 70–71).

155. See pp. 129–30, 224–25. See also pp. 72–73 for a description of Murdoch III.

156. Lallemand (1911); Penck (1910); United Nations (1963, 22–23).

157. Łomnicki (1927b, 1927c); Hinks (1927a, 1927b, 1927c).

158. Deetz and Adams (1934, 63–66); Snyder (1987b, 131–37). See also index in Snyder and Steward (1988), under "International Map of the World (IMW) Projection," for other papers.

159. Maurer (1935, [S179], 55–56, pl. 5; 1968 trans., 109–11, 203). See also Tissat (1881, 256–57; 1887, 161–62).

160. Solov'ev (1949, 165); reference in Graur (1956, 217). Taich (1937); reference in Ginzburg (1952b, 64–65, 137).

161. Ginzburg (1949b, 1951); Salmanova (1951); Ginzburg and Salmanova (1957); references in Maling (1960, 294–95) and Graur (1956, 217, 268).

162. Maling (1960, 295); Graur (1956, 67, 218, 269); Solov'ev (1969, 192–93); Ginzburg and Salmanova (1957, 30, 105–8, 226–33); Ginzburg (1966).

163. Graur (1956, 218). Three coefficients in Maling (1960, 295) should have minus signs.

164. Haines (1981, 1987). Used but not identified in Canada (1974).

165. O. S. Adams (1919a, 167–71). See also Deetz and Adams (1934, 62). For biographical notes about Deetz, see Shalowitz (1946).

166. Steers (1970, 121–22). National Geographic (1990, pl. 74; also earlier eds.).

167. Geographical, Japan (1977, 6th unnumbered page from beginning).

168. National Geographic (1979). Rand McNally (1983, 242–55); although labeled only "oblique conic conformal," it is almost certainly based on the Miller and Briesemeister bipolar oblique conic conformal projection, but in a portion using only one oblique pole.

169. Schjerning (1904), reference in Dahlberg (1962, 47–48). See also Zöppritz and Bludau (1899; 1912 3rd ed., 185–201); Maurer (1935, [S43, S64–69], 22–23, 27–28, tab. pl. 1, 2; 1968 trans., 36, 47–48, 213, 214).

170. See comments by Zöppritz and Bludau (1899; 1912 3d ed., 200–201).

171. James (1943, 1949); references in Wong (1965, 106–7, 155–57, 167).

172. Encyclopaedia Britannica (1911); H.R.M. (1920). Gardiner (1976, 30–44); J. C. Bartholomew (1984; pers. com., 1992). John Bartholomew and Son, Ltd. is now a division of HarperCollins Publishers, moved from Edinburgh to Glasgow in 1995.

173. Maling (1960, 295–96); Solov'ev (1969, 177–78).

174. Lewis and Campbell (1951, 6, VI–XI). In *Shorter Oxford* (1959), the Oxford projection is used for over sixty maps.

175. Van der Grinten (1904b). Also patented in Canada, Great Britain, and France; reference in van der Grinten (1905a, 359).

176. Zöppritz and Bludau (1899; 1912 3d ed., 179–85). Zöppritz died in 1885.

177. For use of Grinten, see Graur (1956, 219–20); Solov'ev (1969, 201–4). The designations I–IV were used in 1935 by Maurer (1935, S166, S167, S156, S161, respectively; see p. 85; or 1968 trans., 192a. This reference was also noted in Royal Society (1966, 66). Reinventions of III were by McKnight and McKnight Map Co., reference in Wong (1965, 118–20), and by Brooks and Roberts (1976), amplified as still original with them by Brooks and Roberts (1986).

178. See van der Grinten (1904a, 156, 158) for geometric construction of I and IV; Snyder (1987b, 239–45) for geometric construction and forward and inverse formulas of I. Inverse formulas were developed by Rubincam (1981). See Snyder and Voxland (1989, 200–205, 237–38) for plotted graticules and forward formulas for all four projections.

179. Caspari (1905); Goode (1905, 373) for quote used; Littlehales (1905); Reeves (1904).

180. Internal correspondence of National Geographic Society, from G.H.G. to Mr. Bumstead, Nov. 14, 1918, courtesy of the National Geographic Society. See also Garver (1990, 132–33).

181. National Geographic (1922). The first three revisions: National Geographic (1932, 773); Nicholas (1943); National Geographic (1951).

182. Wong (1965, 98, 99, 116, 117). Hainsworth (1949); Heintzelman and Highsmith (1955); Highsmith and Jensen (1958); references in Wong (1965, 117).

183. MacFadden and Hall (1940); Geographical Publishing (1942); references in Wong (1965, 86–87, 153). Debenham (1958, 84–85); National Geographic (1990; 1975 4th ed., 10–13). Commercial mapmakers: Esselte, K & F, Ravenstein, Nystrom, as listed by The Map Store Inc., Washington, D.C., 1988, plus National Geographic Society.

184. Graur (1956, 266–67); Solov'ev (1969, 204).

185. Wong (1965, 116, 118–20). Also see n. 177.

186. Maurer (1935, [S180], 56, 58, pl. 2; 1968 trans., 113, 113a, 200).

187. Funkhouser (1938, 301, 334–35, 364); G. Mayr (1874), reference in Arnberger (1966, 72, 487); Robinson (1982).

188. Tobler (1963a, 60–61) reproduces the cartogram "A New Yorker's Idea of the United States of America" by Daniel K. Wallingford.

189. Monmonier (1990, 34). Southworth and Southworth (1982, 116–30) shows numerous subway maps and related cartograms.

190. Cahill (1929, 1934, 1939). For other papers, see Snyder and Steward (1988, 15).

191. Quotation from Cahill (1929). See also Cahill (1937).

192. Maurer (1922; 1935, [S159, S160, S187], 50, 63, tab. pl. 5, 6; 1968 trans., 93, 127, 219, 221).

193. *Surveying and Mapping* (1962); reference and other information supplied by Charles A. Whitten.

194. Raisz (1944), reference in Wong (1965, 86, 125, 153, 183). Bertin (1967, 294–95).

195. For the earlier use of polyhedra as map projection bases, see pp. 43, 143.

196. Fisher (1943, 1948); Fisher and Miller (1944, 102–5).

197. For a reproduction of the flattened icosahedral globe, see Southworth and Southworth (1982, 40).

198. See, for example, Hallinan (1989), wishing that the National Geographic Society in choosing the Robinson projection had considered Fuller's approach "with no visible distortion of the relative size or shape."

199. For other references, see Snyder and Steward (1988, index under "Polyhedral globe").

200. Bradley (1946), with footnote on pp. 103–4 describing Fisher's version.

201. DeLucia and Snyder (1986). In 1989 this projection was adopted as the centerpiece of the new crest of the Royal Geographical Society (1989).

202. For cylindrical and polyconic examples, see nn. 74, 75, and 162. For pseudocylindricals, see Snyder (1985a, 120–31). For general cases, see Canters (1989); Laskowski (1991).

203. Maurer (1935, 12, 84; 1968 trans., 13, 188–92). In German on p. 12, "Für diese Bezeichnungen wurden nach Möglichkeit deutsche Ausdrücke herangezogen und auch für vorhandene fremdsprachliche vielfach Verdeutschungen vorgeschlagen."

204. John D. Hill, pers. com., 1991; Gabler (1943).

205. Snyder and Steward (1988, 103). See also this vol., p. 2.

206. Close and Clarke (1911; 1922 12th ed. and 1926 13th ed., same text, volume, and page nos.; text abridged and revised in 1929 14th ed., 14:846–51, and reprintings to 1960). O. M. Miller (1962).

207. Fiala (1952; 1957 German trans.); Zöppritz and Bludau (1899, 1912 ed.).

208. Snyder and Steward (1988, esp. x–xiii); Stephens (1980).

209. Nyerges and Jankowski (1989); Jankowski and Nyerges (1989); D. G. Smith and Snyder (1989); Kessler (1991).

210. Robinson (1953a, but 1969 3d ed. with Sale, 202).

REFERENCES AND BIBLIOGRAPHY

ABBREVIATIONS

GUGK Glavnoye Upravleniye Geodezii i Kartografii (Chief Administration of Geodesy and Cartography)

NIIGAiK Novosibirskiy Institut Inzhenerov Geodezii, Aerofotos'emki i Kartografii (Novosibirsk Engineering Institute of Geodesy, Aerial Photo Survey and Cartography)

TsNIIGAiK Tsentral'nyy Nauchno-issledovatel'skiy Institut Geodezii, Aerofotos'emki i Kartografii (Central Scientific Research Institute of Geodesy, Aerial Photo Survey and Cartography)

Adams, Byron R. 1984. "Transverse Cylindrical Stereographic and Conical Stereographic Projections." *American Cartographer* 11(1): 40–48.

Adams, Oscar Sherman. 1918. *Lambert Projection Tables for the United States*. Washington: U.S. Coast and Geodetic Survey Special Publication 52.

———. 1919a. *General Theory of Polyconic Projections*. Washington: U.S. Coast and Geodetic Survey Special Publication 57. Reprinted 1934.

———. 1919b. *A Study of Map Projections in General*. Washington: U.S. Coast and Geodetic Survey Special Publication 60.

———. 1921. *Latitude Developments Connected with Geodesy and Cartography with Tables, including a Table for Lambert Equal-Area Meridional Projection*. Washington: U.S. Coast and Geodetic Survey Special Publication 67.

———. 1925. *Elliptic Functions Applied to Conformal World Maps*. Washington: U.S. Coast and Geodetic Survey Special Publication 112.

———. 1927. *Tables for Albers Projection*. Washington: U.S. Coast and Geodetic Survey Special Publication 130.

———. 1929. *Conformal Projection of the Sphere within a Square*. Washington: U.S. Coast and Geodetic Survey Special Publication 153.

———. 1936. "Conformal Map of the World in a Square, Poles in the Middle of Opposite Sides." *Bulletin Géodésique* (52): 461–73.

———. 1945. *General Theory of Equivalent Projections*. Washington: U.S. Coast and Geodetic Survey Special Publication 236.

Airy, George Biddell. 1861. "Explanation of a Projection by Balance of Errors for Maps applying to a very large extent of the Earth's Surface; and Comparison of this projection with other projections." *London, Edinburgh, and Dublin Philosophical Magazine*, 4th ser., 22(149): 409–21.

Aitoff, David. 1889. "Projections des cartes géographiques." In *Atlas de géographie moderne*. Paris: Hachette.

———. 1892. "Note sur le projection zenithale equidistante et sur les canevas qui en est dérivé." *Nouvelles Géographiques* 6 (June): 87–90.

Albers, Heinrich Christian. 1805a. "Beschreibung einer neuen Kegelprojektion." *Zach's Monatliche Correspondenz zur Beförderung der Erd- und Himmels-Kunde* 12(Nov.): 450–59.

———. 1805b. "Über Murdoch's drey Kegelprojectionen." *Zach's Monatliche Correspondenz zur Beförderung der Erd- und Himmels-Kunde* 11(Feb.): 97–114; 11(Mar.): 240–50.

American Cartographer. 1989. 16(3): 222–23.

American Congress on Surveying and Mapping. 1976. *ACSM Bulletin* (54): 33.

———. [1989]. "Introduce Yourself to a New World . . . with Membership in ACSM." Falls Church, Va.

The American Ephemeris and Nautical Almanac for the Year 1943. 1941. Washington: Government Printing Office.

American Geographical Society. 1950–55. *Atlas of Diseases.* New York. Seventeen maps on sheets.

———. 1962–63. *AGS Newsletter* (Winter).

Anatupon ek tōn Technikōn Chronikōn. 1963. *E azimouthiakaē isapechousa probolē tou Hatt.* Athens.

Anderson, Paul S. 1974. "An Oblique, Poly-Cylindrical, Orthographic Azimuthal Equidistant Cartographic Projection: Its Purpose, Construction and Theory." *Cartography* 8(4): 182–86.

Andrews, H. J. 1935. "Note on the Use of Oblique Cylindrical Orthomorphic Projection." *Geographical Journal* 86(5): 446.

———. 1938a. "An Oblique Mercator Projection for Europe and Asia." *Geographical Journal* 92(6): 538.

———. 1938b. "A Round-the-World Map on Airy's Projection." *Geographical Journal* 92(6): 537.

———. 1952. "Aitoff's Projection of the Sphere." *Geographical Journal* 118(2): 236–37.

Anthiaume, A. 1911. "Un pilote et cartographe havrais au XVIe siècle, Guillaume Le Testu." *Bulletin de Géographie Historique et Descriptive*, 135–202. Reprinted in *Acta Cartographica* 19(1974): 10–77.

———. 1916. *Cartes marines.* 1: 125–27; 2: 448–58. Paris.

Apian, Peter. 1524, 1551. *Cosmographicus liber.* Landshut, 1524; Paris, 1551.

Applebaum, Wilbur. 1970–80. "Wilhelm Schickard." In Gillispie 1970–80, 12:162–63.

Appleton, D., and Co. 1872. *General Atlas of the World, containing Maps of Various Countries, and Particularly of the United States.* New York.

Arden-Close, Charles Frederick. 1938. "A Polar-Azimuthal Retroazimuthal Projection." *Geographical Journal* 92(6): 536–37.

———. 1941. "An Oblique Rectangular Cylindrical Projection." *Geographical Journal* 97(6): 349–50.

———. 1942. "The Map of the Pacific." *Geographical Journal* 100(2): 64–72.

———. 1943. "A Combined Projection." *Empire Survey Review* 7(49): 135–36.

———. 1947. "Some Curious Map Projections." In his *Geographical By-Ways and Some Other Essays*, 68–88. London: Edward Arnold.

Arden-Close, Charles Frederick, and Frank George. 1952. "A Forgotten Pseudo-Zenithal Projection." *Geographical Journal* 118(2): 237.

Arnberger, Erik. 1966. *Handbuch der Thematischen Kartographie.* Vienna: Franz Deuticke.

The Astronomical Almanac for the Year 1983. 1982. Washington: Government Printing Office; London: Her Majesty's Stationery Office.

August, Friedrich Wilhelm Oscar. 1874. "Ueber eine conforme Abbildung der Erde nach der epicycloidischen Projection." *Zeitschrift der Gesellschaft für Erdkunde zu Berlin* 9:1–22.

Aujac, Germaine, and eds. 1987a. "Greek Cartography in the Early Roman World." In Harley and Woodward 1987, 161–76.

———. 1987b. "The Growth of an Empirical Cartography in Hellenistic Greece." In Harley and Woodward 1987, 148–60.

Averdunk, H., and J. Müller-Reinhard. 1914. *Gerhard Mercator und die Geographen unter seinen Nachkommen*. Ergänzungsheft (Supplement) no. 182 to *Petermanns Mitteilungen*.

Baar, Edward J. 1947. "The Manipulation of Projections for World Maps." *Geographical Review* 37(1): 112–20.

Babinet, Jacques. 1859. *Géographie nouvelle: Mappemondes et cartes, système homalographique*. Paris: Ernest Bourdin.

———. [1861]. *Atlas universel de géographie*. Paris: Ernest Bourdin.

Bacon, Roger. Ca. 1265–68. *The Opus Majus*. 2 vols. Trans. Robert Belle Burke, 1928. Reprinted, New York: Russell and Russell, 1962.

Bagrow, Leo. 1964. *History of Cartography*. Rev. and enlarged by R. A. Skelton. Trans. D. L. Paisey. Cambridge: Harvard University Press. Reprinted and enlarged, Chicago: Precedent Publishing, 1985.

Baily, Francis. 1835. *An Account of the Rev^d. John Flamsteed, the first Astronomer-Royal; compiled from his own manuscripts, and other authentic documents, never before published*. London.

Baker, J. G. P. 1986. "The 'Dinomic' World Map Projection." *Cartographic Journal* 23(1): 66–67.

Baranyi, János. 1968. "The Problems of the Representation of the Globe on a Plane with Special Reference to the Preservation of the Forms of Continents." *Hungarian Cartographical Studies*, 19–43.

Baranyi, János, and Ervin Földi. 1976. "Megjegyzések egy új térképvetülethez." *Geodézia és Kartográfia* 28(5): 383–886. Published in English as "Remarks on a New Projection," *Hungarian Cartographical Studies 1976*, 7–13.

Baranyi, János, and János Györffy. 1989. "New Form-True Projection in Hungarian Atlases." *Hungarian Cartographical Studies*, 75–85.

Bartholomew, John. 1942. *The Citizen's Atlas of the World*. Edinburgh: John Bartholomew & Son.

———. 1948. *The Regional Atlas of the World*. Edinburgh: Geographical Institute.

———. 1950. *The Advanced Atlas of Modern Geography*. London: Meiklejohn & Son.

———. [1953]. *The Columbus Atlas, or Regional Atlas of the World*. Edinburgh: John Bartholomew & Son.

———, ed. 1958. *The Times Atlas of the World*. Vol. 1. London: Times Publishing.

Bartholomew, John Christopher. 1982. *World Atlas*. 12th ed. Edinburgh: John Bartholomew & Son.

———. 1984. *The Family Atlas of the World*. Edinburgh: John Bartholomew & Son.

Bartholomew, John George. 1889. *Atlas of Commercial Geography*. Cambridge: Cambridge University Press.

———. 1890. *The Library Reference Atlas of the World*. London: Macmillan.

[————]. 1891. "Hermann Berghaus." *Scottish Geographical Magazine* 7:86.

————. 1893? *The Graphic Atlas and Gazetteer of the World.* New York: Thomas Nelson & Sons.

Beaman, W. M. 1928. *Topographic Mapping.* Washington: U.S. Geological Survey, Bulletin 788-E.

Behrmann, Walter. 1910. "Die beste bekannte flächentreue Projektion der ganzen Erde." *Petermanns Mitteilungen* 56-2, (3): 141–44.

————. 1919. "Eine neue Projektion mit geradlinigen grössten Kugelkreisen." *Zeitschrift der Gesellschaft für Erdkunde zu Berlin* (1–2): 86–87.

Bell, Eric Temple. 1937. *Men of Mathematics.* New York: Simon and Schuster.

Berghaus, Heinrich. 1845. *Physikalischer Atlas oder Sammlung von Karten.* Gotha: Justus Perthes.

Berry, Ralph M., and Valdis Bormanis. 1970. *Plane Coordinate Survey System for the Great Lakes Based on the Hotine Orthomorphic Projection.* U.S. Lake Survey Miscellaneous Paper 70-4.

Bertin, Jacques. 1967. "L'identification en situation les projections." In his *Semiologie graphique,* 287–95. Paris, Mouton: Gauthier-Villars. Translated into English as *Semiology of Graphics.* Madison: University of Wisconsin Press, 1984.

Biernacki, Franciszek. 1949. *Teoria odwzorowań powierzchni dla geodetów i Kartografów.* Warsaw: Glówny Urzqd Pomiarów Kraju, Prace Geodezyjnego Instytuto Naukowo-Badawczego 4. Translated into English as *Theory of Representation of Surfaces for Surveyors and Cartographers.* Washington: U.S. Department of Commerce, 1965.

Boaga, Giovanni. 1948. *Trattato di geodesia e topografia, con elementi di fotogrammetria.* Padova: CEDAM (Casa Editrice Dott. A. Milani).

Boerner, H. 1970–80. "Hermann Amandus Schwarz." In Gillispie 1970–80, 12:245–47.

Boggs, Samuel Whittemore. 1929. "A New Equal-Area Projection for World Maps." *Geographical Journal* 73(3): 241–45.

Bolliger, Jakob. 1967. *Die Projektionen der schweizerischen Plan- und Kartenwerke.* Winterthur, Switz.: Druckerei Winterthur AG.

Bol'shoy sovetskiy atlas mira (B.S.A.M.). 1937. Vol. 1. Moscow: Proletarii Vsekh Stran, Soyedinyaytes'.

Bonacker, Wilhelm, and Ernst Anliker. 1930. "Heinrich Christian Albers, der Urheber der flächentreuen Kegelrumpfprojektion." *Petermanns Mitteilungen* 76(9–10): 238–40.

Bonne, Rigobert. 1788. *Atlas encyclopédique contenant la géographie ancienne, et quelques cartes sur la géographie du Moyen Age, la géographie moderne, . . .* Paris: Hotel de Thou.

Boorman, J. Marcus. 1877. *Improvement in Geometric Blocks for Mapping.* U.S. Patent 185,889. Jan. 2, 1877.

Borčić, Branko. 1955. *Matematička kartografija.* Zagreb: Tehnička Knjiga.

Boston University Art Gallery. 1985. *Celestial Images.* Boston. Describing the exhibit Astronomical Charts from 1500 to 1900.

Botley, F. V. 1949. "A Tetrahedral Gnomonic Projection." *Geography* 34(3): 131–36.

————. 1951. "A New Use for the Plate Carrée Projection." *Geographical Review* 41(4): 640–44.

————. 1954. "An Octahedral Gnomonic Projection." *Empire Survey Review* 12(94): 379–81.

Bowditch, Nathaniel (original author). 1966. *American Practical Navigator*. Washington: Government Printing Office.

Bowie, William, and Oscar S. Adams. 1919. *Grid System for Progressive Maps in the United States*. Washington: U.S. Coast and Geodetic Survey Special Publication 59.

Bradford, Thomas Gamaliel. 1835. *A Comprehensive Atlas: Geographical, Historical & Commercial*. Boston: William D. Ticknor; New York: Wiley & Long.

Bradley, A. D. 1946. "Equal-Area Projection on the Icosahedron." *Geographical Review* 36(1): 101–4.

Braun, Carl. 1867. "Ueber zwei neue geographische Entwurfsarten." *Wochenschrift für Astronomie, Meteorologie und Geographie* 10(33): 259–63; 10(34): 269–72; 10(35): 276–78.

Breckman, Jack. 1962. "The Theory and Use of B-Charts." Radio Corporation of America. Manuscript.

Bretterbauer, Kurt. 1989. "Die trimetrische Projektion von W. Chamberlin." *Kartographische Nachrichten* 39(2): 51–55.

Breusing, Friedrich August Arthur. 1892. *Das Verebnen der Kugeloberfläche für Gradnetzentwürfe*. Leipzig: H. Wagner & E. Debes.

Briesemeister, William. 1953. "A New Oblique Equal-Area Projection." *Geographical Review* 43(2): 260–61.

———. 1959. "A World Equal-Area Projection for the Future. The Selection of the Most Suitable Equal-Area Projection for the Purpose of Plotting World Wide Statistics in This Present Day of Super Speed, Jet Planes and Intercontinental Missiles." *Nachrichten aus dem Karten- und Vermessungswesen*, ser. 2, (3): 60–63.

Briggs, J. Morton. 1970–80. "Antoine Parent." In Gillispie 1970–80, 10:319–20.

Bromley, Robert H. 1965. "Mollweide Modified." *Professional Geographer* 17(3): 24.

Brooks, William D., and Charles E. Roberts, Jr. 1976. "An Analysis of a Modified van der Grinten Equatorial Arbitrary Oval Projection." *American Cartographer* 3(2): 143–50. See correction in 6(1)(1979): 81, by John P. Snyder.

———. 1986. "A Mathematical Analysis and Comparison of Three Circular World Graticules." *Cartographica* 23(3): 28–41. See correction in 24(3)(1987): 85, by John P. Snyder.

Brown, B. H. 1935. "Conformal and Equiareal World Maps." *American Mathematical Monthly* 42(4): 212–23.

Brown, Lloyd Arnold. 1949. *The Story of Maps*. Boston: Little, Brown. Reprinted, New York: Dover, 1979.

Burky, Charles. 1934–35. "Un nouveau système de projection cartographique. Projection de la sphère sur un polyédre de 32 faces, icosaédre tronqué par le dodecaédre, selon construction de Mo.O. Dallwigk, à Genève." Le Globe, *Bulletin* 74: 30–33.

Büsching, Anton Friedrich. 1754ff. *Neue Erdbeschreibung*. Translated into English by Patrick Murdoch as *A New System of Geography*. London: A. Millar, 1762.

Bynum, George. 1978. "Chamberlin Trimetric Projection." Manuscript.

Cagnoli, Antonio. 1799. "Della più esalta costruzione delle carte geografiche." *Memorie di Matematica e Fisica della Società Italiana* 2: 658–64.

Cahill, Bernard J. S. 1909. "An Account of a New Land Map of the World." *Scottish Geographical Magazine* 25(9): 449–69.

———. 1913a. *Geographical Globe*. U.S. Patent 1,081,207. Dec. 9, 1913.

———. 1913b. *Map of the World*. U.S. Patent 1,054,276. Feb. 25, 1913.

———. 1929. "Projections for World Maps." *Monthly Weather Review* 57(4): 128–33.

———. 1934. "A World Map to End World Maps." *Geografiska Annaler* 16(2–3): 97–108.

———. 1937. "Die Schmetterlingskarte; das oktaedrische System der Projektion der Erd-kugel." *Petermanns Mitteilungen* 83(5): 129–31.

———. 1939. "Progress of the Butterfly Map: The Three Variants." *Architect and Engineer* 137 (May): 47–49.

Calderera, Gastono. 1910. "Sulla classificazione delle proiezioni geografiche." *Rivista Geografica Italiana* 17(9): 473–87.

Campbell, Tony. 1987. "Portolan Charts from the Late Thirteenth Century to 1500." In Harley and Woodward 1987, 371–463.

Canada. Department of Energy, Mines, and Resources. 1974. *The National Atlas of Canada.* Ottawa.

Canters, Frank. 1989. "New Projections for World Maps: A Quantitative-Perceptive Approach." *Cartographica* 26(2): 53–71.

Canters, Frank, and Hugo Decleir. 1989. *The World in Perspective: A Directory of World Map Projections.* Chichester, Eng.: John Wiley and Sons.

Carey, Mathew. 1818. *Carey's General Atlas, Improved and Enlarged; Being a Collection of Maps of the World and Quarters.* Philadelphia.

Cartographia. 1967. *National Atlas of Hungary.* Budapest.

———. 1987. *Nagy világatlasz* (Great world atlas). Budapest: Kartográfiai Vállalat.

Caspari, C. E. 1905. "Mappemonde circulaire: Projection de M. Alphonse van den [*sic*] Grinten." *Bulletin de la Société de Géographie* 11(1): 58–60.

Chamberlin, Wellman. 1947. *The Round Earth on Flat Paper.* Washington: National Geographic Society. Revised ed., 1950.

Chebyshev, Pafnutiy L'vovich. 1856. "Sur la construction des cartes géographiques." *Bulletin de l'Académie Impériale des Sciences* [St. Petersburg], *classe physico-mathématique* 14: 257–61.

Christensen, Albert H. J. 1992. "The Chamberlin Trimetric Projection." *Cartography and Geographic Information Systems* 19(2): 88–100.

Churski, Zygmunt, and Rajmund Galon. 1968. *Siatki kartograficzne.* Toruń, Poland: Nicolaus Copernicus University. 2d ed., 1974; 3d ed., 1982.

Claire, Charles N. 1968. *State Plane Coordinates by Automatic Data Processing.* Washington: U.S. Coast and Geodetic Survey Publication 62-4.

Clark, R. 1977. "Regional and World Maps Based on an Octahedron." *Classroom Geographer* (Oct.): 17–20.

Clarke, Alexander Ross. 1879. "Geography: Mathematical Geography." *Encyclopaedia Britannica,* 9th ed., 10:197–210.

Clarke, Alexander Ross, and Frederick Robert Helmert. 1911. "Earth, Figure of the." *Encyclopaedia Britannica,* 11th ed., 8:801–13.

Close, Charles Frederick. 1921. "Note on a Doubly-Equidistant Projection." *Geographical Journal* 57(6): 446–48.

———. 1922. *Note on Two Double, or Two-Point, Map Projections.* London: Ordnance Survey, Professional Papers, new ser. (5). See review by Alfred Ernest Young, "Two New Map Projections," *Geographical Journal* 60(4)(1922): 297–99.

———. 1927. *The Transverse Elliptical Equal-Area Projection of the Sphere, Otherwise Transverse Mollweide.* London: Ordnance Survey, Professional Papers, new ser. (11).

———. 1929. "An Oblique Mollweide Projection of the Sphere." *Geographical Journal* 73(3): 251–53; and (discussion), "New Map Projections," *Geographical Journal* 73(3): 248–50.

———. 1934. "A Doubly Equidistant Projection of the Sphere." *Geographical Journal* 83(2): 144–45.

———. 1935. "Two-Point Azimuthal-Equidistant Projection." *Geographical Journal* 86(5): 445–46.

Close, Charles Frederick, and Alexander Ross Clarke. 1911. "Map Projections." *Encyclopaedia Britannica,* 11th ed., 17:653–63.

Cole, J. H. 1943. *The Use of the Conformal Sphere for the Construction of Map Projections.* Giza (Orman), Egypt: Ministry of Finance, Survey of Egypt. Survey Paper 46.

Colles, Christopher. 1789. *A Survey of the Roads of the United States of America, 1789.* Reprinted, Cambridge: Harvard University Press, 1961; Walter W. Ristow, ed.

———. 1794. *Geographical Ledger.* New York: John Buel. Partial publication.

Collignon, Édouard. 1865. "Recherches sur la représentation plane de la surface du globe terrestre." *Journal de l'École Polytechnique* 24:125–32.

Colvocoresses, Alden Partridge. 1974. "Space Oblique Mercator." *Photogrammetric Engineering* 40(8): 921–26.

———. 1977. "Planimetric Mapping from Spacecraft." In Urho A. Uotila, ed., *The Changing World of Geodetic Science,* 352–58. Report 250. Columbus: Ohio State University, Department of Geodetic Science.

Coote, C. H. 1878. *Shakspere's New Map in "Twelfth Night."* London: Dulau & Co. Read before the forty-fourth meeting of the [Shakspere] Society, June 14.

Cornell, Sarah S. 1859. *Cornell's Companion Atlas to Cornell's High School Geography.* New York: D. Appleton.

Craig, James Ireland. 1910. *Map-Projections (Technical Lecture, 1909). The Theory of Map-Projections, with Special Reference to the Projections used in the Survey Department.* Cairo, Egypt: Ministry of Finance, Survey Dept., Paper 13. See review by Charles F. Close, *Geographical Journal* 37(2)(1911): 208–9.

Craig, Thomas. 1882. *A Treatise on Projections.* Washington: U.S. Coast and Geodetic Survey.

Cram Co., George F. 1941. *Cram's Unrivaled Atlas.* Indianapolis.

Craster, John Evelyn Edmund. 1929. "Some Equal-Area Projections of the Sphere." *Geographical Journal* 74(5): 471–74. See also 78(1931): 194–95.

———. 1938. "Oblique Conical Orthomorphic Projection for New Zealand." *Geographical Journal* 92(6): 537–38.

Cumming, William P., Raleigh Ashlin Skelton, and D. B. Quinn. 1972. *The Discovery of North America.* New York: American Heritage Press.

Dahlberg, Richard E. 1961. "Maps without Projections." *Journal of Geography* 60(5): 213–18.

———. 1962. "Evolution of Interrupted Map Projections." *International Yearbook of Cartography* 2:36–54.

d'Aiguillon, François. 1613. *Aguilonii opticorum.* Paris or Antwerp.

d'Ailly, Pierre. Ca. 1483. *Ymago mundi.* Probably Louvain.

D'Alembert, Jean le Rond. 1752. *Essai d'une nouvelle théorie de la résistance des fluides.* Paris.

Dalrymple, Alexander. N.d. *An Account of the Discoveries Made in the South Pacifick Ocean, Previous to 1764.*

Daskalova, M. 1982. ["Deducing the Constants of the Conformal and Equal-Area Conic Projection with the Application of Airy's Criterion"], (in Bulgarian). *Godisnik-Vissiya Institut po Arhitektura i Stroitelstvo, 1981–82* 29(3): 41–52. English abstract in *Geo Abstracts* (5)(1984): 341.

d'Avezac-Macaya, Marie Armand Pascal. 1863. "Coup d'oeil historique sur la projection des cartes de géographie." *Bulletin de la Société de Géographie,* ser. 5, 5(Apr.–May): 257–361, 5(June): 438–85. Reprinted, Paris: E. Martinet, 1863, and in *Acta Cartographica* 25(1977): 21–173.

———. 1866. *Note sur un mappemonde turke du XVIe siècle, conservée à la bibliothèque de Saint-Marc à Venise.* Paris: E. Martinet.

Davies, Arthur. 1949. "An Interrupted Zenithal World Map." *Scottish Geographical Magazine* 65(1): 1–6.

Deakin, R. E. 1990. "The 'Tilted Camera' Perspective Projection of the Earth." *Cartographic Journal* 27(1): 7–14.

Debenham, Frank. 1958. *The Global Atlas: A New View of the World from Space.* New York: Simon and Schuster.

Debes, Ernst. 1895. *Neuer Handatlas über alle Teile der Erde, in 59 Haupt- und 120 Neben-karten.* Leipzig: H. Wagner & E. Debes.

Deetz, Charles Henry. 1918. *The Lambert Conformal Conic Projection with Two Standard Parallels, including a Comparison of the Lambert Projection with the Bonne and Polyconic Projections.* Washington: U.S. Coast and Geodetic Survey Special Publication 47.

Deetz, Charles Henry, and Oscar Sherman Adams. 1934. *Elements of Map Projection with Applications to Map and Chart Construction.* 4th ed. Washington: U.S. Coast and Geodetic Survey Special Publication 68.

de Jode, Gerard. 1578. *Speculum orbis terrarum.* Antwerp. Facsimile atlas, Amsterdam: Theatrum Orbis Terrarum, 1965.

De l'Isle, Guillaume. 1730. *Atlas nouveau, contenant toutes les parties du monde, où sont exactement remarquées les empires, monarchies, royaumes, etats, republiques &c.* Amsterdam: J. Cóvens and C. Mortier.

De l'Isle, Guillaume, and Philippe Buache. 1700–1763. *Atlas géographique & universel, avec la géographie ancienne & moderne, & les détails de la France.* Paris.

———. 1833. *Atlas géographique des quatre parties du monde.* Rev. and enlarged by Dezauche. Paris: Dezauche.

De l'Isle, Joseph Nicolas. 1745. *Atlas' Rossiyskoy, sostoyashchey iz' devyatnattsati spetsi-al'nykh' kart' predstavlyayushchikh' vserossiyskuyu imperiyu s' progranichiymi zemlyami.* St. Petersburg.

DeLucia, Alan A., and John P. Snyder. 1986. "An Innovative World Map Projection." *American Cartographer* 13(2): 165–67.

DeMorgan, Augustus. 1836. *An explanation of the gnomonic projection of the sphere.* London: Baldwin and Cradock.

Dent, Borden D. 1987. "Continental Shapes on World Projections: The Design of a Poly-Centred Oblique Orthographic World Projection." *Cartographic Journal* 24(2): 117–24.

Deutsche Gesellschaft für Kartographie. 1982. "Die sogenannte Peters-Projektion. Eine Stellungnahme der Deutsche Gesellschaft für Kartographie." *Kartographische Nachrichten* 32(1): 33–36. Reprinted in German in at least three other journals and in English in at least two, including *Cartographic Journal* 22(2)(1985): 108–10.

de Vaulx Atlas. N.d. MS Franc. 159, fol. 27. Bibliothèque Nationale, Paris.

Dickinson, H. W. 1949–50. "A Brief History of Draughtsmen's Instruments." *Transactions of the Newcomen Society* 27:73–84. Reprinted in *Bulletin of the Society of University Cartographers* 2(2)(1968): 37–52.

Dilke, O. A. W. 1987. "The Culmination of Greek Cartography in Ptolemy." In Harley and Woodward 1987, 177–200.

Donald, Jay K. 1957. "Donald Elliptic Projection." American Telephone and Telegraph Co., New York. Manuscript. Supplied courtesy Charles W. Stevenson and Ralph E. Waters, 1978.

Dozier, Jeff. 1980. *Improved Algorithm for Calculation of UTM and Geodetic Coordinates*. Washington: National Oceanic and Atmospheric Administration Technical Report NESS 81.

Drecker, Josef. 1917. "Ein Instrument, eine Karte und eine Schrift des Nürnberger Kartographen und Kompastmachers Erhard Etzlaub." *Annalen der Hydrographie und Maritimen Meteorologie* 45(June): 217–24.

Driencourt, Ludovic, and Jean Laborde. 1932. *Traité des projections des cartes géographiques à l'usage des cartographes et des géodésiens*. Paris: Hermann et Cie.

Duchesne, Charles. 1907. "Les projections cartographiques." *Mémoires de la Société Royale des Sciences de Liège*, ser. 3, 7:1–178.

Dudley, Robert. 1646–47. *Arcano del mare*. Florence.

Dumitrescu, Victor. 1966. "Perspectives cosmographiques: Un système utile de projections azimutales." *International Yearbook of Cartography* 6:25–32.

———. 1968. "Cartographic Solution for Deciphering Space-Photographs." *International Yearbook of Cartography* 8:66–74.

Durand, Dana Bennett. 1952. *The Vienna-Klosterneuburg Map Corpus of the Fifteenth Century: A Study in Transition from Medieval to Modern Science*. Leiden: Brill.

Dürer, Albrecht. 1538. *Unterweysung des Messung mit dem Zirckel und Richtscheyt*. Vol. 4.

Dyer, John A., and John P. Snyder. 1989. "Minimum-Error Equal-Area Map Projections." *American Cartographer* 16(1): 39–46.

Eastwood, John. 1489. *Summa anglicana*. Venice.

Eckert, Max. 1906. "Neue Entwürfe für Erdkarten." *Petermanns Mitteilungen* 52(5): 97–109.

———. 1921, 1925. *Die Kartenwissenschaft*. Berlin and Leipzig: Walter de Gruyter.

Eckert-Greifendorff, Max. 1935. "Eine neue flächentreue (azimutaloide) Erdkarte." *Petermanns Mitteilungen* 81(6): 190–92.

Edney, Matthew H. 1991. "The Atlas of India, 1823–1947: The Natural History of a Topographic Map Series." *Cartographica* 28(4): 59–81, esp. 67.

Edwards, Trystan. 1953. *A New Map of the World: The Trystan Edwards Homolographic Projection*. London: B. T. Batsford.

Eggen, Olin J. 1970–80. "George Biddell Airy." In Gillispie 1970–80, 1:84–87.

Eisele, Carolyn. 1963. "Charles S. Peirce and the Problem of Map-Projection." *Proceedings of the American Philosophical Society* 107(4): 299–307.

———. 1970–80a. "Benjamin Peirce." In Gillispie 1970–80, 10:478–81.

———. 1970–80b. "Charles Sanders Peirce." In Gillispie 1970–80, 10:482–88.

Eisenlohr, Friedrich. 1870. "Ueber Flächenabbildung." *Journal für Reine und Angewandte Mathematik* [Crelle's] 72(2): 143–51.

———. 1875. "Ueber Kartenprojection." *Zeitschrift der Gesellschaft für Erdkunde zu Berlin* 10(59): 321–34.

Emerson, William. 1749. *The Projection of the Sphere, Orthographic, Stereographic and Gnomical; both demonstrating the principles, and explaining the practice of these various sorts of projections*. London.

Encyclopaedia Britannica. 1911. "Bartholomew, John." *Encyclopaedia Britannica*, 11th ed., 3:450.

———. 1942. *World Atlas*. Chicago.

———. 1945. *World Atlas*. New York: C. S. Hammond.

———. 1951. *World Atlas*. Chicago.

The English Pilot. The Fourth Book. 1689. London. Reprinted, Amsterdam: Theatrum Orbis Terrarum, 1967.

Environmental Science Services Administration. 1969. *Climates of the World*. Washington.

Érdi-Krausz, György. 1968. "Combined Equal-Area Projection for World Maps." *Hungarian Cartographical Studies*, 44–49.

Euler, Leonhard. 1777. "De projectione geographica de Lisliana in mappa generali Imperii Russici usitata." *Acta Academia Scientiarum Imperialis Petropolitanae* (St. Petersburg), no. 1, 143–53. Translated into German as one of "Drei Abhandlungen über Kartenprojection," in *Ostwald's Klassiker der Exakten Wissenschaften*, no. 93, 53–64. Leipzig: Wilhelm Engelmann, 1898.

Evans, David S. 1970–80. "John Frederick William Herschel." In Gillispie 1970–80, 6:323–28.

Everett, Joseph David. 1904. "On a Flat Model Which Solves Problems in the Use of the Globes." *Geographical Journal* 23(2): 234–42.

Everitt, Charles. 1911. "Algebra: History." *Encyclopaedia Britannica*, 11th ed., 1:616–20.

Fairgrieve, J. 1928. "A New Projection." *Geography* 14, pt. 6 (82): 525–26.

Fang, J. T. 1934. *The Study of Map Projection* (in Chinese). Peking. 1952 ed.; 1957–58 ed.

Fawcett, Charles Bungay. 1949. "A New Net for a World Map." *Geographical Journal* 114(1–3): 68–70.

Fay, Theodore Sedgwick. 1869. *Atlas of Universal Geography for Libraries and Families*. New York: G. P. Putnam.

Ferschke, Hans. 1985. "Karlheinz Wagner." *Kartographische Nachrichten* 35(3): 108–9.

Fiala, František. 1952. *Kartografické zobrazování*. Prague: Státní Pedagocké Naki. Translated from Czech into German as *Mathematische Kartographie*. Berlin: VEB Verlag Technik, 1957.

Finch, Vernor Clifford, and Glenn T. Trewartha. 1936. *Elements of Geography*. New York: McGraw-Hill. 2d ed., 1942; 3d ed., 1949; 4th ed., with Arthur H. Robinson and Edwin H. Hammond, 1957.

Fiorini, Matteo. 1881. *Le projezioni delle carte geografiche.* Bologna: Nicola Zanichelli.

———. 1889. "Le projezioni cordiformi della cartografia." *Bollettino della Società Geografica Italiana* 26:554–79, 676. Reprinted in *Acta Cartographica* 7(1970): 110–35.

———. 1894. "Il mappamondo di Leonardo da Vinci ed altre consimile mappe." *Rivista Geografica Italiana* 1:213–23. Reprinted in *Acta Cartographica* 4(1969): 188–98.

———. 1895. "Sopra tre speciali projezioni meridiane e i mappamondi ovali del secolo XVI." *Memorie della Società Geografica Italiana* 5:165–201. Reprinted in *Acta Cartographica* 1(1967): 37–73.

———. 1900. "Proiezioni cartografiche cicloidali." *Rivista Geografica Italiana* 7(4): 177–86.

Fischer, Joseph, and Franz R. von Wieser. 1903. *The Oldest Map with the Name America of the Year 1507 and the Carta Marina of the Year 1516 by M. Waldseemüller (Ilacomilus).* London: Henry Stevens, Son & Stiles. With facsimile.

Fisher, Irving. 1943. "A World Map on a Regular Icosahedron by Gnomonic Projection." *Geographical Review* 33(4): 605–19.

———. 1948. *Global Map.* U.S. Patent 2,436,860. Mar. 2, 1948.

Fisher, Irving, and Osborn Maitland Miller. 1944. *World Maps and Globes.* New York: Essential Books.

Fite, Emerson D., and Archibald Freeman. 1926. *A Book of Old Maps Delineating American History from the Earliest Days Down to the Close of the Revolutionary War.* Cambridge: Harvard University Press. Reprinted, New York: Dover Publications, 1969.

Flamsteed, John. 1729. *Atlas Coelestis.* London. Published posthumously.

Folkerts, Menso. 1970–80. "Johann(es) Werner." In Gillispie 1970–80, 14:272–77.

Foucaut, H.-C.-A. de Prépetit. 1862. *Notice sur la construction de nouvelles mappemondes et de nouveaux atlas de géographie.* Arras, France.

Fournier, Georges. 1643. *Hydrographie, contenant la théorie et la pratique de toutes les parties de la navigation.* Paris.

Freeman, T. W. 1963. "Rev. James Gall and his Map Projections." *Scottish Geographical Magazine* 79(3): 177.

Freiesleben, H.-Christ. 1970–80. "Karl Brandan Mollweide." In Gillispie 1970–80, 9:463.

Frischauf, Johannes. 1897. "Bemerkungen zu C. S. Peirce Quincuncial Projection." *American Journal of Mathematics* 19(4): 381–82.

Frye, Alexis E. 1895. *Complete Geography.* Boston: Ginn.

Fuller, Richard Buckminster. 1943. "Dymaxion World." *Life,* Mar. 1, p. 41.

———. 1946. *Dymaxion Map.* U.S. Patent 2,393,676. Jan. 29, 1946.

Funkhouser, Howard Gray. 1938. "Historical Development of the Graphical Representation of Statistical Data." *Osiris* 3:269–404.

Gabler, H. 1943. "Hans Maurer 75 Jahre alt." *Annalen der Hydrographie und Maritimen Meteorologie* 71:273–78.

Gall, James. 1855. "On Improved Monographic Projections of the World." In *Report of the Twenty-fifth Meeting of the British Association for the Advancement of Science,* 148.

———. 1871. "On a New Projection for a Map of the World." *Proceedings of the Royal Geographic Society* 15(July 12): 159.

———. 1885. "Use of Cylindrical Projections for Geographical, Astronomical, and Scientific Purposes." *Scottish Geographical Magazine* 1(4): 119–23.

Gardiner, Leslie. 1976. *Bartholomew 150 Years*. Edinburgh: John Bartholomew and Son, Ltd.

Gardner, Martin. 1975. "Mathematical Games: On Map Projections (with Special Reference to Some Inspired Ones)." *Scientific American* 233(5): 120–25.

Garnett, William [Mary Adams, pseud.]. 1914. *A Little Book on Map Projection*. London: G. Philip & Sons. 2d ed., 1920; 3d ed., 1924; 4th ed., 1928.

Garver, John B., Jr. 1988. "New Perspective on the World." *National Geographic* 174(6): 910–13. With supplement "The World," Robinson projection, scale 1:30,840,000, as p. 910A.

———. 1990. "Seventy-five Years of Cartography: A Love Affair with Maps." *National Geographic* 178(5): 130–34.

Gauss, Carl Friedrich. 1825. "Allgemeine Auflösung der Aufgabe: Die Theile einer gegebnen Fläche auf einer andern gegebnen Fläche so abzubilden, dass die Abbildung dem Abgebildeten in den kleinsten Theilen ähnlich wird." Preisarbeit der Kopenhagener Akademie 1822. *Schumachers Astronomische Abhandlungen* (Altona) (3): 5–30. Reprinted in *Ostwald's Klassiker der Exakten Wissenschaften*, no. 55, 57–81. Notes by Albert Wangerin, ed., 97–101. Leipzig: Wilhelm Engelmann, 1894.

———. 1843. "Untersuchungen über Gegenstände der höheren Geodäsie." *Der Königl. Sozietät überreicht*, 1st paper. Oct. 23.

Geisler, Walter. 1939. "Max Eckert-Greifendorffs Bedeutung für die Geographische und Kartographische Wissenschaft." *Petermanns Geographische Mitteilungen* 85(3): 85–89.

Geographical Publishing. 1942. *New International Atlas of the World*.

Geographical Survey Institute, Japan. [1977]. *The National Atlas of Japan*.

Germain, Adrien Adolphe Charles. [1866]. *Traité des projections des cartes géographiques, représentation plane de la sphère et du sphéroïde*. Paris: Arthus Bertrand.

———. 1868. "Note sur deux nouvelles projections géographiques, par le r.p. Braun, de la Compagnie de Jésus. Lettre à M. d'Avezac." *Bulletin de la Société de Géographie*, 5th ser., 16 (Nov.–Dec.): 510–13.

Germany, Federal Republic of. Press and Information Office. 1977. "Peters Projection: To Each Country Its Due on the World Map." *The Bulletin* (Bonn), 25(17): 126–27. Reprinted with credit in *ACSM Bulletin* (59)(1977): 13–15. See also separate comments by Arthur H. Robinson and John P. Snyder, *ACSM Bulletin* (60)(1978): 27.

Gesellschaft für Erdkunde zu Berlin. 1891. *Drei Karten von Gerhard Mercator—Europa—Britische Inseln—Weltkarte, Facsimile-Lichtdruck*. Berlin: W. H. Kühl.

Gilbert, Edgar N. Ca. 1973. "An Atlas of Oddities." Bell Laboratories, Murray Hill, N.J. Manuscript.

Gilfillan, S. C. 1946. "World Projections for the Air Age." *Surveying and Mapping* 6(1): 12–18.

Gillispie, Charles Coulston, ed. 1970–80. *Dictionary of Scientific Biography*. 16 vols. New York: Charles Scribner's Sons.

Gingerich, Owen. 1970–80. "Johannes Kepler." In Gillispie 1970–80, 7:289–312.

Ginzburg, Georgiy Aleksandrovich. 1949a. "Matematicheskaya osnova shkol'nykh kart polushariy" (The mathematical basis of school maps of hemispheres). TsNIIGAiK. *Trudy* (55): 35–55.

———. 1949b. "Psevdotsilindricheskaya proyektsiya TsNIIGAiK" (The pseudocylindrical projection of TsNIIGAiK). *Sbornik Nauchno-Tekhnicheskikh i Proizvodstvennikh Statey GUGK* (24): 68–72.

———. 1951. "O matematicheskoy osnove shkol'nikh kart mira" (The mathematical basis of school maps of the world). TsNIIGAiK. *Trudy* (88): 38–44.

———. 1952a. "Kartograficheskiye proyektsii s izokolami v vide ovalov i ovoid" (Cartographic projections with oval or oval-like lines of constant distortion). *Sbornik statey po kartografii* (1): 39–48.

———. 1952b. "Matematicheskoye obosnovaniye kart kompleksnykh mirovykh geograficheskikh atlasov" (The mathematical basis of maps of comprehensive world geographical atlases). TsNIIGAiK. *Trudy* (91).

———. 1966. "Novyy variant polikonicheskoy proyektsii" (A new variant of the polyconic projection). *Geodeziya i kartografiya* (3): 55–57.

Ginzburg, Georgiy Aleksandrovich, and T. D. Salmanova. 1957. *Atlas dlya vybora kartograficheskikh proyektsiy* (Atlas for the selection of cartographic projections). TsNIIGAiK. *Trudy* (110).

———. 1964. *Posobiye po matematicheskoy kartografii* (Manual of mathematical cartography). TsNIIGAiK. *Trudy* (160).

Glaisher, James Whitbread Lee. 1911. "Logarithms." *Encyclopaedia Britannica*, 11th ed., 16: 868–77.

Gohm, Douglas. 1972. *Antique Maps*. London: Octopus Books.

Goode, John Paul. 1905. "A New Method of Representing the Earth's Surface: The Van Der Grinten Projection." *Journal of Geography* 4(9): 369–73.

———. 1917. "A New Idea for a World Map: A Substitute for Mercator's Projection." *Annals of the Association of American Geographers* 7: 75–76 (abstract only).

———. 1925. "The Homolosine Projection: A New Device for Portraying the Earth's Surface Entire." *Annals of the Association of American Geographers* 15(3): 119–25.

———. 1929a. "A New Projection for the World Map: The Polar Equal Area." *Monthly Weather Review* 57(4): 133–36.

———. 1929b. "The Polar Equal Area, a New Projection for the World Map." *Annals of the Association of American Geographers* 19(3): 157–61.

Goodman, George, Jr. 1979. "Osborn Miller Dies; Noted Mapmaker." *New York Times*, Aug. 2.

Goussinsky, Boris. 1951. "On the Classification of Map Projections." *Empire Survey Review* 11(80): 75–79.

Grassi, Orazio. 1619. *De tribus cometis anni M.DC.XVIII*. Rome.

Graur, Aleksey Vasil'evich. 1956. *Matematicheskaya kartografiya*. 2d ed. rev. by A. A. Pavlov. Leningrad: Leningrad University. Translated into English as *Mathematical Cartography* by U.S. Air Force, Aeronautical Chart and Information Center, St. Louis, Mo., 1970; 1st Russian ed., 1938.

Grave, Dmitry Aleksandrovich. 1896. *Ob osnovnykh zadachakh matematicheskoy teorii postroyeniya geograficheskikh kart*. (Basic problems of the mathematical theory of geographical map construction). St. Petersburg: Academy of Science.

Greenhood, David. 1964. *Mapping*. Chicago: University of Chicago Press.

Gretschel, Heinrich Friedrich. 1873. *Lehrbuch der Karten-Projektion, enthaltened eine An-*

weisung zur Zeichnung der Netze für die verschiedensten Arten von Land- und Himmelskarten. Weimar: B. F. Voigt.

Grienberger, Christoph. 1612. *Catalogus veteres affixarum logitudines ac latitudines conferens cum novis.* Rome.

Grosvenor, Gilbert H. 1937. "The Society's Map of South America." *National Geographic* 72(6): 809–10.

Grynaeus. 1532. *Novus orbis regionum ac insularum veteribus incognitarum.* Paris.

Gunther, Robert William Theodore. 1932. *The Astrolabes of the World.* Oxford: Oxford University Press.

Gurba, Stefania. 1970. "Ortodroma na globusie dwunastościennym" (The orthodrome on a dodecahedron globe). *Polski Przegląd Kartograficzny* 2(4): 160–68.

Guthorn, Peter J. 1984. *United States Coastal Charts, 1783–1861.* Exton, Pa.: Schiffer Publishing.

Guyou, Émile. 1886. "Sur un nouveau système de projection de la sphère." *Comptes Rendus de l'Académie des Sciences* 102(6): 308–10.

———. 1887. "Nouveau système de projection de la sphère: Généralisation de la projection de Mercator." *Annales Hydrographiques,* 2d ser., 9:16–35. Also in *Revue Maritime et Coloniale* 94(1887): 228–47.

Haas, William H., and Harold B. Ward. 1933. "J. Paul Goode." *Annals of the Association of American Geographers* 23:241–46.

Hägerstrand, Torsten. 1957. "Migration and Area." In "Migration in Sweden." *Lund Studies in Geography* (Lund, Sweden, Royal University of Lund, Dept. of Geography), ser. B, Human Geography (13): 27–158.

Hagiwara, Yasuchi. 1984. "A Retro-Azimuthal Projection of the Whole Sphere." In *Technical Papers* 1:840–48. International Cartographic Association, 12th Conference, Perth, Australia, Aug. 6–13.

Haines, G. V. 1981. "The Modified Polyconic Projection." *Cartographica* 18(1): 49–58.

———. 1987. "The Inverse Modified Polyconic Projection." *Cartographica* 24(4): 14–24.

Hainsworth, R. G. 1949. *A Graphic Summary of World Agriculture.* Washington: U.S. Department of Agriculture, Miscellaneous Publication 705.

Hakluyt, Richard. 1582. *Divers Voyages Touching the Discoverie of America.* London.

———. 1598–1600. *The principal navigations, voyages, traffiques and discoveries.* Reproduced Glasgow.

Hall, H. Duncan. 1948. "Zones of the International Frontier." *Geographical Review* 38(4): 615–26.

Halley, Edmond. 1695–96. "An easy Demonstration of the Analogy of the Logarithmic Tangents, to the Meridian Line, or Sum of the Secants: with various Methods for computing the same to the utmost Exactness." *Philosophical Transactions* 19:202–14. Title as included with reprint in abridged *Philosophical Transactions,* 4:68–77. London: C. and R. Baldwin, 1809.

Hallinan, Bill. 1989. Letter to the editor. *National Geographic* 175(4): 8th unnumbered page.

Hammer, Ernst Hermann Heinrich. 1889. "Über Projektionen der Karte von Afrika." *Zeitschrift der Gesellschaft für Erdkunde zu Berlin* 24(4): 222–39.

————. 1892. "Über die Planisphäre von Aitow und verwante Entwürfe, insbesondere neue flächentreue ähnlicher Art." *Petermanns Mitteilungen* 38(4): 85–87.

————. 1900. "Unechtcylindrische und unechtkonische flächentreue Abbildungen. Mittel und Auftragen gegebener Bogenlängen auf gezeichneten Kreisbögen von bekannten Halbmessern." *Petermanns Mitteilungen* 46(2): 42–46.

————. 1910. "Gegenazimutale Projektionen." *Petermanns Mitteilungen* 56-1(3): 153–55.

————. 1917. "Die Mercator-Projektion und—Erhart [*sic*] Etzlaub." *Petermanns Mitteilungen* 63(Oct.): 303–4.

Hammond, C. S., & Co. 1945. *Encyclopaedia Britannica World Atlas.* New York.

————. 1946. *Hammond's Historical Atlas.* New York.

————. 1952. *Universal World Atlas.* New York.

Hammond Inc. 1957. *Hammond's Historical Atlas.* Maplewood, N.J.

————. 1966. *Hammond Citation World Atlas, New Perspective Edition.* Maplewood, N.J.

————. 1993. *Hammond Atlas of the World.* Maplewood, N.J.

Harding, Charles Louis. 1808–22. *Atlas novus coelestis XXVII tabulis continens stellas inter polum borealem et trigesimum gradum declinationis australis adhuc observatas.* Göttingen.

Harley, John Brian. 1975. *Ordnance Survey Maps: A Descriptive Manual.* Southampton, Eng.: Ordnance Survey.

Harley, John Brian, and David Woodward, eds. 1987. *The History of Cartography.* Vol. 1, *Cartography in Prehistoric, Ancient, and Medieval Europe and the Mediterranean.* Chicago: University of Chicago Press.

————. 1992. *The History of Cartography.* Vol. 2, bk. 1, *Cartography in the Traditional Islamic and South Asian Societies.* Chicago: University of Chicago Press.

Harrison, Richard Edes. 1944. *Look at the World: The Fortune Atlas for World Strategy.* New York: Alfred A. Knopf.

Hassler, Ferdinand Rudolph. 1825. "On the mechanical organisation of a large survey, and the particular application to the Survey of the Coast." *American Philosophical Society Transactions,* n. ser., 2:385–408.

Hatano, Masataka. 1972. "Consideration on the Projection Suitable for Asia-Pacific Type World Map and the Construction of Elliptical Projection Diagram." *Geographical Review of Japan* 45(9): 637–47. English translation.

Hatt, Philippe. 1886. "Définition et emploi des coordonées azimutales." *Annales Hydrographiques* 8, ser. 2, "semestre" 2:477–504.

Hazay, István. 1954. *Földi vetületek* (Earth projections). Budapest: Akadémiai Kiadó.

————. 1964. *Vetülettan* (Study of projections). Budapest: Tankönyvkiadó.

Heinfogel. Ca. 1515. *Het Boek* 17.

Heintzelman, D. H., and Richard M. Highsmith. 1955. *World Regional Geography.* New York: Prentice-Hall.

Hendrikz, D. R. 1946–47. *The Theory of Map Projections.* Pretoria, Union of South Africa: Trigonometric Survey Special Publication 4.

Hennig, Richard. 1948. "The Representation on Maps of the Magalhães Straits before Their Discovery." *Imago Mundi* 5:33–37.

Herausgegeben von der Gesellschaft der Wissenschaften zu Göttingen in Kommission bei Julius Springer in Berlin. 1929. 12:1–9.

Herschel, John Frederick William. 1859. "On a New Projection of the Sphere." *Proceedings of the Royal Geographical Society* 3:174–77. Same title in *Journal of the Royal Geographical Society* 30(1860): 100–106.

Herz, Norbert. 1885. *Lehrbuch der Landkartenprojektionen.* Leipzig: B. G. Teubner.

Hevelius, Johannes. 1687. *Firmamentum Sobiescianum sive Uranographia.*

Highsmith, Richard M., and J. G. Jensen. 1958. *Geography of Commodity Production.* Chicago: Lippincott Co.

Hill, Gillian. 1978. *Cartographical Curiosities.* London: British Library.

Hinks, Arthur Robert. 1912. *Map Projections.* Cambridge: Cambridge University Press. 2d ed., 1921.

———. 1921. "The Projection of the Sphere on the Circumscribed Cube." *Geographical Journal* 57(6): 454–57.

———. 1927a. "The Projection of the International Map." *Geographical Journal* 70(3): 289–91.

———. 1927b. "A Reply to the Criticisms of Dr. Antoni Łomnicki on the Projection of the International Map of the World on the Scale of 1/M." *Wiadomości Sluzby Geograficznej* 1:33–38.

———. 1927c. "W sprawie projekcji mapy miedzynarodowej 1/M; A Reply to the Criticisms of Dr. Antoni Łomnicki on the Projection of the International Map of the World on the Scale of 1/M" (in English with Polish introduction). *Polski Przegląd Kartograficzny* 3(18–20): 33–38.

———. 1929. "A Retro-Azimuthal Equidistant Projection of the Whole Sphere." *Geographical Journal* 73(3): 245–47.

———. 1940. "Maps of the World on an Oblique Mercator Projection." *Geographical Journal* 95(5): 381–83.

———. 1941a. "More World Maps on Oblique Mercator Projections." *Geographical Journal* 97(6): 353–56. See also comments on p. 362.

———. 1941b. "Murdoch's Third Projection." *Geographical Journal* 97(6): 358–62.

Hojovec, Vladislav. 1977. "Přibližná metoda transformace daného zobrazeni podle Airyho kritéria" (Approximate method of transformation of a given projection using the Airy criterion). *Geodetický a Kartografický Obzor* 23(7): 159–64.

Homann, Johann Baptist. [1750]. *Atlas novus terrarum orbis imperia, regna et status exactis tabulis geographicè demonstrans.* Nuremberg.

Hopfner, E. 1938. "Die beiden Kegelprojectionen des Ptolemaios." In "Theorie und Grundlagen der darstellenden Erdkunde," with commentaries by Hans von Mžik. *Des Klaudios Ptolemaios Einführung in die darstellende Erdkunde,* pt. 1. Vienna.

Hoschek, Josef. 1969. *Mathematische Grundlagen der Kartographie.* Mannheim, Zürich: Bibliographisches Institut. 2d ed., 1984.

Hotine, Martin. 1946–47. "The Orthomorphic Projection of the Spheroid." *Empire Survey Review* 8(62)(1946): 300–311; 9(63)(1947): 25–35; 9(64): 52–70; 9(65): 112–23; 9(66): 157–66.

H.R.M. 1920. "Obituary, John George Bartholomew." *Geographical Journal* 55(6): 483–84.

Hunt, Edward Bissell, and Charles A. Schott. 1854. "Tables for Projecting Maps, with Notes on Map Projections." In *Report of the Superintendent of the Coast Survey . . . 1853.* Appendix 39, 96–163. Washington, D.C.: Robert Armstrong, Public Printer.

Huxley, G. 1970–80. "Henry Briggs." In Gillispie 1970–80, 2:461–63.

Immler, W. 1919. "Ein doppelazimutaler gnomonischer Kartenentwurf und seine Anwendung auf Kreuzpeilungen für grosse Entfernungen." *Annalen der Hydrographie und Maritimen Meteorologie* 47(1–2): 22–36.

India. National Atlas and Thematic Mapping Organisation. 1979. *National Atlas of India.* Calcutta.

Jackson, John Eric. 1968. "On Retro-Azimuthal Projections." *Survey Review* 19(149): 319–28.

Jäger, G. 1865a. "Der Nordpol, ein thiergeographisches Centrum." *Das Ausland* (37): 865–67.

———. 1865b. "Der Nordpol, ein thiergeographisches Centrum." *Mittheilungen aus Justus Perthes' Geographischer Anstalt . . . von Dr. A. Petermann.* Ergänzungsheft 16:67–70.

James, Henry. 1860. "Description of the Projection used in the Topographical Department of the War Office for Maps Embracing large portions of the Earth's Surface." *Journal of the Royal Geographical Society* 30:106–11. Communicated by James; preparation attributed to A. R. Clarke.

James, Henry, and Alexander Ross Clarke. 1862. "On Projections for Maps applying to a very large extent of the Earth's Surface." *London, Edinburgh, and Dublin Philosophical Magazine,* 4th ser., 23(154): 306–12.

James, Preston Everett. 1943. *An Outline of Geography.* Boston: Ginn.

———. 1949. *A Geography of Man.* Boston: Ginn. 2d ed., 1958.

James, Preston Everett, and Nelda Davis. 1959. *The Wide World.* New York: Macmillan.

Jankowski, Piotr, and Timothy L. Nyerges. 1989. "Design Considerations for MaPKBS-Map Projection Knowledge-Based System." *American Cartographer* 16(2): 85–95.

Jervis, Thomas Best. 1895. *New cycloidal projection, by which entire continents may be represented with the least distortion of any projection hitherto known.* Turin.

Jervis, William P. 1898. *Thomas Best Jervis, as Christian Soldier, Geographer, and Friend of India, 1796–1857.* London: Elliot Stock.

Jordan, Wilhelm, and Otto Eggert. 1939–41. *Handbuch der Vermessungskunde.* Stuttgart: J. B. Metzlersche. 1939, vol. 3, first half; 1941, vol. 3, 2d half. Translated into English by Martha W. Carta as *Jordan's Handbook of Geodesy.* U.S. Army Corps of Engineers, Army Map Service, 1962.

Justus Perthes, pub. 1892. *Berghaus' Physikalischer Atlas, Dritte Ausgabe (herausgegeben von Dr. Herm. Berghaus).* Gotha.

Kadmon, Naftali, and Eli Shlomi. 1978. "A Polyfocal Projection for Statistical Surfaces." *Cartographic Journal* 15(1): 36–41.

Kaiser, Ward L. 1987. *A New View of the World—A Handbook to the World Map: Peters Projection.* New York: Friendship Press.

Kavrayskiy, Vladimir Vladimirovich. 1933. "Issledovaniya po matematicheskoy kartografii" (Investigation in mathematical cartography). TsNIIGAiK. *Trudy* (6).

————. 1934. *Matematischeskaya kartografiya* (Mathematical cartography). Moscow-Leningrad.

Kellaway, George P. 1946. *Map Projections*. London: Methuen. 2d ed., 1949. Reprinted 1962, 1970.

Kepler, Johannes. 1606. *De stella nova in pede serpentarii*. Prague: Ex Officino Calcographica Pauli Sessii.

Kessler, Fritz C. 1991. "The Development and Implementation of MaPPS: An Expert System Designed to Assist in the Selection of a Suitable Map Projection." M.A. thesis. Penn State University, State College, Pa., Geography Department.

Keuning, Johannes. 1947. "The History of an Atlas: Mercator-Hondius." *Imago Mundi* 4: 37–40.

————. 1954. "[Globe Gores Engraved by] Nicolaas Geelkerken." *Imago Mundi* 11:174–77.

————. 1955. "The History of Geographical Map Projections until 1600." *Imago Mundi* 12: 1–24.

Khanikof, Nicolas de. 1862. "Sur la nouvelle carte de l'Empire de Russie." *Bulletin de la Société de Géographie*, 5th ser., 4:185–91.

Kharchenko, A. S. 1951. "Novaya perspectivno-tsilindricheskaya proyektsiya" (A new perspective cylindrical projection). NIIGAiK. *Trudy* (3).

Khristov, Vladimir K. 1949. *Kartii proyektsiy matematicheska kartografie* (Map projections in mathematical cartography). Sofia.

————. 1961. *Matematicheska kartografiya* (Mathematical cartography). Sofia: Tekhtsmka.

Kish, George. 1957. *The Suppressed Turkish Map of 1560*. Ann Arbor, Mich.: William L. Clements Library.

————. 1965. "The Cosmographic Heart: Cordiform Maps of the 16th Century." *Imago Mundi* 19:13–21.

————. 1970–80a. "Gerardus Mercator." In Gillispie 1970–80, 9:309–10.

————. 1970–80b. "Peter Apian." In Gillispie 1970–80, 1:178–79.

Konvitz, Josef. 1987. *Cartography in France, 1660–1848: Science, Engineering, and Statecraft*. Chicago: University of Chicago Press.

Krasovskiy, Fedodosiy Nikolayevich. 1922. "Novye kartograficheskiye proyektsii" (New cartographic projections). *Vyshevo Geodezicheskovo Upravleniya*.

————. 1925. "Vychisleniye konicheskoy ravnopromezhutochnoy proyektsii, nailuchshe prisposoblennoy dlya izobrazheniya dannoy strany" (Computation of a conic equidistant projection with the best adjustment for representation of a given country). *Geodezist* (6–7).

Krüger, Louis. 1912. *Konforme Abbildung des Erdellipsoids in der Ebene. Veröffentlichung.* (Königlich Preussisches Geodätisches Institut, Potsdam), n. ser. (52).

————. 1922. *Zur stereographischen Projektion. Veröffentlichung.* (Preussisches Geodätisches Institut, Berlin), n. ser. (60).

Kuska, František. 1960. *Matematická kartografia*. Bratislava: Slovenské Vydavateľstvo Technickej Literatúry.

Laborde, Jean. 1928. *La nouvelle projection du Service Géographique de Madagascar.* Tananarive: Cahiers du Service Géographique de Madagascar (1).

Lagrange, Joseph Louis de. 1779. "Sur la construction des cartes géographiques." In *Nouveaux mémoires de l'Académie Royale des Sciences et Belles-lettres*, 161–210. Also in *Oeu-*

vres de Lagrange, 4:635–92. Paris: Gauthier-Villars, 1869. Also in German as *Ueber die Construction geographischer Karten*, in *Ostwald's Klassiker der Exakten Wissenschaften* (55), 1–56. Notes by Albert Wangerin, ed., 82–97. Leipzig: Wilhelm Engelmann, 1894.

La Hire, Philippe de. 1704. "Construction d'un nouvel astrolabe universel." In *Histoire de l'Académie Royale des Sciences, année MDCCI, avec les mémoires de mathématique et de physique pour la même année*, 97–101. Paris.

Lallemand, Charles. 1911. "Sur les déformations résultant du mode de construction de la carte internationale du monde au millionième." *Comptes Rendus des Séances de l'Académie des Sciences* 153(12): 559–67.

Lambert, Johann Heinrich. 1772. "Anmerkungen und Zusätze zur Entwerfung der Land- und Himmelscharten." In *Beiträge zum Gebrauche der Mathematik und deren Anwendung*, pt. 3, sec. 6. Translated into English and introduced by Waldo R. Tobler as *Notes and Comments on the Composition of Terrestrial and Celestial Maps (1772)*. Ann Arbor: University of Michigan, 1972. Also in German in *Ostwald's Klassiker der Exakten Wissenschaften* (54). Ed. Albert Wangerin. Leipzig: Wilhelm Engelmann, 1894.

Lanman, Jonathan Trumbull. 1987. *On the Origin of Portolan Charts*. Chicago: Newberry Library.

Larrivée, Léo. 1988. "Une création québécoise projetée dans le monde." *Revue Carto-Québec* 9(1): 16–17.

Laskowski, Piotr. 1991. "On a Mixed Local-Global Error Measure for a Minimum Distortion Projection." In *Technical Papers* 2:181–86. American Congress on Surveying and Mapping—American Society for Photogrammetry and Remote Sensing, Annual Conference, Baltimore, Md.

Lauf, G. B. 1983. *Geodesy and Map Projections*. Collingswood, Victoria, Australia: TAFE Publications Unit.

Lee, Laurence Patrick. 1944. "The Nomenclature and Classification of Map Projections." *Empire Survey Review* 7(51): 190–200.

———. 1965. "Some Conformal Projections Based on Elliptic Functions." *Geographical Review* 55(4): 563–80. See also 58(3)(1968): 490–91.

———. 1974. "A Conformal Projection for the Map of the Pacific." *New Zealand Geographer* 30(1): 75–77.

———. 1976. *Conformal Projections Based on Elliptic Functions*. Cartographica Monograph 16, supp. 1, to *Canadian Cartographer* 13.

Leighly, John B. 1955. "Aitoff and Hammer: An Attempt at Clarification." *Geographical Review* 45(2): 246–49.

Letoval'tsev, I. G. 1968. "O proyektsii Russilya." *Geodeziya i Aerofotos'emka* (2): 51–55. Translated into English as "The Roussilhe Projection," *Geodesy and Aerophotography* (2)(1968): 92–94. See correction by V. P. Morozov in *Geodeziya i aerofotos'emka* (6)(1968): 142; translated into English in *Geodesy and Aerophotography* (6)(1968): 418.

Lewis, Clinton, and J. D. Campbell. 1951. *The American Oxford Atlas*. New York: Oxford University Press.

Li, Guozao. 1981. "Cylindrical Transverse Conformal Projection with Two Standard Meridians" (in Chinese with English abstract). *Acta Geodetica et Cartographica Sinica* 10(4): 309–12.

Lilius, Zacharius. 1493. *Orbis breviarium*. Florence.

Littlehales, G. W. 1905. "Van der Grinten's Circular Projection." *Bulletin of the American Geographical Society* 37(1): 14–18.

Littrow, Joseph Johann von. 1833. *Chorographie; oder Anleitung, alle Arten von Land- See- und Himmelskarten zu verfertigen.* Vienna: F. Beck.

Łomnicki, Antoni. 1927a. *Kartografja matematyczna.* Lwow-Warsaw. 2d ed., 1956.

———. 1927b. "Matematyczna analiza projekcji mapy miedzynarodowej w skali 1:1,000,000" (in Polish with French abstract). *Wiadomości Sluzby Geograficznej* (1): 3–31.

———. 1927c. "Projekcja miedzynarodowej mapy świata: The International Map Projection 1/M" (in English with Polish introduction). *Polski Przegląd Kartograficzny* 3(17): 1–14.

Loomer, Scott A. 1986. "Mathematical Analysis of Medieval Sea Charts." American Congress on Surveying and Mapping—American Society for Photogrammetry and Remote Sensing, Annual Convention, March 16–21, Washington. *Technical Reports* 1:123–32.

Lorgna, Anton-Mario. 1789. *Principi di geografia astronomico-geometrica.* Verona.

[Lowry, John, projection, description of]. 1825. "Construction d'une mappemonde." *Bulletin de la Société de Géographie* 4(27–32): 127–30.

Lud, Gualterius [Walther Ludd]. [1507]. *Speculi orbis succintissima sed poenitenda, neque inelegans declaratio et canon.* Strasbourg.

Lunny, Robert M. 1961. *Early Maps of North America.* Newark: New Jersey Historical Society.

McBryde, Felix Webster. 1978. "A New Series of Composite Equal-Area World Map Projections." In *Abstracts,* 76–77. International Cartographic Association, 9th International Conference on Cartography, College Park, Md.

———. 1982a. *Homolinear Composite Equal-Area World Projections.* U.S. Patent 4,315,747. Feb. 16, 1982. Covers process for developing projections, including but not only S3.

———. 1982b. "World Oceans and Seas: Maritime Claims by Countries and Categories: 200-Nautical-Mile Exclusive Economic Zones." Approx. scale 1:53,080,000. Washington, D.C.: Transematics.

McBryde, Felix Webster, and Paul D. Thomas. 1949. *Equal-Area Projections for World Statistical Maps.* Washington: U.S. Coast and Geodetic Survey Special Publication 245.

McCaw, George Tyrrell. 1917. *Report on the Trigonometric Survey of Vanua Levu and Taveuni.* London: Stanford.

———. 1921. "A Modified Rectangular Polyconic Projection." *Geographical Journal* 57(6): 451–54.

Macdonald, R. R. 1968. "An Optimum Continental Projection." *Cartographic Journal* 5(1): 46–47.

McDonnell, Porter W., Jr. 1979. *Introduction to Map Projections.* New York: Marcel Dekker, Inc. 2d ed., 1991; Rancho Cordova, Calif.: Landmark Enterprises.

MacFadden, C. H., and R. B. Hall. 1940. *An Atlas of World Review.* New York: Thomas Y. Crowell.

McKinney, William M. 1969. "The Wright Projection." *Journal of Geography* 68(8): 472.

[MacLeod, N. M.?] 1953. "Colonel Sir Charles F. Arden-Close, K.B.E., C.B., C.M.G., F.R.S." *Geographical Journal* 119(2): 251–52.

Macrobius. 1483. *In somnium scipionis expositio.* Brescia.

Mainwaring, James. 1942. *An Introduction to the Study of Map Projection.* London: Macmillan.

Maling, Derek Hylton. 1960. "A Review of Some Russian Map Projections." *Empire Survey Review* 15(115): 203–15; 15(116): 255–66; 15(117): 294–303.

———. 1962. "The Hammer-Aitoff Projection and Some Modifications." In *Proceedings*, 41–57. Geography Department, Cartographic Symposium, University of Glasgow, Edinburgh.

———. 1965. "Suitable Projections for Maps of the Visible Surface of the Moon." *Cartographic Journal* 2(2): 95–99.

———. 1966. "Some Notes about the Trystan Edwards Projection." *Cartographic Journal* 3(2): 94–97.

———. 1973. *Coordinate Systems and Map Projections*. London: George Philip and Son Ltd. 2d ed., Oxford: Pergamon Press, 1992.

———. 1974a. "Personal Projections." *Geographical Magazine* 46(11): 599–600.

———. 1974b. "Peters' Wunderwerk" (in English). *Kartographische Nachrichten* 24(4): 153–56.

———. 1983. "'A little, round, fat, oily man of God': Rev Patrick Murdoch and His Contributions to Eighteenth-Century Cartography and Geography." *Cartographic Journal* 20(2): 110–18.

Marquis Who's Who. 1981–82. *Who's Who in the East*. 18th ed. Chicago.

Martinovich, Sándor. 1989. "The Great World Atlas of Cartographia." *Hungarian Cartographical Studies*, 29–36.

Maurer, Hans. 1905. "Eine neue graphische Azimut- und Kurs-Tafel und eine winkeltreue Kartenprojektion." *Annalen der Hydrographie und Maritimen Meteorologie* 33(3): 125–30; see also 33(7): 323.

———. 1914. "Die Definitionen in der Kartenentwurfslehre im Anschluss an die Begriffe zenital, azimutal und gegenazimutal." *Petermanns Mitteilungen* 60-2 (Aug.): 61–67; (Sept.): 116–21.

———. 1919a. "'Doppelbüschelstrahlige, ortodromische' statt 'doppelazimutale, gnomonische' Kartenentwürfe. Doppel-mittabstandstreue Kartogramme. (Bemerkungen zu den Aufsätzen von W. Immler und H. Thorade. Ann. d. Hydr. usw 1919, S. 22 und 35.)." *Annalen der Hydrographie und Maritimen Meteorologie* 47(3–4): 75–78.

———. 1919b. "Das winkeltreue gegenazimutale Kartennetz nach Littrow (Weirs Azimutdiagramm)." *Annalen der Hydrographie und Maritimen Meteorologie* 47(1–2): 14–22.

———. 1922. "Über Bezeichnungen von Kartenentwürfen, insbesondere über doppelsymmetrische unecht-abstandstreue Weltkarten mit Pol-Punkten." *Zeitschrift der Gesellschaft für Erdkunde zu Berlin* (3–4): 115–26.

———. 1931. "Johann Heinrich Lambert." *Hydrographic Review* 8:69–82. Translated into English by Waldo R. Tobler and included in his 1972 translation of Lambert 1772.

———. 1935. *Ebene Kugelbilder. Ein Linnésches System der Kartenentwürfe*. Published as *Petermanns Mitteilungen*, Ergänzungsheft (supplement) 221. Translated into English by Peter Ludwig as *Plane Globe Projection: A Linnean System of Map Projection*. Ed. William Warntz. Harvard Papers in Theoretical Geography, Geography and the Properties of Surfaces series, paper 23, Cambridge, Mass., 1968.

Mayr, Franz. 1964. "Flächentreue Plattkarten. Eine bisher vernachlässigte Gruppe unechter Zylinderprojektionen" (in German with English abstract). *International Yearbook of Cartography* 4:13–19.

Mayr, Georg. 1874. "Gutachten über die Anwendung der graphischen Methode in der Statistik." *Zeitschrift des Koeniglich Bayerischen Statistischen Bureau* 6(1–2): 36–44.

[Mead, Bradock, alias John Green]. 1717. *The Construction of Maps and Globes. In Two Parts. First, Contains the various Ways of projecting Maps, exhibited in fifteen different Methods, with their Uses; Second, Treats of making diverse Sorts of Globes.* London: T. Horne [et al.].

Melluish, R. K. 1931. *An Introduction to the Mathematics of Map Projections.* Cambridge: Cambridge University Press.

Mendeleev, Dmitri Ivanovich. 1907. "K poznanyu Rossi." Akademiya Nauk [St. Petersburg]. *Izvestiya*, ser. 6, 1:141–57.

Mercator-Hondius-Janssonius Atlas or A Geographicke Description of the World. 1636. Amsterdam. Reprinted, Amsterdam: Theatrum Orbis Terrarum, 1968.

Merriman, Arthur Douglas. 1947. *An Introduction to Map Projections.* London: G. G. Harrup.

Meshcheryakov, G. A. 1968. *Teoreticheskiye osnovy matematicheskoy kartografii* (Theoretical foundations of mathematical cartography). Moscow: Nedra.

Mikhaylov, Aleksandr Aleksandrovich. 1911–12. "O proyektsiyakh na sekushchem konuse, sokhranyayushchikh razmery po meridianam" (Projections on secant cones, preserving scale along meridians). *Otdel'nyy ottisk iz otcheta Moskovskogo o-va Iyubiteley astronomii za 1911–12,* 5–6.

Miller, Osborn Maitland. 1941. "A Conformal Map Projection for the Americas." *Geographical Review* 31(1): 100–104.

———. 1942. "Notes on Cylindrical World Map Projections." *Geographical Review* 32(3): 424–30.

———. 1953. "A New Conformal Projection for Europe and Asia [*sic;* should read Africa]." *Geographical Review* 43(3): 405–9.

———. 1955. *Specifications for a Projection System for Mapping Continuously Africa, Europe, Asia, and Australasia on a Scale of 1:5,000,000.* New York: American Geographical Society. Contract report to Army Map Service, May 1955. Also corrections in letter by Miller, June 2, 1955, and Supplement, Jan. 31, 1956.

———. 1962. "Map Projections." *Encyclopaedia Britannica*, 15th ed., 14:841–44. Reprintings to 1972.

———. 1965. "Some Equivalent Map Projection Transformations in the Plane." *Survey Review* 18(136): 73–77.

Miller, Ronald. 1948. "Map Projections of the Entire Sphere. *Scottish Geographical Magazine* 64(3): 113–22. See also *Geography* 36(4)(1951): 270.

Mitchell, Hugh C., and Lansing G. Simmons. 1945. *The State Coordinate Systems (A Manual for Surveyors).* Washington: U.S. Coast and Geodetic Survey Special Publication 235.

Mitchell, Samuel Augustus. 1866a. *Mitchell's Modern Atlas.* Philadelphia: E. H. Butler and Co.

———. 1866b. *Mitchell's New General Atlas.* Philadelphia: S. Augustus Mitchell, Jr.

Mittelstaedt, Otto. 1976. "Dr. Karl Heinrich Wagner zum 70. Geburtstag." *Kartographische Nachrichten* 26(4): 148–49.

Moccia, A., and S. Vetrella. 1986. "An Integrated Approach to Geometric Precision Processing of Spaceborne High-Resolution Sensors." *International Journal of Remote Sensing* 7(3): 349–59.

Modelski, Andrew M. 1984. *Railroad Maps of North America: The First Hundred Years.* Washington: Library of Congress.

Mollweide, Karl Brandan. 1805a. "Mappirungskunst des Claudius Ptolemaeus, ein Beytrag zur Geschichte der Landkarten." *Zach's Monatliche Correspondenz zur Beförderung der Erd- und Himmels-Kunde* 11(Apr.): 319–40; 11(June): 504–14; 12(July): 13–22.

———. 1805b. "Ueber die vom Prof. *Schmidt* in Giessen in der zweyten Abtheilung seines Handbuchs der Naturlehre S. 595 angegebene *Projection der Halbkugelfläche.*" *Zach's Monatliche Correspondenz zur Beförderung der Erd- und Himmels-Kunde* 12(Aug.): 152–63.

———. 1807. "Einege Projectionsarten der sphäroidischen Erde." *Zach's Monatliche Correspondenz zur Beförderung der Erd- und Himmels-Kunde* 16(Sept.): 197–210.

Monmonier, Mark Stephen. 1985. *Technological Transition in Cartography.* Madison: University of Wisconsin Press.

———. 1990. *How to Lie with Maps.* Chicago: University of Chicago Press.

Monmonier, Mark, and George A. Schnell. 1988. *Map Appreciation.* Englewood Cliffs, N.J.: Prentice-Hall.

Morrison, G. J. 1901. *Maps, Their Uses and Construction. A Short Popular Treatise on the Advantages and Defects of Maps on Various Projections.* London: Edward Stanford. Rev. ed., 1911.

Morse, Jedediah. 1814. *A Compendium and Complete System of Modern Geography or a View of the Present State of the World.* Boston: Thomas and Andrews.

Morskogo General'nogo Shtaba. 1950. *Morskoy atlas.*

Mugnier, Clifford J. [1985]. "Grids and Datums of the World." Manuscript.

Münster, Sebastian. 1540. *Geographia.* Basel.

Murchison, Roderick Impey. 1857. "New Geometrical Projection of Two-Thirds of a Sphere." In his "Address to the Royal Geographical Society of London, delivered at the Anniversary meeting on the 25th May, 1857." *Journal of the Royal Geographical Society* 27:cxli–cxlii. Also in *Proceedings of the Royal Geographical Society* 1(1857): 421–22.

Murdoch, Patrick. 1758. "On the best form of Geographical Maps." *Philosophical Transactions* (of the Royal Society) 50(2): 553–62. Abridged and reprinted in *Philosophical Transactions* 11(1755–63): 215–18. London: C. and R. Baldwin, 1809.

NASA Tech Briefs. 1991. "Program Computes Universal Transverse Mercator Projection." 17(7): 62–63.

National Geographic Society. 1922. "The Society's New Map of the World." *National Geographic Magazine* 42(6): 690, with world map supplement.

———. 1932. "The Story of the Map." *National Geographic Magazine* 62(6): 759–74, with world map supplement.

———. 1951. "Our Narrowing World." *National Geographic Magazine* 100(6): 751–54, with world map supplement.

———. 1959. "Portrait of Earth's Largest Continent." *National Geographic* 116(6): 751, with supplement map of Asia.

———. 1971. *National Geographic* 139(3): supplement map of Asia.

————. 1979. *National Geographic* 156(2): 154A, supplement map of the Americas, scale 1:20,000,000.

————. 1988. *The World.* World map on Robinson projection; various sizes.

————. 1990. *Atlas of the World.* 6th ed. Washington: National Geographic Society.

Nell, Adam Maximilian. 1852. *Vorschlag zu einer neuen charten Projektion.* Mainz: Inaugural-schrift, F. Kupferberg.

————. 1885–86. "Fischers perspektivische Projektion zur Darstellung der Kontinente." *Petermanns Mitteilungen* 31(8)(1885): 295–97; 32(8)(1886): 247–48.

————. 1890. "Äquivalente Kartenprojektionen." *Petermanns Mitteilungen* 36(4): 93–98.

Neugebauer, Otto. 1959. "Ptolemy's *Geography,* Book VII, Chapters 6 and 7." *Isis* 50, pt. 1 (159): 22–29.

————. 1975. *A History of Ancient Mathematical Astronomy.* New York: Springer-Verlag.

New York Times. 1976. "Wellman Chamberlin Dies at 68; Geographic Unit's Map Expert." July 7, p. 36.

New Zealand Cartographic Journal. 1985. "Obituary: Laurence Patrick Lee." 15(1): 32.

Nicholas, William H. 1943. "The World's Words." *National Geographic Magazine* 84(6): 689–700, with world map supplement.

Nicolls, Michael. 1986. *Wellman Chamberlin and the Chamberlin Trimetric Map Projection.* George Mason University, Fairfax, Va. Manuscript for course GECA 661.

Nicolosi, Giovanni Battista. 1660. *Hercole siciliano, studio geografico.* Rome.

Nordenskiöld, Adolf Erik. 1889. *Facsimile-Atlas to the Early History of Cartography with Reproductions of the Most Important Maps Printed in the XV and XVI Centuries.* English translation of Swedish original by Johan Adolf Ekelöf and Clements R. Markham. Stockholm, 1889. Reprinted, New York: Dover Publications, 1973.

————. 1897. *Periplus.* English translation of Swedish original by Francis A. Bather. Stockholm: P. A. Norstedt. Reprinted, New York: B. Franklin, 1962, 1967.

Nowicki, Albert L. 1962. "Topographic Lunar Mapping." *Canadian Surveyor* 16(3): 141–48.

Nyerges, Timothy L., and Piotr Jankowski. 1989. "A Knowledge Base for Map Projection Selection." *American Cartographer* 16(1): 29–38.

Odyssey Books. 1966. *The Odyssey World Atlas.* New York.

O'Keefe, John A. 1948. "The New Military Grid of the Department of the Army." *Surveying and Mapping* 8(4): 214–16.

————. 1952. "The Universal Transverse Mercator Grid and Projection." *Professional Geographer* 4(5): 19–24.

Ordnance Survey. 1991. *Ordnance Survey, 1791–1991.* [Southampton, Eng.].

Ormeling, Ferdinand J., comp. 1988. *ICA 1959–1984: The First Twenty-five Years of the International Cartographic Association.* New York: Elsevier.

Osley, Arthur S. 1969. *Mercator.* New York: Watson-Guptill Publications.

Parent, Antoine. 1704. "Détermination des projections cylindriques et sphériques les plus parfaites pour la construction des cartes géographiques." In *Histoire de l'Académie Royale des Sciences, année MDCCII, avec les mémoires de mathématique et de physique pour la même année,* 2:613–30. Paris. Additions on pp. 917–21: "Sur les projections de la sphere le plus égales qu'il se puisse" (presented Dec. 20, 1702).

Pattison, William D. 1960. "The Star of the AAG." *Professional Geographer* 12(5): 18–19.

Pearson, Frederick II. 1977. *Map Projection Equations*. Dahlgren, Va.: Naval Surface Weapons Center. Revised as *Map Projection Methods*. Blacksburg, Va.: Sigma Scientific, Inc., 1984. Revised as *Map Projections: Theory and Applications*. Boca Raton, Fla.: CRC Press, 1990.

Peirce, Charles Sanders. 1879. "A Quincuncial Projection of the Sphere." *American Journal of Mathematics* 2(4): 394–96. Also in *U.S. Coast Survey Report* for the year ending June 1877, pub. 1880, Appendix 15, pp. 191–92.

Penck, Albrecht. 1910. "Die Weltkartenkonferenz in London 16–22. November 1909," *Petermanns Mitteilungen* 56-1(1): 33–35.

Peter Heiler Ltd. [Before 1988]. *The World/Le Monde/Die Erde*. Whitby, Ontario. Scale 1: 35,000,000.

Petermanns Mitteilungen 63(1917): 221.

Peters, Aribert B. 1984. "Distance-related Maps." *American Cartographer* 11(2): 119–31.

Peters, Arno. [1983]. *Die Neue Kartographie/The New Cartography* (in German and English). Klagenfurt, Austria: Carinthia University; New York: Friendship Press.

———. 1989–90. *Peters Atlas of the World*. England: Longman House, 1989; New York: Harper and Row, 1990, with minor textual changes.

Petri, Henri. 1541. *Cl. Ptolemaeus, omnia, quae extant opera, geographia excepta*. Basel.

Philbrick, Allen K. 1953. "An Oblique Equal Area Map for World Distributions." *Annals of the Association of American Geographers* 43(3): 201–15.

Pierpont, James. 1896. "Note on C. S. Peirce's Paper on 'A Quincuncial Projection of the Sphere.'" *American Journal of Mathematics* 18(2): 145–52.

Poole, H. 1934. "An Oblique Cylindrical Equal-Area Map." *Geographical Journal* 83(2): 142–43.

———. 1935. "A Map Projection for the England-Australia Air Route." *Geographical Journal* 86(5): 446–48.

Portinaro, Pierluigi, and Franco Knirsch. 1987. *The Cartography of North America*. New York: Facts on File.

Potter, A. J. 1925. "The Tetrahedral Gnomonic Projection." *Geographical Teacher* 13(pt. 1) (71): 52–56.

Potter, Jonathan. 1989. *Country Life Book of Antique Maps*. Secaucus, N.J.: Chartwell Books.

Prentice-Hall. 1983. *The Prentice-Hall Universal Atlas*. Englewood Cliffs, N.J.

Putniņš, Reinholds V. 1934. "Jaunas projekci jas pasaules kartēm" (in Latvian with extensive French résumé). *Geografiski Raksti, Folia Geographica 3 and 4*, 180–209.

[Pye, Norman?]. 1955. Review of *A New Map of the World* by T. Edwards, 1953. *Geography* 40(1): 71.

Quilette/Hachette. 1984. *Atlas mondial*. Paris.

Quill, Humphrey. 1966. *John Harrison—The Man Who Found Longitude*. London: John Baker.

Raisz, Erwin Josephus. 1934. "The Rectangular Statistical Cartogram." *Geographical Review* 24(2): 292–96.

———. 1938. *General Cartography*. New York: McGraw-Hill. 2d ed., 1948.

————. 1943. "Orthoapsidal World Maps." *Geographical Review* 33(1): 132–34.

————. 1944. *Atlas of Global Geography*. New York: Global Press Corp., 1944.

————. 1962. *Principles of Cartography*. New York: McGraw-Hill.

Rand McNally & Co. 1896. *New Pictorial Atlas of the World*. Chicago.

————. 1916. *The Rand McNally Imperial Atlas of the World*. Chicago.

————. 1932. *Rand McNally World Atlas, Premier Edition*. Chicago.

————. 1949a. *Cosmopolitan World Atlas*. Chicago.

————. 1949b. *Rand McNally World Atlas, Readers Edition*. New York.

————. 1968. *Around the World—A View from Space*. Chicago.

————. 1971. *Rand McNally Cosmopolitan World Atlas*. Chicago.

————. 1983. *The New International Atlas*. Chicago.

————. 1986. *Goode's World Atlas*. 17th ed. Chicago.

————. 1989. "Rand McNally World Portrait Map, Based on the Robinson Projection, 1989 Edition."

Ravenhill, William. 1978. "John Adams, His Map of England, Its Projection, and His *Index Villaris* of 1680." *Geographical Journal* 144(3): 424–37.

————. 1981. "Projections for the Large General Maps of Britain, 1583–1700." *Imago Mundi* 33:21–32.

Reeves, Edward A. 1904. "Van der Grinten's Projection." *Geographical Journal* 24(6): 670–72.

————. 1929. "A Chart Showing the True Bearing of Rugby from All Parts of the World." *Geographical Journal* 73(3): 247–48.

Reichard, Christian Gottlieb. 1803. *Atlas der ganzen Erdkreises, nach den besten astronomischen Bestimmungen, neuesten Entdeckungen in eigenen Untersuchungen in der Central-Projection auf 6 Tafeln entworfen*. Weimar: Lander Industrie Comptoirs.

Reignier, François. 1957. *Les systèmes de projection et leurs applications*. Vol. 1. Paris: Institut Géographique National.

Reilly, W. I. 1973. "A Conformal Mapping Projection with Minimum Scale Error." *Survey Review* 22(168): 57–71.

Reilly, W. I., and H. M. Bibby. 1975–76. "A Conformal Mapping Projection with Minimum Scale Error, Part 2: Scale and Convergence in Projection Coordinates." *Survey Review* 23(176)(1975): 79–87. "Part 3: Transformation of Coordinates, and Examples of Use." *Survey Review* 23(181)(1976): 302–15.

Reinganum. 1839. *Geschichte der Erd- und Himmelsabbildungen der Alten I*. Jena.

Reisch, Gregor. 1512. *Margarita philosophica nova*. Strasbourg. Also later ed.

Richardson, Robert T. 1989. "Area Deformation on the Robinson Projection." *American Cartographer* 16(4): 294–96.

Richardus, Peter, and Ron K. Adler. 1972. *Map Projections for Geodesists, Cartographers, and Geographers*. Amsterdam: North Holland Publishing.

Riel, C. 1875. *Das Sonnen- und Sirius-Jahr der Rammessiden*. Leipzig.

Ristow, Walter William, ed. 1972. *A la carte*. Washington: Library of Congress.

————. 1979. "Aborted American Atlases." *Quarterly Journal of the Library of Congress* (Summer): 320–45.

————. 1985. *American Maps and Mapmakers, Commercial Cartography in the Nineteenth Century.* Detroit: Wayne State University Press.

Ritter, M. Franciscus. 1610. *Speculum orbis.* Nuremberg: Paul Fürstens.

Robinson, Arthur Howard. 1953a. *Elements of Cartography.* New York: John Wiley and Sons. 2d ed., 1960; 3d ed., with Randall D. Sale, 1969; 4th ed., with Sale and Joel L. Morrison, 1978; 5th ed., with Sale, Morrison, and Phillip C. Muehrcke, 1984.

————. 1953b. "Interrupting a Map Projection: A Partial Analysis of Its Value." *Annals of the Association of American Geographers* 43(3): 216–25.

————. 1970. "Erwin Josephus Raisz, 1893–1968." *Annals of the Association of American Geographers* 60(1): 189–93.

————. 1974. "A New Map Projection: Its Development and Characteristics." *International Yearbook of Cartography* 14: 145–55.

————. 1982. *Early Thematic Mapping in the History of Cartography.* Chicago: University of Chicago Press.

————. 1985. "Arno Peters and His New Cartography." *American Cartographer* 12(2): 103–11.

[————]. 1986. Committee on Map Projections of the American Cartographic Association. *Which Map Is Best? Projections for World Maps.* Falls Church, Va.: American Congress on Surveying and Mapping, Special Publication 1 of ACA.

[————]. 1988. Committee on Map Projections of the American Cartographic Association. *Choosing a World Map: Attributes, Distortions, Classes, Aspects.* Falls Church, Va.: American Congress on Surveying and Mapping, Special Publication 2 of ACA.

————. 1989. "Flattening the Round Earth: Fashions and Fads in Map Projections." Annual meeting, State Historical Society of Wisconsin, Manitowoc, Wis. June 16. Manuscript.

————. 1990. "Rectangular World Maps—No!" *Professional Geographer* 42(1): 101–4.

Ronan, Colin A. 1983. *The Cambridge Illustrated History of the World's Sciences.* Cambridge: Cambridge University Press.

Rosen, Edward. 1970–80. "Johannes Schöner." In Gillispie 1970–80, 12:199–200.

Rosenmund, Max. 1903. *Die Änderung des Projektionssystems der schweizerischen Landesvermessung.* Bern, Switz.: Militärdepartements, Abteilung für Landestopographie, Verlag.

Rotz, Jean. N.d. *Booke of Hydrography.* British Museum Royal MS. 20EIX.

Roussilhe, Henri. 1922. "Emploi des coordonnées rectangulaires stéréographiques pour le calcul de la triangulation dans un rayon de 560 kilomètres autour de l'origine." Section of Geodesy, International Union of Geodesy and Geophysics. May.

Royal Geographical Society. 1989. *Newsletter* (Nov.).

Royal Society. 1966. "Appendix 1: Named Map Projections." In *Glossary of Technical Terms in Cartography,* 41–67. London.

Royer, Augustin. 1679. *Cartes du ciel reduites en quatre tables contenant toutes des constellations.* Paris: J. B. Coignard.

Rubincam, David P. 1981. "Latitude and Longitude from van der Grinten Grid Coordinates." *American Cartographer* 8(2): 177–80.

Ruscelli, Girolamo. 1561–74. *Geographia.* Venice. 1561, 1562, 1574.

Salmanova, T. D. 1951. "Vidoizmenennaya polikonicheskaya proyektsiya dlya stennoy uchebnoy karty Sovetskogo Soyuza" (Modified polyconic projection for an educational wall map of the Soviet Union). TsNIIGAiK. *Trudy* (88): 32–37.

Sanchez, Pedro C., and Octavio Bustamente. 1928. *Apuntes sobre cartografia.* Mexico City: Secretaria de Agriculture y Fomento de Mexico, Talleres Graficos de la Secretaria de Agriculture y Fomento.

Sanchez-Saavedra, E. M. 1975. *Description of the Country: Virginia's Cartographers and Their Maps, 1607–1881.* Richmond: Virginia State Library.

Sanson, Nicolas. 1675. *Cartes generales de la géographie ancienne et nouvelle.* Paris.

——. 1683. *L'Europe* [with subsections *l'Asie, l'Afrique,* and *l'Amerique*] *en plusieurs cartes, et en divers traittés de géographie et d'histoire.* Paris.

Savage-Smith, Emilie. 1992. "Islamic Cartography: Celestial Mapping." In Harley and Woodward 1992, 12–70.

Saxton, Christopher. 1579. *An Atlas of England and Wales.* London.

Scarborough, James B. 1930. *Numerical Mathematical Analysis.* Baltimore, Md.: Johns Hopkins Press.

Schickard, Wilhelm. 1623. *Astroscopium pro facillima stellarum cognitione excogitatum & commentariolo illustratum.* Tübingen. Also Nordlingae, 1655; Stuttgart, 1687.

Schjerning, Wilhelm. 1904. "Über mittabstandstreue Karten." *Mitteilungen der Österreichischen Geographischen Gesellschaft* 3(4).

Schmid, Erwin. 1962. *The Earth as Viewed from a Satellite.* Washington: U.S. Coast and Geodetic Survey Technical Bulletin 20.

——. 1974. *World Maps on the August Epicycloidal Conformal Projection.* Washington: National Oceanic and Atmospheric Administration, NOAA Technical Report NOS 63.

Schmidt, Georg Gottlieb. 1801–3. "Projection der Halbkugelfläche." In his *Handbuch der Naturlehre.* Giessen.

Schöner, Johannes. 1533. *Opusculum geographicum.* Nuremberg.

——. 1551. "Planispherium, seu meteoroscopium." In *Opera mathematica.* Nuremberg.

Schott, Charles A. 1882. "A comparison of the relative value of the Polyconic projection used on the Coast and Geodetic Survey, with some other projections." In *Report of the Superintendent of the U.S. Coast and Geodetic Survey . . . June 1880,* Appendix 15, pp. 287–96. Washington: Government Printing Office.

Schoy, Carl. 1913a. "Azimutale und gegenazimutale Karten mit gleichabständigen parallelen Meridianen." *Annalen der Hydrographie und Maritimen Meteorologie* 41(1): 33–42.

——. 1913b. "Die gegenazimutale mittabstandstreue Karte in konstruktiver und theoretischer Behandlung." *Annalen der Hydrographie und Maritimen Meteorologie* 41(9): 466–73.

——. 1915. "Nochmals 'Azimutale und gegenazimutale Karten.'" *Petermanns Mitteilungen* 61-1(Apr.): 137–38.

Schreiber, Oskar. 1866. *Theorie der Projektionsmethode der hannoverschen Landesvermessung.* Hannover: Hahnsche Hofbuchhandlung.

——. 1897. *Die konforme Doppelprojektion der trigonometrischen Abteilung der Königl. Preussischen Landesaufnahme.* Berlin.

——. 1899–1900. "Zur konformen Doppelprojektion der preussischen Landesaufnahme. Sphäroid und Kugel. Gauss'sche Projection." *Zeitschrift für Vermessungswesen* (1899)28: 491–502, 593–613; (1900)29: 257–81, 289–310.

Schwartz, Seymour I., and Ralph E. Ehrenberg. 1980. *The Mapping of America.* New York: Harry N. Abrams.

Schwarz, Hermann Amandus. 1869. "Ueber einege Abbildungsaufgaben." *Journal für die Reine und Angewandte Mathematik* (Crelle's) 70(2): 105–20.

Seller, John. 1679. *Atlas maritimus*. London.

Serapinas, B. B. 1984. "O poluchenii ravnougol'nykh kartograficheskikh proyektsiy trekhosnogo ellipsoida" (The calculation of conformal cartographic projections of the triaxial ellipsoid). *Geodeziya i Kartografiya* (8): 48–50.

Shabanova, A. I. 1952. "K voprosu o razrabotke novykh tsilindricheskikh proyektsiy" (On the question of development of new cylindrical projections). *Sbornik Statey po Kartografii* 2: 67–72.

Shalowitz, Aaron L. 1946. "Charles Henry Deetz." *Surveying and Mapping* 6(1): 19.

———. 1964. *Shore and Sea Boundaries*. Vol. 2. Washington: U.S. Coast and Geodetic Survey Publication 10-1.

Shirley, Rodney W. 1983. *The Mapping of the World: Early Printed Maps, 1472–1700.* London: Holland Press.

Shlomi, Eli. 1977. "A Polyfocal Projection: Development and Applications." Master's thesis. Tel-Aviv University.

The Shorter Oxford Economic Atlas of the World. 1959. 2d ed. London: Oxford University Press.

Siemon, Karl. 1935. "Wegtreue Ortskurskarten." *Mitteilungen des Reichsamts für Landesaufnahme* 11(2): 88–95.

———. 1937. "Flächenproportionales Umgraden von Kartenentwürfen." *Mitteilungen des Reichsamts für Landesaufnahme* 13(2): 88–102.

Smith, Benjamin E. 1897. *The Century Atlas of the World.* New York: Century.

Smith, Doyle G., and John P. Snyder. 1989. "Expert Map Projection Selection System." In *U.S. Geological Survey Yearbook, Fiscal Year 1988*, 14–15.

Smith, James Addison. 1939. *Globe*. U.S. Patent 2,153,053. Apr. 4, 1939.

Smith, James R. 1986. *From Plane to Spheroid: Determining the Figure of the Earth from 3000 B.C. to the 18th Century Lapland and Peruvian Survey Expeditions.* Rancho Cordova, Calif.: Landmark Enterprises.

Snyder, John Parr. 1977. "A Comparison of Pseudocylindrical Map Projections." *American Cartographer* 4(1): 59–81. See corrections in 6(1)(1979): 81.

———. 1978a. "Equidistant Conic Map Projections." *Annals of the Association of American Geographers* 68(3): 373–83.

———. 1978b. "The Space Oblique Mercator Projection." *Photogrammetric Engineering and Remote Sensing* 44(5): 585–96.

———. 1979. "Calculating Map Projections for the Ellipsoid." *American Cartographer* 6(1): 67–76. See correction in 8(2)(1981): 160.

———. 1981a. "Map Projections for Satellite Tracking." *Photogrammetric Engineering and Remote Sensing* 47(2): 205–13.

———. 1981b. "The Perspective Map Projection of the Earth." *American Cartographer* 8(2): 149–60. See corrections in 9(1)(1982): 84.

———. 1981c. *Space Oblique Mercator Projection: Mathematical Development.* Washington: U.S. Geological Survey Bulletin 1518. See corrections in review by Joseph C. Loon, in *Photogrammetric Engineering and Remote Sensing* 48(10)(1982): 1581.

———. 1982. *Map Projections Used by the U.S. Geological Survey.* Washington: U.S. Geological Survey Bulletin 1532. 2d ed., 1983.

———. 1984a. "A Low-Error Conformal Map Projection for the 50 States." *American Cartographer* 11(1): 27–39. See correction in Snyder 1987b, 207.

———. 1984b. "Minimum-Error Map Projections Bounded by Polygons." *Cartographic Journal* 21(2): 112–20. See corrections in 22(1)(1985): 73.

———. 1985a. *Computer-Assisted Map Projection Research.* Washington: U.S. Geological Survey Bulletin 1629.

———. 1985b. "Conformal Mapping of the Triaxial Ellipsoid." *Survey Review* 28(217): 130–48.

———. 1985c. "The Transverse and Oblique Cylindrical Equal-Area Projection of the Ellipsoid." *Annals of the Association of American Geographers* 75(3): 431–42.

———. 1986. "A New Low-Error Map Projection for Alaska." In *Technical Papers,* 307–14. American Society for Photogrammetry and Remote Sensing—American Congress on Surveying and Mapping, Fall Convention, Anchorage, Alaska.

———. 1987a. "'Magnifying-Glass' Azimuthal Map Projections." *American Cartographer* 14(1): 61–68.

———. 1987b. *Map Projections: A Working Manual.* Washington: U.S. Geological Survey Professional Paper 1395.

———. 1988. "New Equal-Area Map Projections for Non-Circular Regions." *American Cartographer* 15(4): 341–55.

Snyder, John Parr, and Harry Steward, eds. 1988. *Bibliography of Map Projections.* Washington: U.S. Geological Survey Bulletin 1856.

Snyder, John Parr, and Philip M. Voxland. 1989. *An Album of Map Projections.* Washington: U.S. Geological Survey Professional Paper 1453.

Solov'ev. Mikhail Dmitriyevich. 1937. *Kartograficheskiye proyektsii.* Moscow: Geodezicheskoy i Kartograficheskoy dit-ry. 2d ed., 1946. Rev. ed. (Moscow: Geodezizdat), 1964.

———. 1949. "Atlas kartograficheskikh proyektsiy TsNIIGAiK" (Atlas of cartographic projections for the TsNIIGAiK). TsNIIGAiK. *Trudy* (61).

———. 1952. *Prakticheskoye posobiye po matematicheskoy kartografii* (Practical manual for mathematical cartography). Moscow: Geodezizdat.

———. 1969. *Matematicheskaya kartografiya* (Mathematical cartography). Moscow: Nedra.

Southworth, Michael, and Susan Southworth. 1982. *Maps: A Visual Survey and Design Guide.* Boston: Little, Brown.

Spilhaus, Athelstan F. 1942. "Maps of the Whole World Ocean." *Geographical Review* 32(3): 431–35.

Spilhaus, Athelstan F., and John P. Snyder. 1991. "World Maps with Natural Boundaries." *Cartography and Geographic Information Systems* 18(4): 246–54.

Spinnaker Press. 1983. *World Ocean Map (Spilhaus Projection).*

Sprinsky, William H., and John P. Snyder. 1986. "The Miller Oblated Stereographic Projection for Africa, Europe, Asia and Australasia." *American Cartographer* 13(3): 253–61.

Stamp, Dudley, ed. 1966. *Longmans Dictionary of Geography.* London: Longman, Green.

Stanley, Albert A. 1946. "A Quincuncial Projection of the World." *Surveying and Mapping* 6(1): 19.

Starostin, F. A., L. A. Vakhrameyeva, and L. M. Bugayevskiy. 1981. "Obobshchennaya klassi-fikatsiya kartograficheskikh proyektsiy po vidu izobrazheniya meridianov i paralleley" (Generalized classification of cartographic projections in regard to the representation of meridians and parallels). *Geodeziya i Aerofotos'emka* (6): 111–16.

Steck, Max. 1970–80. "Albrecht Dürer." In Gillispie 1970–80, 4:258–61.

Steers, James Alfred. 1970. *An Introduction to the Study of Map Projections.* 15th ed. London: University of London Press.

Steinhauser, Anton. 1883. "Ueber die Anwendung der Kegelprojektion auf Darstellungen der ganzen Erde." *Zeitschrift für Wissenschaftliche Geographie* 4:34–36.

Stephens, John D. 1980. "Current Cartographic Serials: An Annotated International List." *American Cartographer* 7(2): 123–38.

Stevenson, Edward Luther. 1903–6. *Maps Illustrating the Early Discovery of America, 1502–1530.* New Brunswick, N.J.

———. 1921. *Terrestrial and Celestial Globes: Their History and Construction, Including a Consideration of Their Value as Aids in the Study of Geography and Astronomy.* Hispanic Society of America, Publication 86. New Haven: Yale University Press. Reprinted, New York and London: Johnson Reprint Corp., 1971.

———. 1932. *Geography of Claudius Ptolemy.* New York: New York Public Library. English translation from Greek, Latin, and other sources. Reprinted as Claudius Ptolemy, *The Geography.* (New York: Dover Publications, 1991).

Stieler, Adolf. [1855]. *Hand-Atlas über alle Theile der Erde nach dem neuesten Zustande.* Gotha: Justus Perthes.

Stooke, Philip J. 1989. "Sizing up Phobos." *Sky and Telescope* 77(5): 477–79.

Stooke, Philip J., and C. Peter Keller. 1990. "Map Projections for Non-Spherical Worlds/The Variable-Radius Map Projections." *Cartographica* 27(2): 82–100.

Struik, Dirk J. 1948. *A Concise History of Mathematics.* 2d ed. New York: Dover Publications.

Surveying and Mapping. 1962. "Deaths." 22(2): 324.

Sylvano, Bernardo. 1511. *Geographia.* Venice. Reprinted, Amsterdam: Theatrum Orbis Terrarum, 1969.

Taich, V. D. 1937. "Uslovnaya proyektsiya c polyusnoy dugoy i ravnorazdelennymi parallely-ami" (A conventional projection with a polar arc and equally divided parallels). *Geodezist* (2).

Taton, René. 1970–80. "Philippe de La Hire." In Gillispie 1970–80, 7:576–79.

Thomas, Paul D. 1952. *Conformal Projections in Geodesy and Cartography.* Washington: U.S. Coast and Geodetic Survey Special Publication 251.

[———]. 1970. *Spheroidal Geodesics, Reference Systems, & Local Geometry.* Washington: U.S. Naval Oceanographic Office SP-138.

Thomson, John. 1816. *New General Atlas.*

Thorade, Hermann. 1919. "Zur Umwandlung von Kartenprojektion (Bemerkungen zu der vorangehenden Abhandlung Immlers)." *Annalen der Hydrographie und Maritimen Meteorologie* 47(1–2): 36–38.

Thornthwaite, C. Warren. 1927. "The Cylindrical Equal-Area Projection for a New Map of Eurasia and Africa." *University of California Publications in Geography* (Berkeley) 2(6): 211–30.

Thrower, Norman J. W. 1972. *Maps and Man.* Englewood Cliffs, N.J.: Prentice-Hall.

Tibbetts, Gerald R. 1992. "Islamic Cartography: Later Cartographic Developments." In Harley and Woodward 1992, 137–155.

Times Books. 1985. *The Times Atlas of the World.* 7th ed. London.

Tissot, Nicolas Auguste. 1859. "Sur les cartes géographiques." *Comptes Rendus des Séances de l'Académie des Sciences* 49(19): 673–76.

———. 1860. "Sur les cartes géographiques." *Comptes Rendus des Séances de l'Académie des Sciences* 51:964–69. Reprinted in Germain 1866, 230–34.

———. 1878. "Sur la représentation des surfaces et les projections des cartes géographiques." *Nouvelles Annales de Mathématiques,* 2d ser., 17:49–55, 145–63, 351–66.

———. 1881. *Mémoire sur la représentation des surfaces et les projections des cartes géographiques.* Paris: Gauthier Villars.

———. 1887. *Die Netzentwürfe Geographischer Karten.* Stuttgart: J. B. Metzlersche Buchhandlung. Tissot 1881 translated into German with additions by Ernst Hammer.

Tobler, Waldo Rudolph. 1962a. "A Classification of Map Projections." *Annals of the Association of American Geographers* 52(2): 167–75.

———. 1962b. "The Polar Case of Hammer's Projection." *Professional Geographer* 14(2): 20–22. See correction in Tobler 1964, 242.

———. 1963a. "Geographic Area and Map Projections." *Geographical Review* 53(1): 59–78.

———. 1963b. "Map Projection Research by Digital Computer." Paper presented to Association of American Geographers, Denver. September.

———. 1964. "Some New Equal Area Map Projections." *Survey Review* 17(131): 240–43.

———. 1966a. "Medieval Distortions: The Projections of Ancient Maps." *Annals of the Association of American Geographers* 56(2): 351–61.

———. 1966b. "Notes on Two Projections." *Cartographic Journal* 3(2): 87–89.

———. 1973a. "A Continuous Transformation Useful for Districting." *Annals of the New York Academy of Sciences* 219:215–20.

———. 1973b. "The Hyperelliptical and Other New Pseudo Cylindrical Equal Area Map Projections." *Journal of Geophysical Research* 78(11): 1753–59.

———. 1974a. *Cartogram Programs.* University of Michigan, Department of Geography, Cartographic Laboratory Report 3.

———. 1974b. "Local Map Projections." *American Cartographer* 1(1): 51–62.

———. 1977. "Numerical Approaches to Map Projections." In Ingrid Kretschmer, ed., *Beiträge zur Theoretischen Kartographie* (Studies in theoretical cartography), 51–64. Vienna: Franz Deuticke.

Tolstova, T. I. 1969. "Kriteriy Eyri v primenenii k azimutal'nym proyektsiyam." *Geodeziya i Aerofotos'emka* (6): 115–18. Translated into English as "The Airy Criterion as Applied to Azimuthal Projections (onto the Tangent Plane)." *Geodesy and Aerophotography* (6)(1969): 427–29.

Tooley, Ronald Vere. 1949. *Maps and Map-makers.* New York: Bonanza Books.

Tooley, Ronald Vere, and Charles Bricker. 1968. *Landmarks of Mapmaking: An Illustrated Survey of Maps and Mapmakers.* Amsterdam: Elsevier.

Toomer, G. J. 1970–80. "Ptolemy." In Gillispie 1970–80, 11:186–206.

True, David O. 1954. "Some Early Maps Relating to Florida." *Imago Mundi* 11:73–84.

Tsinger, Nikolay Yakovlevich. 1916. "O naivygodneyskikh vidakh konicheskikh proyetsiy" (The most suitable-appearing conic projections). Akademiya Nauk (St. Petersburg). *Izvestiya*, ser. 6, 10(17): 1693.

"Uebersicht der Literatur für Vermessungswesen auf 1876." *Zeitschrift für Vermessungswesen* 6(1877): Ergänzungsheft (supplement), pp. (78)–(79).

Uhden, Richard. 1935. "Die antiken Grundlagen der mittelalterlichen Seekarten." *Imago Mundi* 1:1–19.

———. 1937. "An Equidistant and a Trapezoidal Projection of the Early Fifteenth Century." *Imago Mundi* 2:8.

Union of Soviet Socialist Republics. Chief Administration of Geodesy and Cartography under the Council of Ministers. 1954. *Atlas mira.* Moscow. 2d ed., in both Russian and English (*The World Atlas*) editions, 1967.

United Nations. 1954. "Resolutions of the Second International Map Conference (Paris, 1913)." *World Cartography* 4:33–42.

———. 1963. *United Nations Technical Conference on the International Map of the World on the Millionth Scale,* vol. 2: *Specifications of the International Map of the World on the Millionth Scale (IMW).* New York.

United States. Department of the Army. 1973. *Universal Transverse Mercator Grid.* Washington: U.S. Army Technical Manual TM 5-241-8.

[United States Central Intelligence Agency]. 1973. *Projection Handbook.* Washington.

United States Coast and Geodetic Survey. 1951. *The World on the Azimuthal Equidistant Projection Centered at New York City.* Map 3,042. Washington.

United States Coast Survey. 1867. *Report of the Superintendent of the U.S. Coast Survey . . . 1865.* Appendix 20, pp. 176–86. Washington: Government Printing Office.

United States Environmental Science Services Administration 1969. *Climates of the World.* Washington.

United States Geological Survey. 1916. *United States Relief Map.* Washington. Scale 1:7,000,000. "Compiled by Henry Gannett. . . . Polyconic projection."

———. 1970. *National Atlas of the United States.* Washington.

Universum-Verlag. Ca. 1980. *The Peters Projection—World Press—Professional Press.* München-Solln.

Urmayev, Nikolay Andreyevich. 1941. *Matematicheskaya kartografiya* (Mathematical cartography). Moscow: Voyennoinzhenernaya Akademiya.

———. 1947. *Metody izskaniya novykh kartograficheskikh proyektsiy* (An approach to establishment of new cartographic projections). Moscow: VTU GSh VSSSR.

———. 1950. "Izyskaniye nekotorykh novykh tsilindricheskikh, azimutal'nykh i psevdotsilindricheskikh proyektsiy" (The search for some new cylindrical, azimuthal, and pseudocylindrical projections). Nauchno-Tekhnicheskiye i Proizvodstvennye Statey GUGK. *Sbornik* (29).

———. 1962. "Osnovy matematicheskoy kartografii" (The fundamentals of mathematical cartography). TsNIIGAiK. *Trudy* (144). Rev. ed., 1969.

van de Gohm, Richard. 1972. *Antique Maps for the Collector.* New York: Macmillan.

van der Grinten, Alphons J. 1904a. "Darstellung der ganzen Erdoberfläche auf einer kreisförmigen Projektionsebene." *Petermanns Mitteilungen* 50(7): 155–59. See also corrections: 50(10): 250; 51(2)(1905): 48.

———. 1904b. *Map.* U.S. Patent 751,226. Feb. 2, 1904. Also patented in Canada, Great Britain, and France (van der Grinten 1905a, 359).

———. 1905a. "New Circular Projection of the Whole Earth's Surface." *American Journal of Science,* ser. 4, 19(113): 357–66.

———. 1905b. "Zur Verebnung der ganzen Erdoberfläche." *Petermanns Mitteilungen* 51(10): 237. See further comments in 52(2)(1906): 46.

Varenius, Bernardus. 1650. *Geographia generalis, in qua affectiones generales Telluris explicantur.* Amsterdam. Translated into English by Richard Blome as *Cosmography and Geography in Two Parts: The First, Containing the General and Absolute Part of Cosmography and Geography; being a Translation from that Eminent and much Esteemed Geographer, Varenius, Wherein are at large handled All such Arts as are necessary to be Understood, for the True Knowledge thereof. To which is added the much wanted Schemes, omitted by the Author.* London: 1693. Esp. bk. 3, "Comparative Part of the Affections from Comparing of Places"; chap. 32, "General Geography"; proposition 6, "To Compose Geographical Maps," 316–34.

Velten, A. W. 1898. "Neue Methode, eine in azimutaler Projection." *Zeitschrift für Vermessungswesen* 27(4): 103–13.

Vietor, Alexander O. 1963. "A Pre-Columbian Map of the World, circa 1489." *Imago Mundi* 17:95–96.

Vitkovskiy, Vasily Vasil'evich. 1907. *Kartografiya: Teoriya kartograficheskikh proyektsiy* (Cartography: The theory of cartographic projections). 14. St. Petersburg: Yu.N. Erlikh.

Vitruvius Pollio, Marcus. N.d. *De architectura.*

Volodarsky, A. I. 1970–80. "Dmitry Aleksandrovich Grave." In Gillispie 1970–80, 5:508–9.

von Braunmühl, A. 1900. *Vorlesungen über die Geschichte der Mathematik.* Leipzig.

VonderMühll, Carl. 1868. "Ueber die Abbildung von Ebenen auf Ebenen." *Journal für die Reine und Angewandte Mathematik* (Crelle's) 69(3): 264–85.

Wagner, Heinrich. 1915. "Kartometrische Analyse der Weltkarte G. Mercators vom Jahre 1569." *Annalen der Hydrographie und Maritimen Meteorologie* 43(9): 377–94. Reprinted in *Acta Cartographica* 25(1977): 435–52.

Wagner, Hermann. 1925. "Ernst Hammer." *Petermanns Mitteilungen* 71(11/12): 254–56.

Wagner, Karl-Heinrich (Karlheinz). 1932. "Die unechten Zylinderprojektionen." *Aus dem Archiv der Deutschen Seewarte* 51(4).

———. 1941. "Neue ökumenische Netzentwürfe für die kartographische Praxis." *Jahrbuch der Kartographie,* 176–202.

———. 1949. *Kartographische Netzentwürfe.* Leipzig: Bibliographisches Institut. 2d ed., 1962.

Waldseemüller, Martin. 1507. *Universalis cosmographia secundum Ptholomaei traditionem et Americi Vespucii aliorumque lustrationes.* Strasbourg.

Wallis, P. J. 1970–80. "Edward Wright." In Gillispie 1970–80, 14:513–15.

Ward, H. B., ed. 1943. *Octovue Map of the World.* Milwaukee: L. E. Pitner.

Warner, Benjamin, and M. Carey & Son, pub. 1820. *A General Atlas: Being a Collection of Maps of the World and Quarters.* Philadelphia.

Warner, Deborah Jean. 1979. *The Sky Explored: Celestial Cartography, 1500–1800.* New York: Alan R. Liss.

Watts, D. G. 1970. "Some New Map Projections of the World." *Cartographic Journal* 7(1): 41–46.

Werenskiold, Werner. 1945. "A Class of Equal Area Map Projections." Norske Videnskaps-Akademi i Oslo, I. Matematisk-Naturvidenskapelig Klasse 1944, *Avhandlinger* (11).

Werner, Johannes. 1514. *Noua translatio primi libri geographiae Cl. Ptolemaei*. Nuremberg.

Westfall, John E. 1970. "The Ecumenical Projection." *Canadian Cartographer* 7(1): 42–47.

Wiechel, H. 1879. "Rationelle Gradnetzprojectionen." *Civilingenieur*, n. ser., 25, cols. 401–22.

Wilford, John Noble. 1972. "Mars Variety Shown in First Detailed Map." *New York Times*, Nov. 27, pp. 1, 30.

——. 1981. *The Mapmakers*. New York: Alfred A. Knopf.

William-Olsson, William. 1968. "A New Equal Area Projection of the World." *Acta Geographica* (Helsinki) 20:389–93.

Williams, Joseph E., ed. [1958]. *World Atlas*. Englewood Cliffs, N.J.: Prentice-Hall. 2d ed., 1964.

Winkel, Oswald. 1909. "Flächentreue, schiefachsige Zylinderprojektion mit längertreuem Grundkreis für eine Karte von Nord-, Mittel- und Südamerika." *Petermanns Mitteilungen* 55(11): 329–30.

——. 1921. "Neue Gradnetzkombinationen." *Petermanns Mitteilungen* 67(Dec.): 248–52.

Wong, Frank Kuen Chun. 1965. "World Map Projections in the United States from 1940 to 1960." Master's thesis. Syracuse University.

Woodward, David, ed. 1975. *Five Centuries of Map Printing*. Chicago: University of Chicago Press.

——. 1979. "Early Gnomonic Projection." *Mapline* (13): [1–2].

——. 1987. "Medieval *Mappaemundi*." In Harley and Woodward 1987, 286–370.

World Book. 1982. *The World Book Great Geographical Atlas*. Chicago.

Wraight, A. Joseph, and Elliott B. Roberts. 1957. *The Coast and Geodetic Survey, 1807–1957: 150 Years of History*. Washington: U.S. Coast and Geodetic Survey.

Wright, Edward. 1610. *Certaine Errors in Navigation*. London. 1st ed., 1599; later ed., 1657.

[Wright, John K.] 1955. "Obituary: Samuel Whittemore Boggs." *Geographical Review* 45(1): 130.

Yang, Qihe. 1990. *The Principles and Methods of Map Projection Transformation* (in Chinese). Chinese People's Liberation Army Press.

Young, Alfred Ernest. 1920. *Some Investigations in the Theory of Map Projections*. London: Royal Geographical Society, Technical Series 1.

——. 1923. "Note on Professor J. D. Everett's Application of Murdoch's Third Projection." *Geographical Journal* 62(5): 359–61.

Youschkevitch, A. P. 1970–80. "Pafnuty Lvovich Chebyshev." In Gillispie 1970–80, 3: 222–32.

Zetterstrand, Seth J. H., and Karl D. P. Rosén. 1926. *Nordisk världsatlas*. Stockholm.

Zöppritz, Karl J., and Alois Bludau, 1899. "Die Projectionslehre." In their *Leitfaden der Kartenentwurfslehre*. Leipzig: B. G. Teubner. 2d ed., vol. 1. 3d ed., 1912.

INDEX